# TOURISM PLANNING

Second Edition ▪ Revised and expanded

## Clare A. Gunn

Taylor & Francis

New York • Philadelphia • London

| USA | Publishing Office: | Taylor & Francis • New York |
| | | 3 East 44th St., New York, NY 10017 |
| | Sales Office: | Taylor & Francis • Philadelphia |
| | | 242 Cherry St., Philadelphia, PA 19106-1906 |
| UK | | Taylor & Francis Ltd. |
| | | 4 John St., London WC1N 2ET |

**Tourism Planning** Second Edition, Revised and Expanded

First Edition published 1979. Second Edition 1988
Printed in the United States of America

**Library of Congress Cataloging In Publication Data**

Gunn, Clare A.
  Tourism planning.

  Includes bibliographies and index.
  1. Tourist trade.  1. Title.
G155.A1G86  1988      380.1'4591      87-18074
ISBN 0-8448-1538-1
ISBN 0-8448-1540-3 (pbk.)

# Contents

# Foreword

*Tourism Planning,* which is based on academic training and practical experience, charts a course for all institutions and people involved in tourism. As such, it is heavily involved in policy and now that there is an increasing need for better training and better planning in tourism, new information and concepts are badly needed. The text is based on worldwide experience. Its principles apply internationally and will provide a solid background for policy makers, planners, entrepreneurs wherever and whenever tourism is under consideration.

Tourism has been and will continue to be one of the fastest growing social and economic phenomena of the twentieth century, and there is no sign of any slowdown as we look ahead to the twenty-first century. The number of tourists, both those who travel internationally and domestically, will continue to increase; and they will be drawn from a wider range of socio-economic groups than at present. The needs, desires, and expectations of each segment of the tourism market has changed rapidly and will continue to change. New segments will arisc; existing ones will wither away. Good, current market data is essential for tourism planning.

The geographic spread of tourism will widen, aided by the increased propensity to travel and new technology to make it possible. Nearly all parts of the world will be made accessible to visitors and will be sought by visitors.

But this trend of the geographic spread of tourism and its consequent impact have given rise to rumblings that indicate that all is not well. Tourism creates its unique problems. The concerns of the 1960s and 1970s as to whether tourism would be a "blessing or blight" and whether tourists would be described as the "Golden Horde" still exist. Equally important is the desire of regions and countries to give their tourism a distinctive character. There is a real danger of tourism products looking the same no matter where they are located.

If tourism is allowed to grow willy-nilly, it may well be seen as a blight and the tourists will be unwelcome. The desire to travel to many different destinations may wither if tourism opportunities become too homogenous.

That situation need not arise—not with planning. A planned approach to its present and, more importantly, to its future is essential if the full benefits of tourism to places and people are to be realized.

Although planning for tourism is considered to be almost a contradiction in terms, if it is done properly, it is the only way to ensure the future of tourism and the fundamental right of people to travel. What is being planned is not the travel experience itself, but the opportunity to achieve the desired experience that lies at the base of all tourism.

Thus, a proper policy stance should be to ensure that every resident has the opportunity to take part in a wide variety of activities, active or passive, in suitable surroundings, and that a minimum of restrictions apply to those people crossing borders for touristic purposes.

It is increasingly evident that policy for planning must be more sensitive to developers and development. The interface between the many developers of tourism has not been treated well in the past. Planning is seen as the only means by which land use conflicts, contradictions, and inefficiencies can be avoided.

Tourism planning is essential if the appropriate tourism product is to be available for the many and varied segments of the market that will be looking for a place to go. The tourism product is an amalgam of services, facilities and location. It is the totality of these elements that provide a destination with its particular attractiveness. Some parts of the product are common to all market segments, and others are unique to specific segments. Planning is the key to getting the right combination at the right place, and to ensuring that all of the players on the supply side understand what their role is on the product team.

Physical planning, the primary focus of this book, also implies policy that recognizes possible limits. What are the limits of natural resources to withstand use? What are the number limits tourists will place on themselves for quality experiences? Although precise standards are not known, planning shows the only promise of solution to avoid chaos. It is also important to recognize that planning is not the prerogative of any one level of government but that all levels are involved.

Tourism requires planning—now, more than ever before. If it is carried out in the spirit of this book, tourism's future will be far happier than it might be if left to grow uncoordinated and, all too often, with little thought for the consequences.

*Gordon D. Taylor*

# Preface

The first edition of *Tourism Planning* opened with the statement that the planning of tourism was "an uncommon, an unpopular, and even an unwanted idea." The decade since has proven this to no longer apply. Although not yet commonplace, tourism planning is being discussed and implemented more widely than ever before. Not yet a popular topic of general conversation, the concept of tourism planning is no longer limited to esoteric treatises and governmental documents. Nations, provinces, states, and communities the world over are now interested in learning more about how better tourism planning can be practiced. And the acceptance of the first edition of this book as a university text and reference for the teaching of tourism suggested that it deserved revision and updating. These were part of the promptings for issuing a second edition. There were other reasons.

In the last decade, tourism has grown to encompass total travel rather than only its traditional scope of pleasure travel. Whereas vacationing and personal travel remain a large portion of tourism, planning must encompass the entire phenomenon, not just one part. This new definition is used throughout this revision to reflect contemporary thinking.

The linkage between planning, policy, and politics was not seen as clearly a decade ago. Preoccupation with private sector services, the "tourism industry," tended to omit the vital role of government, not so much for governance as for guidance, direction, and public policy. Consideration of the political and policy aspects of tourism planning have been incorporated into this revision.

In this last decade, serious writings about tourism and planning implications have multiplied greatly. Books and journal articles, good and bad, have proliferated. The earlier edition could not benefit by these findings and opinions of scholars, government agents, and practitioners. Major changes and additions to the text reflect these new insights.

The first edition was provincial, limited primarily to the U.S. experience. In the last decade it has been my privilege to have worked professionally in several countries outside the U.S.A. In addition I made requests to regions, nations, and provinces worldwide for their documents relating to policy and planning. The results are substantive additions to the text.

Planning as a concept has changed in the last decade. Once considered primarily physical and practiced by a highly trained and skilled elite, planning is now considered in much more comprehensive terms. It is political. It is no longer the sole prerogative of a select few, but requires input from many constituencies. Planning must be continuous and integrative. These newer approaches and processes are incorporated here.

Other opportunities for improving and updating the text were seized upon. Increased interest in tourism development at the community level prompted the writing of an entirely new chapter on this topic. The original elaboration of one process of tourism planning has been truncated to allow discussion of other approaches. And to provide the reader with more detailed descriptions of policies and plans around the world, an appendix of summaries of examples has been added.

In spite of all these changes and additions, study over the last decade has shown that many basic tenets of the original edition still deserve strong support and coverage in the text. Considering tourism as a functioning system, understanding the several components, planning with a regional perspective, mixing project with continuous planning, and striving towards more than economic goals are as valid today as a decade ago.

It is hoped that the reader will agree with the author that the cuts from the original text are justified by the addition of new, more universal, better documented, and more pertinent ideas and information. As before, the reader is encouraged to use this text as a guide to even greater depth of reading and understanding of the many facets important to the planning of a better tourism world.

## Acknowledgments

Many individuals from academia as well as the several tourism sectors have assisted the author with information and ideas for this revision. Their contributions are genuinely appreciated. In addition to the many persons who responded to requests for information from throughout the world, the following deserve special thanks: Dr. Robert B. Ditton, Dr. Dan Fesenmaier, Dr. Carlton S. VanDoren, and Dr. Carson Watt, Recreation and Parks Department, Texas A&M University; Prof. Brian Archer, Department of Management Studies for Tourism and Hotel Industries, University of Surrey, England; Dr. Jan Auyong, Sea Grant Extension Service, University of Hawaii; Dr. Rene Baretje, Centre des Hautes Etudes Touristiques, Aix-en-Provence, France; Manuel Baud-Bovy, planning consultant, Geneva, Switzerland; Jorge Casasempere, Ministry of Tourism, Madrid; Chu Ta-rong, Tourism Bureau, Taiwan, Republic of China; Dr. David L. Edgell, Sr., U.S. Travel and Tourism Administration; Dr. Jose

Miguel Fernandez Guell, Regional Government of Madrid, Spain; Dr. Douglas C. Frechtling, formerly, U.S. Travel Data Center; Dr. Charles R. Goeldner, Bureau of Business Research, University of Colorado; Prof. Jafar Jafari, Department of Habitational Resources, University of Wisconsin-Stout; Dr. Atid Kaplan, School of Hotel Administration, University of Denver; Karl Knechtel, Regional Operations, Tourism Canada, Ottawa; Dr. Gary E. Machlis, College of Forestry, Wildlife and Range Sciences, University of Idaho; Kamel O. Mahadin, landscape architect, Jordan; Prof. Douglas Pearce, Geography, University of Canterbury, Christchurch, New Zealand; Dr. J. R. Brent Ritchie, Faculty of Management, University of Calgary; Dr. Linda Richter, Department of Political Science, Kansas State University; Ronald L. Schmeid, National Marine Fisheries, St. Petersburg, Florida; Gordon Shamir, Physical Planning, Ministry of Tourism, Israel; Mark Sparrow, Research and Planning, Western Australian Tourism Commission; Kim Stubbs, Planning and Research, Travel Alberta, Edmonton; Rabbi Peter Tarlow, College Station, Texas; Gordon D. Taylor, Special Research Projects, Tourism Canada, Ottawa; Dr. Josef W. M. Van Doorn, consultant, Amsterdam; Dennis N. Williamson, tourism planner, South Melbourne, Australia; and Prof. Arch G. Woodside, A. B. Freeman School of Business, Tulane University. The assistance of graduate students for the report on tourism potential in Central Texas (Ian Reid, Bud Battles, Larry Mutter, Cindy Danicourt, Karen Watkins, Lynn Arneson, and Dave Polakowski) is appreciated. Special gratitude goes to my wife, Mary Alice, for many tireless hours of word processing for the manuscript.

*Clare A. Gunn*

# List of Figures

# List of Tables

# Chapter 1

# The Nature and Scope of Tourism Planning

Any book on the topic of tourism planning runs the risk of confused communication, the very objective it is intended to clarify. This dilemma stems from the abstract concept of tourism and also from what is variously interpreted as planning. For some, tourism means frivolous hedonism that need not be taken seriously. For others it means commerce—jobs, economic impact. And for still others it means tolerating masses of foreigners for periods of unusual congestion. Equally confusing are images of planning. The entrepreneur believes planning is business planning and oriented only to his own site. The architect sees planning as external and internal design. The environmentalist views planning as the means of alleviating resource degradation. In some parts of the world, planning means the centralized creation and implementation of governmentally mandated economic plans. These impressions and many more beg the question of what a book on tourism planning is about. A beginning point is to place some boundaries on what is meant here as concepts of tourism and planning.

## Definition of Tourism

In this book, tourism is defined as encompassing *all travel* with the exception of commuting. This very broad definition seems necessary from a planning perspective even though it does not agree with many other views. Some specialists restrict tourism to trip distance—more than 50 or 100 miles from home. Some definitions require that a person stay overnight to be counted as a tourist. Other definitions, more traditional, include only vacation or pleasure travel. Whereas the negative image of pleasure travelers is no longer as prevalent, a pejorative element remains. For some, tourists are funny, stupid, unsure, ugly, philistine, rich, exploitive, and environmentally insensitive (Krippendorf: 1986, 132).

Today the trend is toward the use of tourism and travel as synonymous terms. Some organizations and publications combine the terms—"travel and tour-

ism"—to make clear that both business and pleasure travel are included. In both, all the support services for travelers are included. Perhaps the best working definition of tourism is that of Mathieson and Wall (1982, 1):

> Tourism is the temporary movement of people to destinations outside their normal places of work and residence, the activities undertaken during their stay in those destinations, and the facilities created to cater to their needs.

Such a definition dramatizes the complexity of the task of planning tourism. If one accepts this broad definition, many facets become implicit parts of the overall responsibility of planning. The most conspicuous parts, of course, are hotels, restaurants, and airlines. For many regions, automobile travel is dominant, encompassing highways as well as automobiles, scheduled buses, rental cars, and motor coach tours. Some destinations are available only by water—cruise ship, ferry, personal craft. The hundreds of things to see and do in cities and countrysides are also very important parts of tourism.

Parks, theme parks, beaches, resorts, camps, entertainment centers, convention centers, casinos, ski slopes, and homes of friends and relatives are among the many places important to travel destinations. Magazine advertising, the travel guide book, and the variety of maps and schedules are also part of the tourism complex. Certainly, the traditions, customs, and policies of host communities are a part of tourism. Federal policy and political decisions are critical to tourism. Regulations, controls, and standards have much to do with the quality of what is offered to travelers and how satisfying the travel experiences really are. All these and more must come within the scope of tourism planning concern.

## Impacts of Tourism

For many years, the positive impacts of tourism—incomes, jobs, taxes—have been well promoted. Until recently, these were considered bounties that could come to any area with only a little development and a lot of promotion. The costs—social, economic, and environmental—were of no concern. However, in recent years, thoughtful observers, environmentalists, social scientists, and even developers and managers of tourism have begun to raise questions about the erosion of resource assets, congestion, social conflict, and opportunities foregone. Because tourism was viewed narrowly and as primarily the concern of only hotel developers, transportation interests, and promoters, the many negative impacts began to show a great need for planning. The following review of both positive and negative impacts shows that whereas tourism can be of great value, it does have its costs. Most of its problems can be ameliorated or eliminated through planning.

## Positive Impact of Tourism

*Economic.* Worldwide, tourism is looked upon as a smokeless industry with strong and stable economic impacts on the host areas. Tourism continues to be cited as a major item of world trade and of considerable economic importance to domestic economies. In 1985, worldwide domestic and international travel expenditures were estimated at $1,800 billion (Edgell: 1987, iii). Tourism is viewed as the new wave of economic opportunity and is promoted heavily for this reason alone. For the United States in 1985, domestic and international tourism receipts were estimated at $260 billion by the U.S. Travel Data Center.

Whereas economic impact has been variously estimated in the past, better methods of measurement are being applied today. The greatest problem of economic measurement is the extreme complexity of the concept of tourism. Tourism does not lend itself to traditional industrial measurements of "end-use activities" (Frechtling: 1987a, 326). As a result many theoretical models of economic impact have been developed and applied around the world.

The positive impact of tourism, no matter how measured, shows that tourism does strengthen the economy of many areas. This economic benefit is best understood as "a gross increase in the wealth or income, measured in monetary terms, of people located in an area over and above the levels that would prevail in the absence of the activity under study" (Frechtling: 1987a, 328). Economic benefits can be expressed by both primary and secondary affects as shown in Table 1-1.

### TABLE 1-1
### ECONOMIC BENEFITS OF TRAVEL AND TOURISM

A. PRIMARY or DIRECT BENEFITS
1. Business receipts
2. Income
   a. Labor and proprietor's income
   b. Corporate profits, dividends, interest, and rent
3. Employment
   a. Private employment
   b. Public employment
4. Government receipts
   a. Federal
   b. State
   c. Local

B. SECONDARY BENEFITS
  1. Indirect benefits generated by primary business outlays, including investment
     a. Business receipts
     b. Income
     c. Employment
     d. Government receipts
  2. Induced benefits generated by spending of primary income
     a. Business receipts
     b. Income
     c. Employment
     d. Government receipts

(Source: Frechtling: 1987, 330)

Just now being recognized in the U.S.A. (and well accepted in other countries) is the inducement of foreign tourism through expansion of domestic tourism. Evidence of the economic impact of tourism is so overwhelming that it is no wonder that undeveloped countries seek it and industrialized nations wish to protect it.

*Social and personal.* Both research and professional opinion support the many social and personal values derived from tourism.

Plog's (1987, 203) adaptation of Murray's (1938) psychological needs suggests that several needs are satisfied by pleasure travel. The need for achievement may be expressed in competitive sports on vacation (tennis, golf, sailing regattas). Personal achievement through jogging, fishing, hunting, and other individual outdoor sports may be important rewards for some travel market segments. The need for affiliation may be satisfied by motorcoach tours of like interest groups. Visiting historic sites, attending festivals, or meeting with resident hosts can satisfy a need for intraception. Increasingly, the need for succorance is satisfied by complete travel packages where most activities, food service, lodging, and transportation are arranged and provided by others.

Duty and business travel provides for the social and commercial needs of many travelers. The many seminars, conferences, meetings, and conventions called by more and more organizations satisfy the objectives of direct contact between participants.

Cross-cultural exchange is probably the greatest social value of tourism. Seeing and experiencing another region or country exposes the traveler to different political, religious, and economic systems. How these different patterns influence the lives of residents can make lasting impressions on astute and observing travelers. Observing land use and development can provide excellent lessons in geography, economy, and life-styles. Some tourists become so attached to a travel locale that they finally settle there.

The intangible benefits of travel were cited as important as the economic benefits by the 107 state delegations and 91 observer delegations at the World

Tourism Conference in Manila, 1980. In its Manila Declaration on World Tourism, it was stated in part that:

> Through its effects on the physical and mental health of individuals practising it, tourism is a factor that favours social stability, improves the working capacity of communities and promotes individual as well as collective well-being.
>
> With respect to international relations and the search for peace, based on justice and respect of individual and national aspirations, tourism stands out as a positive and ever-present factor in promoting mutual knowledge and understanding and as a basis for reaching a greater level of respect and confidence among all the peoples of the world (Records: 1981, 118).

Often, destination areas are enriched by newcomers with new ideas, new social interests, and even greater respect for their own cultural qualities and traditions. Lancaster, Pennsylvania, is an example of an area receiving heavy tourist use and major positive tourism social impact. Although some of the Old Order Amish resent this invasion of their privacy, on balance the tourist has fostered increased cohesiveness and stimulated even stronger protection of their way of life. ". . . Tourist enterprisers may find they have a significant positive role to play in environmental protection and ecological balance" (Buck: 1977, 29).

In many areas, tourism has stimulated a renaissance of traditional art forms with increased development of ancient crafts—carvings, fabrics, paintings, pottery, jewelry (Mathieson and Wall: 1982, 166).

*Environmental.* Viewed broadly, improvements in manufactured as well as natural environments can be attributed to tourism.

In spite of legitimate criticism of some aspects of modern transportation, tourists today are far better off than in the past. Early road, train, and plane travel were undependable because breakdowns were frequent. The dusty gravel roads, open automobiles, and sooty train coaches of the past demanded that camera equipment and picnic supplies be well insulated and that faces and clothing be washed frequently. Ansel Adams, noted photographer of natural scenery for over half a century, made the following observations regarding Yosemite National Park in California, since 1916. "The roads weren't paved. Dust all over the place. The concessionaires were very bad, fighting with each other, letting people camp in the meadows, turning animals loose in the meadows. Gradually it got better. It's now more beautiful than it's ever been" (Forgey: 1977).

Marsh (1986, 217) has cited environmental enhancements due to visitors to national parks. Tourists from all over the world have become friends of conservation and increasingly provide financial support for parks. National parks, especially inland and remote, have helped disperse tourists, spreading economic input to less developed parts of a country.

Modern highways, and particularly automobiles that provide temperature control and other comfort features, reduce riding fatigue and allow greater opportunity for pleasurable and meaningful participation than in the 1930s. The increased speed of travel has provided the chance to spend many more hours at tourist destinations. The newer expressway systems offer a much greater number of interesting urban and rural travel vistas than did old highways. Creative landscaping eases the driving chore, thus enabling the driver as well as passengers to enjoy the scenery.

Improvements in technology have dramatically influenced both the number and the quality of tourism environments. Few factors are as important in explaining increased tourism in the South as that of air conditioning. The modification of man's travel microclimate has virtually ensured polydirectional dispersal of tourism, once confined to the cooler areas of the North in summer.

Engineering improvements in central heating have made fall, winter, and spring travel more acceptable in the North. Anyone old enough to recall the very uncomfortable cabins of the 1940s, heated by wood stoves, oil space heaters, or the questionable unvented gas space heaters, is very grateful for modern central heating-cooling systems. The innovative use of outdoor air conditioning in the theme parks, such as Six Flags Over Texas and Disneyland, contributed much to their success.

Climate control has not been limited to the commercial attractions, however. Application of new fabrics and insulation materials to the development of camping equipment has provided even the most rustic backpacker with a life-support system rivaling that used on moonwalks. Camping has further been revolutionized and popularized by technological innovations in recreational vehicles that enable the tourist to carry along the amenities of modern urban development.

Because of expanding tourist demands, some species of plant and animal life have increased in volume rather than dwindled. Hunting and fishing organizations have lobbied for and obtained governmental investments and operating budgets that have increased the opportunities for seeing and hunting more fish and wildlife by investments in improved habitat. Many streams and lakes have been modified to increase the yields and diversity of species. Artificial reef development has multiplied many times the sports fishing yields (Ditton: 1985).

Hunting pressure has often been increased in order to maintain better balance between certain species, such as deer, and their natural food supply. In some cases more animals have been produced and their habitat improved for special hunting preserves. Many game forest management programs have been implemented that have increased both animal and tree production. In recent years, a great many cultural assets have been identified, restored, and interpreted. Increased tourism has provided the motivation and economic support for many of these programs (Cohen: 1978, 218; Pigram: 1980, 580).

The modern tourist has at his disposal the greatest multitude of opportunities to view and experience cultural and physical environments than ever before in the history of man. The number of "developed" destination areas and attractions has been multiplied greatly in recent years. Hundreds more parks, preserves, recreation areas, playgrounds, refuges, forests, historic sites, and other areas have been designated by governmental agencies for tourist use.

The nonprofit sector—women's societies, youth organizations, churches, quasi-educational organizations, historical societies, foundations, and conservation organizations—has increased greatly the bank of special environmental settings for tourists. Few laymen realize the importance of such organizational sponsorship of environmental enhancement. Examples are abundant. Some in the U.S.A. include Mt. Vernon in Virginia; absence of billboards in Hawaii; the Polynesian Culture Center; Mystic Seaport in Connecticut; Williamsburg in Virginia; Boy Scout camps; and the Arizona-Sonora Desert Museum.

The private sector has also added greatly to the multiplicity of environmental settings. Improved air travel has stimulated the introduction of many alternative destinations for persons previously out of the market. Thousands more vacation home sites, affording millions of people an opportunity of getting close to mountain lake scenic settings are now available.

The prospects of adapting the hostile environment of space to accommodate extraterrestrial tourists of the future are already being discussed. Such a development would be in keeping with the historic truism that discovery, exploration, and associated transportation innovations have a direct effect on tourism by making travel possible to previously inaccessible destinations.

Visually and esthetically, the developed tourism landscape is in many instances an improvement over years past. Today more of the environmental modification of landscapes and structures for tourists is being influenced by professional designers. Parks, motels, resorts, recreation areas, theme parks, airports, highways, and many other physical and land developments are more pleasing and functionally much better designed than the crude cabins, play areas, and roads of the past. More and more the environmental modifications made for tourists come from highly trained, talented, and experienced teams representing specialists from many fields, such as economics, marketing, sociology, geography, ecology, biological science, horticulture, archeology, and history, as well as the more traditional design professions of landscape architecture, architecture, planning, and engineering. These efforts are producing new beauty as environmental modifications are made for tourist use.

## Negative Impact of Tourism

*Economic.* There is increasing recognition that the positive economic impacts of tourism have their costs. Although accounting systems do not yet allow adequate assessment of the extent of such costs, it seems that there are many. For one, new tourism attractions, facilities, and services place new stress on existing infrastructure. Expansions of water supply, waste disposal, electrical power, and fuel systems and supplies are usually demanded by tourism development. As these are not usually funded by the developers, the community is saddled with an added burden. Expanded police, fire protection, and medical service (especially in high-risk sports and visitor activities) are often needed. Increased maintenance and repair of street systems and other public services are required.

Ownership of the various segments of tourism has much to do with the extent of economic gain for an area. If a foreign airline acts as tour operator, carrier, and hotelier, the economic benefit to the host area is extremely limited (Winpenny: 1982, 6). Traveler expenditure statistics, often cited as the value of tourism to an area, are misleading if most money is expended elsewhere.

In some destination areas, a significant percentage of tourists actually become vagrants, demanding welfare support in host communities. Often cited by areas resisting tourism expansion are the negative aspects of high-risk and seasonality of much of the tourism enterprise (Mathieson and Wall: 1982, 88). Even though one or two seasons might be beneficial, the residual effects of underemployed or unemployed people for the remaining seasons create an economic drain on the community.

Many public amenities, such a parks, convention centers, arenas, museums, game areas, recreation areas, libraries, theaters, and historic restorations require additional capital and maintenance costs to meet the needs of visitors as well as residents. The added restroom facilities, control features, and manpower required are often not calculated in the initial stages of these local developments. Frechtling (1987b, 355) has added several indirect fiscal costs associated with visitors, who, in turn, create greater resident populations: education, hospitals, housing, public welfare, and overall economic development.

Finally, tourism expansion, especially of certain types such as vacation homes, demands greater quantities of land, which may compete with existing land uses and other economic development. Industry and agriculture may describe tourism land use as pre-empting prime industrial sites and yet providing relatively less impact.

A comparison several years ago in Hawaii raises some question about tourism's economic value as compared to agriculture. One study (Tourism in Hawaii: 1972, 117) reported the following changes as the major economy shifted from agriculture to tourism:

1. Indirect employment generated by primary tourism investments is at least 23 percent lower than that generated by agricultural investments.
2. Median and average family cash incomes may increase as per employee cash income falls.
3. Tourism requires more employees per household than agriculture in order to obtain similar incomes.
4. Tourism requires a less skilled labor force than does agriculture.
5. As employment opportunities move from agriculture to tourism, employees must bear more costs of housing, transportation, and recreation.
6. Tourism appears to require more public infrastructural investments per employee than industrialized agriculture. No data are available to compare revenue production of tourism with that of agriculture.
7. If nonagricultural and nonrural labor is attracted to tourism employment within rural communities, noncash income-supplementing activities may not be sustained by these in-migrants. At present, over 75 percent of the tourism-based households report some form of noncash income-supplementing activities.

However, the extent to which this example of Hawaii can be generalized to other areas is subject to question.

*Social and personal.* The invasions of masses of tourists can disrupt existing cultures and subcultures. "It is perfectly legitimate to compare tourists with barbarian tribes. Both involve the mass migration of people who collide with cultures far removed from their own" (Turner and Ash: 1975, 1). Tourism impact is both by direct interface between hosts and tourists and indirect influence of outside investors, developers, managers, and labor. New ideas and physical changes can be disrupting as well as helpful. Furthermore, masses of tourists can produce congestion and competition for local services, both detrimental to the host society. The competition for parking space and for the purchase of goods and services often meets with local resistance. Finally, the social patterns of leadership and political power can be shifted away from traditional locals to newcomers.

For example, in the Pacific, tourist development, spurred by Hawaii's spectacular growth in the 1960s, caused one leader to view tourism more cautiously:

> When we examine other island tourist industries and the reappraisals which they are undergoing, we are inclined to hold back lest we are drawn into their trap of development for the benefit of developers whose sole purpose is the maximization of profits with only token regard for local participation and the social upheaval caused to the community by their developments.

So stated Baron Vaea, Minister of Tourism, Kingdom of Tonga (Ross and Farrell: 1975, 153).

Local populations, often the poor, have been displaced by new development of hotels, convention centers, and food services, exacerbating their already difficult plight (Crandall: 1987, 373). Other negative social effects of tourism have been observed: growth in crime and prostitution, conflicts in values, loss of local culture, growth of frustration, resentment, and hostility. Negative social impacts of tourism are being felt in many destinations, particularly Hawaii. "The well-financed efforts of the Visitors' Industry Educational Council and other propaganda arms of the tourism industry to maintain a 'good public attitude' toward tourism (the so-called aloha spirit) have not stopped the change in the attitude of Hawaii's people from one of benevolent neutrality to one of resentment and rage" (Kent: 1983, 184). This results primarily from the low wages and socially demeaning jobs in most of tourism enterprises. Kent calls this an "awakening of an antideveloper, antitourism consciousness, a desire to reassert local control and local integrity among large numbers of people." Jafari (1973) observed what he calls "premature departure to modernization," a too-swift and abrupt social upheaval.

The degree of local opposition to tourism by local residents varies with the type of development, vacation home development being more disruptive than other forms. "Several holiday homes of absentee English proprietors have been burnt down as a protest against the incursion of second home ownership" (Duffield: 1982, 11).

A study of the impacts of tourism on Sri Lanka (Thahir: 1983) revealed several negative effects. About 71% of local residents believed that indiscriminate entry of tourists does cause or enhance existing social problems. Among these were the introduction of new drugs, increased homosexuality, nudism; most jobs were given to outsiders; and young people now try to imitate tourists, who generally are beyond their socioeconomic status.

Not all these negative social impacts occur immediately upon tourism development. Butler (1975) has identified several characteristics that influence the rate of change: For visitor factors, he includes volume, length of stay, racial characteristics, economic characteristics, and tourist activities. Among destination area factors, he lists degree of economic development (less impact if nontourism development is strong), spatial characteristics and land absorption capacity, degree of local involvement, strength of local culture, and general history of stability.

A few researchers are beginning to identify social impacts in terms of irritation to residents, shifts from euphoria to xenophobia, cultural dislocation, introduction of conflicting ideologies, promise of unattainable goals, and increased community divisiveness (Mathieson and Wall: 1982, 137).

*Environmental.* At the same time that some tourism environments have been enhanced, it must be admitted that tourism frequently has negative environmental impact. It may work against other human activities and long-range protection of natural resource assets. As reported in a United Nations study, the environmentally erosive qualities of tourism are now a subject of worldwide concern. It emphasizes the damage being done to "coastlines, aquatic ecosystems, islands, mountain regions, the countryside in general, unique scenic spots, historical sites and monuments. . . ." (Planning and Development: 1976, 36).

In some instances, touristic development is actually causing pollution of water and air. Some remote resort communities pour all raw sewage directly into nearby waters. The volume of tourists who travel by automobile certainly must contribute to air pollution even though exhaust emissions have been reduced in recent years. The tourist segment of air travel contributes its share of increasing the pollution from airplane exhaust. Some equipment, such as houseboats and cruisers, not only spill oil and gasoline and emit toxicants, but also dump raw sewage directly into the waters.

Insecticides, herbicides, and fertilizer additives used around resort and vacation home complexes sometimes create pollutants in runoff and percolation waters. It has been found that the practice of salting highways for prevention of ice can produce groundwater contamination nearby.

Some touristic uses, particularly from great masses of uncontrolled visitors, apply such pressure on the natural resource base to cause its deterioration. Foot trampling in picnic and camping areas can erode natural ground cover, exposing the surface to erosion from rains. Use of dunebuggies and excessive foot traffic can open up severe wind erosion "blowouts" in beach dune areas. Compaction of soils can kill trees and other plants.

Williams (1987, 386) has observed that the ecosystems most sensitive to all development including tourism are coastal systems, mountain habitats, and landscapes with shallow topsoils. Because these are often the very environmental settings that become attractive to tourists, they are especially vulnerable.

One of the most serious aspects of mass tourist use is the wear and tear on historic sites and buildings. The patina of age is essential to visitor understanding and yet the construction of durable walkways, stairs, lighting, lookout points, and information areas for masses of tourists sometimes destroys the image. For example, the ancient Mayan Indian marketplace of Chichicastenango, Guatemala, owes its tourist appeal to brilliantly colored and artistic native tapestries and crafts, displayed primarily for their own trade. However, the doubling in recent years (and projected redoubling) of tourists, with the accompanying shift from native to tourist market, from quaint native village architecture to international hotel box is gradually eroding its original appeal (Reid et al.:

1975; Hudman: 1977). Cohen (1978, 232) emphasizes that environmental threats from tourism development are greatest in undeveloped countries where fewer human and economic resources are available to cope.

Other impacts threaten. The removal of plants, coral, animals, artifacts, semiprecious or precious stones, and other collectors' items produces severe wear and tear upon many rare environments. Technological changes, such as the manufacture of snowmobiles, trail bikes, and hovercraft, have increased environmental wear-and-tear, particularly on fragile resources.

Becker (1986) proposes a technique—Cross-impact Assessment Process (CAP)—to identify and analyze threats to national park environments. The method is a participatory approach and builds a constituent network of information. Several panels representing diverse interests are chosen. An application of CAP was made on coastal parks in the eastern U.S.A. Panels included land developers, investment bankers, engineers, planners, ecologists, biologists, sociologists, area managers, foresters, water resource specialists, landscape architects, private citizens, and local and state governmental officials. They were chosen also because of known expertise and understanding of the park being studied. The panel listed biological, physical, and social threats to the park. Another panel reduced the list to an arbitrary 20 by assigning priorities. Relationships between threats were also identified. Greatest threats to Gulf islands national parks were found to be mineral development, visual quality, traffic, acid rain, park user conflicts, surface water quality, off-road vehicle erosion, shoreline alteration, public opinion, exotic species encroachment, and number of visitors (Becker et al.: 1986).

It is hoped that the function of planning can provide an answer to the following question: "For as long as most of us shall live, the race between despoilers and improvers of the landscape will be a hard one, and who is to say which is the 'winner'?" (Clay: 1977, 309).

## Relationship to Recreation and Conservation

It seems that the greatest forces that bind or divide peoples of the world are ideological. Doctrines and beliefs, because they are not completely provable in fact, often coalesce people into movements that they will vigorously defend, sometimes even with their lives. Two ideologies whose doctrines are believed in by a great many people and whose tenets often conflict (or appear to) with tourism are *recreation* and *conservation*.

### Recreation

Recreation is defined in many ways, but most definitions include such terms as: activity engaged in during leisure, activity for pleasure and enjoyment, or activity that enriches the lives of people. In many countries, however, the word does not even exist. Instead, the several components, such as sports, physical training, dance, hunting, and fishing, are actively engaged in but are not under control of recreation agencies or a discipline of recreation.

In North America, as soon as recreation became a role of government, definitions became more important. As it was formalized and institutionalized, recreation became whatever the proponents and agencies created as policy. Some recreational professionals draw a strong distinction between that which is an end in itself and that which is purposeful, claiming that the former is negative, whereas the latter is positive. Leisure, engaged in for its own sake, provides no focus; those recreational activities accepted by society as wholesome, creative, and uplifting are worthy of public support (Rodney: 1964, 4).

Recreation agencies having land, facilities, and programs are now well institutionalized at all levels of government in both Canada and the U.S.A. They vary from those that are resource-oriented (intensive parks that accept a minimum of people-use) to those that are user-oriented (marinas, beaches, picnic areas, and playgrounds).

Because recreation is value-loaded (healthful, purposeful), its proponents often view tourism, with its emphasis on consumerism and commercialism, as an adversary. Recreation proponents believe that theirs is a higher ideal and should not be distorted into only those activities that can be made profitable. Tourism leaders increasingly recognize that recreational motivations are only one part of the many purposes of travel, that business and duty activities must also be considered. These devisive influences tend to create barriers between organizations and agencies of tourism and recreation.

## Conservation

Conservation, as a concept, began in North America and grew from several independent and even conflicting roots.

The park movement, often associated with conservation, actually grew out of social concern. The ills of expanding industrialization and industrial cities gave rise to a demand for parks and open space (French: n.d.). The dedication of public park land was (and still is) seen as an antidote to delinquency, crime, illness, and the drudgery of work. In the eyes of many, the moral and ethical values associated with parks are strong components of conservation.

Early conservation efforts, both in the U.S.A. and Canada, were expressions of efficiency of resource use (Allen: 1966; Nelson: 1968). It was less wasteful to consider long-range programs, especially for renewable resources such as timber, but the emphasis was on utilization, not preservation. Water resources were to be harnessed and soils were to be stabilized and made more productive. Much of modern agriculture production is based on this concept of conservation.

The idea of land conservation in an esthetic sense came relatively late, historically. A very popular recreation activity today, sightseeing depends heavily on a strong contemporary definition of conservation. Wilderness beauty is described as timeless, dimensionless, all-encompassing, dynamic, uncluttered by the artist's conception, and a form of beauty that gratifies all the senses (Marshall: 1972, 79).

A modern concept of conservation is that of science, particularly ecology. An important principle is the protection of natural resource characteristics. Conservation, accordingly, means the exercise of rigid controls to prevent habitat destruction, habitat homogenization, reduction of species, and natural resource pollution (Darnell: 1973).

The idea that conservation is the preservation of the cultural heritage is popularly supported today. Many man-made artifacts have a scarcity value that becomes as important to society's well-being as do the natural resources. Therefore, conservation means their protection, restoration, and interpretation.

But mass tourist developments have frequently been destructive of the very resource base that attracted visitors, placing conservationists and tourist promoters in adversarial positions. Even the death in 1985 of Dian Fossey, conservationist and mountain gorilla protectionist, is attributed to controversy over tourist tours (60% of Rwanda's tourism) to visit the gorilla habitat at the Karisoke Research Center (McGuire: 1987, 32). Environmental impacts on water, forest, soil, and wildlife resources have at times been erosive, especially for the esthetics of these resources.

## The Powerful Mixes of Recreation, Conservation, and Tourism

Even with some clearly identifiable conflicts between the ideologies of recreation, conservation, and tourism, there is stronger interdependence. A large segment of tourism does depend upon the attractiveness of destinations fostered by the conservation of natural and cultural resources such as national parks. Throughout the world, national parks are striking examples of the compatible mix of recreational, conservation, and tourism values for millions of visitors. U.S. national parks are significant to foreign as well as domestic visitors. In 1984, ten percent of all overseas visitors visited at least one U.S. national park (Baker: 1986, 53). Parks near gateway cities experience the greatest number of foreign visitors. The importance of cities to these natural and cultural attractions is further dramatized by the fact that two-thirds of the visits to national park systems occur at sites near the standard metropolitan statistical areas.

Conservation movements depend on public financial and political support, dominantly fostered by tourists who have visited protected resource areas.

The long list of outdoor recreational activities away from home represents a substantial amount of tourism worldwide. Health, fitness, relaxation, and rehabilitation as well as cultural and physical enrichment—the catechism of recreationists—are also the gospel of tourism. And so, there is a strong case for functional interdependence between the ideologies of recreation, conservation, and tourism, suggesting that more is to be gained by cooperation, and at times even collaboration, than by conflict.

Tourism is not only rounded out by the addition of commercialism to recreation and conservation; it is different, stronger, and more penetrating because of the conservation and recreation components within its makeup. It would not survive without them. Tourism planning, as considered here, recognizes the differences but also the very important overlap and complementarity of recreation and conservation as contemporary ideologies influencing land use. For example, tourism in the Galapagos Islands National Park, off the coast of Ecuador, has been so well planned and developed "that the ecosystem is protected, tourists are satisfied, and economic development results" from the 20,000 visitors each year (Marsh: 1986, 236.)

The Okefenokee Swamp Park, a dramatic wildlife area in Georgia, that is visited by thousands of tourists each year, illustrates resource protection and mass tourist use (Elkin: n.d.). This park of 1,200 acres is leased by a nonprofit organization from the state of Georgia. The main body of the swamp covers some 400,000 acres, largely owned and managed by the U.S. Fish and Wildlife Service and the U.S. Conservation Service. Boardwalks, observation tower, and interpretive centers cater to the needs of visitors, while the unusual swamp wildlife continues to proliferate. A Serpentarium and Wildlife Observatory features native bear and interpretation of native reptiles. A Swamp Creation Center provides pertinent information on the history and content of the Okefenokee headwaters of the Suwannee and St. Mary's rivers. Some 180 species of birds, 20 species of frogs and toads, 45 species of mammals, 30 species of fish, 28 species of snakes, and 12 species of lizards have been identified there. The sphagnum bogs, muck, and jungle form lush floating islands in pure freshwater lakes, creating a dramatic scene as well as a rare wildlife sanctuary. Conservation and tourists *can* mix—where properly planned and managed.

## Planning as a Concept

Planning is predicting. Prediction requires some estimated perception of the future. Absence of planning or short-range planning that does not anticipate a future can result in serious malfunctions and inefficiencies. Levels of planning experience vary across the world, but some official planning has taken place almost everywhere.

Town planning has been practiced in the United Kingdom for two centuries (Cherry: 1984, 187) and physical layout planning reaches back to early Greek and Roman times. For England, interest in planning was stimulated by the physical and social ills resulting from industrialization. Visionaries and philanthropists dreamed of utopian cities. The bias for many years was on physical planning—the visual appearance of architecture and patterns of land use. This concept was followed by trends toward comprehensive planning set into law. In recent years two dimensions have been added to planning—social and economic.

> Planning is a multidimensional activity and seeks to be integrative. It embraces social, economic, political, psychological, anthropological, and technological factors. It is concerned with the past, present and future (Rose: 1984, 45).

Although such lofty goals are at the heart of the planning concept, carrying them out has not been simple or easy. For many reasons, including the complexity of thousands of decisions made by individuals, corporations, and governments all over the world, planning has not been as effective as planners might have wished. Professional planners have agreed upon some general directions— a better place to live and the like—but there is no neat body of theory for planning as can be found in other disciplines. In fact, planners generally agree that planning is not a distinct discipline but an amalgam of many.

Throughout this book, the term *planning* does not include what could better be defined as design—the physical and esthetic design of land and structures at the site scale usually carried out by professions of architecture, landscape architecture, interior design, and fine art. Instead, the emphasis is on planning that integrates these professional activities with social, economic, and political action.

Planning relies heavily on values even though it may depend on science in the form of research conclusions (Rose: 1984, 49). The values of the community are critical to planning and they vary from community to community and over time. Whereas technical planners profess objectivity, more and more planners admit that they cannot remain neutral and that their own values are a legitimate part of their activities.

Planning is political because in order for plans to be implemented, some governance is required. Cherry (1984, 187) has observed that planning is interplay between the value systems of professionals, bureaucracy, community (and its various subsections), and politicians. Much more depends on the key actors in each of these sectors than upon logic or rightness. Thus, a new advocacy role is replacing a purely technical advisory and elitist role.

In recent years, the concept of planning has shifted from making a *plan* (noun) to *planning* (verb). The making of plans was a "batch" process akin to completing the manufacture of one product at a time. Planning, on the other hand, could be compared to an assembly line operation that is continuous. The problem with plans was that by the time the experts had completed their research and made recommendations, the plans were already obsolete. Furthermore, this process assumed several factors that frequently do not exist: consensus on objectives, lack of uncertainty, known alternatives, a high degree of centralized control, and ample time and money to prepare a plan (Lang: 1986, 15). Now, more and more municipalities and nations are continually planning so that new conditions can be incorporated as a process.

Perhaps the greatest change in planning in recent years is recognition of the great complexity of planning and the limitations of even the most apt professional planners. This has led to the concept of "planning for integrated development" (Lang: 1986). The difficult problems facing managers, politicians, and developers generally are those of interconnectedness, complicatedness, uncertainty, ambiguity, pluralism, and societal constraints (Mason and Mitroff: 1981).

Planning that is practical as a process is now frequently called *strategic* planning, implying action rather than a static state when a plan has been accomplished. Lang (1986) has compared key characteristics between strategic and conventional planning in Figure 1-1.

Now, well into the age of information technology, older forms of planning, more dependent on professional expertise, are being challenged. Computers now allow much greater volume of data collection and storage as well as more rapid retrieval. This new ability may stimulate stronger centralized planning because automated analysis procedures will allow centralized data base formation (Openshaw: 1986, 69). "Planning machines" could be devised to meet almost every planning situation. Decision-making would be reinforced and explained quickly and accurately. On the other hand, networking with local planning computer systems could foster the opposite—greater decentralization but meeting national as well as local objectives.

## The Case for Tourism Planning

The review of planning and tourism as concepts, the positive and negative aspects of tourism development, and the influence of conservation and recreation ideologies establish the need for better planning of tourism throughout the world. But recognizing need is one thing; taking remedial action is quite another.

| STRATEGIC PLANNING | CONVENTIONAL PLANNING |
|---|---|
| Action-oriented; planning and implementation as a single process | Plan-oriented; planning separated from implementation |
| Oriented to the organization's mandate and its internal/external environment | Oriented to substantive issues; organizational issues are suppressed |
| Focused and selective | All-encompassing |
| Situational analysis includes examination of organization's values critique of its performance | Organization's values not considered and its performance not examined critically |
| Environmental scan considers factors in external environment affecting achievement of objectives | Environmental scan rarely done |
| Explicit mission statement, fully cognizant of implementation capability | Vague goals, not tested for consistency or implementability in a shared action space |
| Proactive, with contingency planning | Preactive and reactive; no contingency planning |
| Strongly oriented to allocation of organizational resources; budget is the key integrator | Planning often separated from budgeting; land becomes the key integrator |
| Planning process is ongoing | Planning process is periodic |
| Builds capacity for planning and organizational learning | Capacity-building not an explicit objective |
| Values intuition and judgement highly | Values analysis highly |

**Figure 1-1.** Strategic planning vs. conventional planning. A comparison of key characteristics of conventional planning with those of a strategic and integrated approach (Lang: 1985, Fig 3, 28).

In the decade of the 1970s, several embryonic moves toward improved development and planning of tourism appeared.

## Site Scale Need

Perhaps the first step in recognizing the need for better planning has occurred at the site scale by individual businesses and governmental agencies. From Germany came a complaint about how tourist facilities were being developed:

> . . . The result obtained is extremely unsatisfactory; firstly, because the facilities are freely developed in a haphazard way and this does not result in a highly efficient,

overall structure; secondly, competitive investment ventures are made and are unprofitable because each one ignores (or knows too well) what the other one is doing and plans his investments accordingly. Cooperation is necessary, because only a well-organized structure which answers a definite need is valid (AIT: 1974, 57).

Resort developers along waterfronts have long used the rule of thumb that the establishment of highways and accommodations tight to the water's edge was a planning fundamental. Whereas use was low and underdeveloped natural resources were relatively abundant, the principle was workable. Now several problems are increasingly apparent. With high-density, high-rise towers creating a veritable concrete wall along the beach, many other visitors are denied two of the most important use advantages of beach—access and views. Furthermore, as buildings and highways usurped finite stretches of beach, backlands were stripped of much of their utility, further compounding the pressure on small strips adjacent to the water. Again, this is a worldwide problem:

> Hundreds of miles of coastline have been ruined irremediably by virtually uncontrolled building of hotels, restaurants, bars and houses. Beaches have been divided into unsightly allotments, and noise from juke-boxes, fumes from traffic and sheer human overpopulation pay witness to the chaos man has made of the organization of his leisure. These evil consequences are not *inherent* in the development of tourism; they just happen when tourism is developed in a thoughtless and casual way (Young: 1973, 157).

Another example of the lack of overall planning of tourism is that of the coastal pollution problem in Mississippi in 1973–1974. Bickering over jurisdiction of the beaches and over how to define pollution overshadowed the issue of cleaning up the beaches. No planning coordination brought the tourism interests into support for protection of beaches. On the contrary, tourist businesses opposed action by the state health and pollution agencies that required the posting of beach pollution signs in their lobbies (Cartee: 1975).

It seems that only in the last two decades were there the beginnings of concern over planning at the site scale. These concerns began to show need for better location analysis, financial feasibility, and compliance with codes and other development issues. Business planning became more complex. The older rules of thumb and conventional wisdom no longer seemed to work. The coming of a new kind of tourist, one with greater expectations and sharper discrimination, caught the business establishment by surprise. The demand for larger complexes with a wider range of services, and at much greater capital investment and sophisticated management, rendered many business enterprises obsolete almost overnight.

## Broad Scale Need

Much slower to comprehend was the need for better planning at a larger scale. Lundberg observed in 1972 that the sequence of tourism development in the Caribbean was through six phases.

Phase one includes governmental incentives for development, promiscuous location of projects, no identification of land qualities requiring protection, and highly optimistic feasibilities. Phase two is short-run success and is a "halcyon period for all concerned and may last five to ten years." Phase three begins to show some realities; less economic impact than anticipated, labor unrest, local resistance, environmental errors. Phase four is a tourism recession resulting from overbuilding, high labor costs, and a backlash because of poor service. Phase five exposes even deeper difficulty: local decline in visitor popularity. The final phase is a reflective one in which investors, developers, managers, local society, and political leadership reassess the entire tourism development pattern and wish they had planned (Lundberg: 1972, 192).

The Aspen, Colorado, region reacted from explosive growth of winter sports tourism that revealed the many weaknesses of no early overall collaborative planning. Arbitrarily, the growth of 20 percent a year was stopped; a proposal for a highway expansion was stopped; plans for a future population increment of 200,000 was downgraded to less than 50,000, and eight lawsuits were initiated claiming damages of $32 million (Blomquist: 1975, 21). All this happened because a local electorate, which includes those who work in winter sports tourism, questioned the proposals and narrow plans of promoters, developers, and planners. They discovered that no one was asking "whether it should be done at all" (Blomquist: 1975, 21).

Because of the lack of overall tourism planning, research findings that related to tourism were seldom introduced. For example, few cities with heavy tourist traffic have implemented ordinances that reflect research proving the impotency of sign clutter in business sections. Research showed that the multiplicity of messages from signs and businesses was not even comprehended by the tourist. The "average observer cannot distinguish between more than seven different sights or sounds presented to him simultaneously" (Ewald: 1971, 29).

A tourism interface problem between governmental resource development and nearby commercial development was cited:

> The national government designates a park and makes it clear that it will accept responsibility for only the area inside the boundaries. State governments generally have said the area around the park is not our responsibility—that is a local matter. Most often the local units of government have not been equipped to meet the inordinate pressure put upon them by these developments (Christenson: 1974).

In addition, state tourism organizations did not include tourism planning as a responsible function. Instead, mandates were usually limited to advertising and promotion. The case of Hilton Head Island, South Carolina, illustrates this point. It took two years and over a million dollars for environmentalists, fishing and oyster interests, and owners of a major resort complex to defeat the establishment of a German petrochemical plant. The state industrial development agency had lured the plant proposal and the state tourism agency took no part in the issue. It had no plans to handle the situation (McCaskey: 1972). Hunt (1976, 3) observed in Utah that the development of a comprehensive tourist plan is paramount to the future of tourism in that state. It is basic to building awareness, credibility, and viability. Unfortunately, most tourist endeavors are characterized by a lack of planning.

## Conclusions

Well established now is the fact that tourism is of great economic significance worldwide. Domestic and international travel provides many jobs, wealth, and tax support for many governmental services. This fact has stimulated nations and businesses to establish promotional staffs and spend many millions of dollars to attract more travel.

But preoccupation with promotion has masked the equally important fact that in order to obtain and maintain tourism, many other critical factors demand attention. Promotion alone is not enough.

If the desirable impacts of tourism are to be enjoyed, certainly the negative impacts must be dealt with. Once identified, the problems of tourism can be divided into two groups—those that can be solved with better planning and those that must be accepted as inherent in tourism development. For example, increased congestion, erosion of fragile resource assets, and added burden on infrastructure can be resolved through better circulation planning and development, better location and design of facilities to avoid fragile resources, and planning for larger capacity infrastructure. However, the social and cultural fact of greater volumes of visitors in a community must be accepted by residents if tourism's benefits are to be obtained.

An important quirk of the geography of tourism is the reality of North-South tourism (English: 1986). Most of the developed countries, those that send out the most tourists, are located in the North. Many of the amenities and also the economically expanding countries are located in the South. The dramatic impacts of the first wave of northern tourists on newly developed southern destinations have been documented.

English provides a series of concepts for lessening the negative and enhancing the positive results of tourism development in the South. Measures can be taken to strengthen local rather than foreign ownership and control of tourism services and facilities. Increased use of indigenous design and local absorption of transportation, tour companies, accommodations, and food services could gradually increase appeal to visitors and reduce dependency upon outsiders. This improvement in local control should reflect positive change in employment. Seasonal tourism, for example, can be handled better by local part-time businesses than by large international corporations. Whereas tourism should not be viewed as the only answer to better income distribution, it can assist economically depressed areas with proper guidance. Gradual and cautious growth in areas with resource advantages can turn a local economy around. The problem of too narrow a dependency upon a few tourism developers can be offset by strong policies of diversification. When a local area maintains control, not given up to a few multinational firms, dependency can be spread, reducing the negative impact of economic shift. Finally, sociocultural shock from tourism can be lessened with concerted application of policy. Host governments and enterprises could do a more effective job of monitoring those tour operators and developers who bring in the tourists. Better promotion-product match can alleviate much host-guest conflict. Better education of both travelers and local residents regarding the functions of tourism can prepare both for tourism's consequences.

Planning as a concept of viewing the future and dealing with anticipated consequences is the only way that tourism's advantages can be obtained. The following conclusions are important foundations upon which better tourism planning concepts can be built.

*Tourism is of major worldwide significance.* Although well-understood among tourism proponents, it bears repeating that tourism has become a vital force throughout the world. This relatively recent phenomenon is increasingly demanding attention by governments and citizens who previously regarded tourism with indifference. Developed countries now cite tourism as a significant contributor to the economy, and developing countries are rushing into tourism for its economic potential. It must have increasing personal priority because the number of travelers continues to grow. Some see tourism as an effective instrument of world peace. When tourism is defined to include total travel, its implications become very potent. There is little question that tourism has strong positive economic and social import throughout the world.

*Only planning can avert negative impacts.* Increasingly, scholars, observers, and practitioners are recognizing that tourism exacts some costs. The slogan that tourism is a "smokeless" industry is only half true. Improperly developed, tourism can erode the physical and social environment. Tourism requires infrastructure, dominantly provided by government. If tourism is to yield the benefits that are possible and proven, the concept of planning is essential. But such planning must be implemented by all actors involved in tourism, not delegated to professional planners alone.

*Tourism is symbiotic with conservation and recreation.* In the past, the movements and ideologies of recreation and conservation have often polarized into open conflict. Deeper examination of these ideologies shows greater compatibility and interdependence than conflict, as is now being recognized to a greater degree. Most of mass tourism today depends upon clean physical environments and protected resources, the very foundations of conservation. Conserved lands gain greater public support for protection when publics have some knowledge of and access to protected resources. All elements of leisure and recreation are great stimulants to travel. Increasingly, health and fitness provide travel motivations. Needed is greater cooperation of the many agencies of tourism, recreation and conservation so that the symbiotic qualities of these forces can flourish.

*Planning today is pluralistic—social, economic, physical.* Although the concept and practice of planning had its origins in physical planning, the economic and social dimensions are now being considered in most countries. This broadening of objective is seen as essential to potent planning. However, planning with such increased breadth complicates all planning processes and techniques. Physical planning is much more easily measured. Planning that involves social, conservation, and economic consequences seemed to be antithetic to market-economy philosophies. But many businesses related to tourism are now recognizing that business success demands greater sensitivity to the need for the broad and pluralistic concept of planning.

*Planning is political.* Past emphasis in both training and practice on the physical and technical sides of planning ignored the body politic. Now being added to the technical and professional aspects of planning is the political side. Politics in any country, of particular significance to tourism, has two important dimensions. Regulation, a crucial function of all governments, has both positive and negative implications for tourism. Too much regulation can stifle innovation and entrepreneurship and even disallow the realization of many tourism objectives. But in some areas, more and better regulation may be needed for health, safety, and resource protection—all vital to tourism. Second, governments reflect the value systems of constituencies. These values must enter into all planning for tourism development and management. Increasingly, the need for tourism development policy at all levels is being recognized throughout the world.

*Tourism planning must be strategic and integrative.* In the past, all planning has been faulted on its visionary, intangible, and elitist qualities. Whereas no one questions the need for vision and creativity that conceives of new and better tourism, plans must be implementable. Conventional planning has too often been oriented only to a plan, too vague and all-encompassing, reactive, sporadic, divorced from budgets and extraneous data-producing. Needed is tourism planning that is action-oriented, focused, explicit in mission, proactive, continuing, integrative, and values intuition and judgment. These qualities are particularly important for tourism because it is far more complicated an economic, social, and environmental activity than most.

*Tourism must have a regional planning perspective.* Because tourism geography is far more extensive than most social and economic activities, it demands a larger scale concern. Traditionally the focus of tourism development has been at the site scale—hotels, restaurants, historic sites, beaches, mountains. Site scale design and management must be of high quality. But organizations, governments, and all sectors involved in tourism are not recognizing how interdependent are all the separate entities and fragments that make up the whole. Tourism markets and supply are more widely scattered across the earth than for any other economic activity. It is at the interface of the many parts that many problems arise, requiring a more complex but a more broad-scale planning horizon than has been practiced in the past.

# Bibliography

*AIT Congress on Leisure and Touring* (Fifth International). Algrave, Portugal: Alliance Internationale de Allen, Tourisme, (1974).

Allen, Shirley W. and Justin W. Leonard. *Conserving Natural Resources.* New York: McGraw-Hill, (1966).

Baker, Priscilla. "Tourism and the National Parks," *Parks & Recreation,* 21 (10), October 1986, 50+.

Becker, Robert H. et al. "Threats to Coastal National Parks: A Technique for Establishing Management Priorities," *Leisure Sciences,* 8 (3), 1986, 241–256.

Blomquist, Allen. "County Government's View of Ski Area Development and its Impact," *Man, Leisure and Wildlands: A Complex Interaction.* Eisenhower Consortium Bulletin No. 1. Springfield, VA: National Technical Information Service, 1975.

Buck, Roy C. "Tourism Containment and Culture Preservation: The Case of 'Amish Country' in Lancaster County, Pennsylvania," presentation, Travel and Tourism Research Association Annual Meeting, June 12–15, 1977, Scottsdale, AZ.

Butler, R. "Tourism as an Agent of Social Change," occasional paper 4, Trent University, Peterborough, Ont.: Department of Geography, 1975.

Cartee, Charles P. et al. *A Case Study of Estuarine Pollution and Local Agency Interactions.* Water Resources Institute, Hattiesburg: University of Southern Mississippi, 1975.

Cherry, Gordon E. "Town Planning: An Overview." In: *The Spirit and Purpose of Planning,* 2nd ed., M.J. Bruton, ed. London: Hutchinson, 1984, pp. 170–188.

Christenson, Gerald. "Voyageurs," *Duluth News-Tribune,* June 23, 1974.

Clay, Grady. "Winners in the 1977 Professional Design Competition of ASLA, *Landscape Architecture,* 67 (4), July 1977, 309.

Cohen, Erik. "The Impact of Tourism on the Physical Environment," *Annals of Tourism Research,* 5 (2), April/June 1978, 215–237.

Crandall, Louise. "The Social Impact of Tourism on Developing Regions and Its Measurement," *Travel, Tourism and Hospitality Research,* Ritchie and Goeldner, eds. New York: John Wiley & Sons, 1987, pp. 373–384.

Darnell, Rezneat M. *Ecology and Man.* Dubuque, IA: Wm. C. Brown, 1973.

Ditton, Robert B. and Lynne Bonner Burke. *Artificial Reef Development for Recreational Fishing: A Planning Guide.* Washington, DC: Sports Fishing Institute, 1985.

Duffield, Brian S. "Tourism: The Measurement of Economic and Social Impact," conference on Trends in Tourism Planning and Development, September 1–3, 1982, University of Surrey, Department of Hotel, Catering and Tourism Management, Guildford, Surrey, England.

Edgell, David L. *International Trade in Tourism,* A Manual for Managers and Executives. U.S. Department of Commerce. Washington, DC: U.S. Travel and Tourism Administration, 1985.

Edgell, David L. *International Tourism Prospects, 1987–2000.* Washington, DC: U.S. Travel and Tourism Administration, 1987.

Elkin, Liston. *Story of the Okefenokee.* Waycross, GA: Okefenokee Association, n.d.

English E., Philip. *The Great Escape? An Examination of North-South Tourism.* Ottawa, Canada: The North-South Institute, 1986.

Ewald, William K. and Daniel R. Mandelker. *Street Graphics.* Washington, DC: American Society of Landscape Architects, 1971.

Forgey, Benjamin. "Q and A," *The Washington (D.C.) Star,* January 5, 1977.

Frechtling, Douglas C. "Assessing the Impacts of Travel and Tourism—Introduction to Travel Impact Estimation," Chap. 27, *Travel, Tourism and Hospitality Research,* Ritchie and Goeldner, eds. New York: John Wiley & Sons, 1987a, pp. 325–332.

Frechtling, Douglas C. "Assessing the Impacts of Travel and Tourism—Measuring Economic Costs." Chap. 29, *Travel, Tourism and Hospitality Research,* Ritchie and Goeldner, eds. New York: John Wiley and Sons, 1987b, pp. 353–362.

French, Jere. *History of the Public Park Movement.* Pomona, CA: California Polytechnical Institute, n.d.

Hudman, Lloyd E. "Tourist Impacts: The Need for Regional Planning," unpublished paper, Travel Research Association Meeting, Scottsdale, AZ, June 12–15, 1977.

Hunt, John D. "What Direction Tourism? A Plea for Planning," *Utah Tourism and Recreational Review,* 5 (3) July 1976, 3.

Jafari, Jafar. "Role of Tourism in the Socio-Economic Transformation of Developing Countries," M.A. Thesis, Ithaca, NY: Cornell University, 1973.

Kent, Noel J. *Hawaii: Islands Under the Influence.* New York: Monthly Review Press, 1983.

Krippendorf, Jost. "The New Tourist—Turning Point for Leisure Travel," *Tourism Management,* 7 (2) June 1986, 131–135.

Lang, Reg. "Planning for Integrated Development," conference on Integrated Development Beyond the City, June 14–16, 1986, Rural and Small Town Research and Studies Programme, Mount Allison University, Sackville, New Brunswick, Canada.

Lundberg, Donald E. *The Tourist Business.* Boston: Cahners, 1972.

McCaskey, Glen E. "The Travel Industry's Stake in the Environment," presentation, Durham, North Carolina Tourism Commission, 1972.

McGuire, Wayne. "I Didn't Kill Dian, She Was My Friend," *Discover,* 8 (2), February 1987, 28+.

Marsh, J.S. "National Parks, Tourism and Development: Easter Island and the Galapagos Islands." In: *Canadian Studies of Parks, Recreation and Tourism in Foreign Lands,* Department of Geography, J.S. Marsh, ed. Peterborough, Ont.: Trent University, 1986, pp. 215–240.

Marshall, Robert. "The Unplanted Garden: Wilderness Esthetics." In: *A Documentary History of Conservation in America,* Robert McHenry and Charles VanDorne, eds., New York: Praeger, 1972.

Mason, R. and I. Mitroff. *Challenging Strategic Planning Assumptions.* New York: John Wiley & Sons, 1981.

Matheison, Alister and Geoffrey Wall. *Tourism—Economic, Physical and Social Impacts.* London: Longman, 1982.

Murray, Henry A. *Explorations in Personality.* New York: John Wiley & Sons, 1938.

Nelson, J.G. and R.C. Scace, eds. *The Canadian National Parks: Today and Tomorrow.* Calgary, Alb.: University of Calgary, 1968.

Openshaw, Stan. "Towards a New Planning System for the 1990s and Beyond," *Planning Outlook,* 29 (2), 1986, 66–70.

Pigram, John J. "Environmental Implications of Tourism Development," *Annals of Tourism Research,* 7 (4), 1980, 554–583.

*Planning and Development of the Tourist Industry in the ECE Region.* Proceedings, Symposium on . . . New York: United Nations, 1976.

Plog, Stanley C. "Understanding Psychographics in Tourism Research," Chap. 17, *Travel, Tourism and Hospitality Research,* Ritchie and Goeldner, eds. New York: John Wiley & Sons, 1987, pp. 325–332.

*Records of the World Tourism Conference.* Madrid: World Tourism Organization, 1981.

Reid, Leslie M., Clare A. Gunn, and Mario R. Dary. *Prefeasibility Study for a Master Plan of the Renewable Natural Resources of Guatemala,* Vol 7, Natural Parks. Houston: Bovay Engineers, 1975.

Rodney, Lynn S. *Administration of Public Recreation.* New York: Ronald Press, 1964.

Rose, Edgar A. "Philosophy and Purpose in Planning." In: *The Spirit and Purpose of Planning,* 2nd ed., M.J. Bruton, ed. London: Hutchinson, 1984, pp. 31–65.

Ross, Dianne Reid and Bryan H. Farrell, eds. *Source Materials for Pacific Tourism.* Center for South Pacific Studies. Santa Cruz: Univ. of California Press, 1975.

Thahir, M.Y.M. "Integrated Planning for Tourism Development—Sri Lanka Case Study." *Workshop on Environmental Aspects of Tourism,* pp. 79–113. Madrid: World Tourism Organization, 1983.

*Tourism in Hawaii,* Tourism Impact Plan, Vol. 1, Statewide. Hololulu: Department of Planning and Economic Development, 1972.

Turner, Louis and John Ash. *The Golden Hordes.* London: Constable, 1975.

Williams, Peter W. "Evaluating Environmental Impact and Physical Carrying Capacity in Tourism." In: *Travel, Tourism and Hospitality Research,* Chap. 32, Ritchie and Goeldner, eds. New York: John Wiley & Sons, 1987, pp. 385–398.

Winpenny, J.T. "Some Issues in the Identification and Appraisal of Tourism Projects in Developing Countries," conference on Trends in Tourism Planning and Development, September 1982, University of Surrey, Department of Hotel, Catering and Tourism Management, Guildford, Surrey, England.

Young, George. *Tourism: Blessing or Blight?* Baltimore: Penguin, 1973.

# Chapter 2

# Tourism Planning Progress

At a tourism seminar in Yugoslavia in 1972, Dr. Kresa Car, deputy president, Republican Council of Tourism, summed up the attitude of that period: "Is planning necessary? Is planning possible? We don't know how to plan for tourism." But even at that time others were showing that tourism planning was necessary and possible. Gradually, academics, professional planners, and governments began to look toward planning as the only way to solve the several issues that had become more apparent. Preoccupation with promotion had diverted interest away from planning. Now the complexity and enormity of tourism suggest that it must be taken more seriously and more comprehensively. Now developing countries seek its rewards and developed nations seek to maintain their market share. And everywhere there is the desire to ameliorate some of the problems tourism has created.

This chapter sketches evidence of tourism policy and planning around the world. By means of literature review and contact with several countries, some insight into policies and programs of policy and planning for development was obtained. These examples are not necessarily representative of all nations and areas but were selected to show some of the accomplishment issues and diversity. These are edited and condensed versions without editorial comment by the author. The emphasis is on policy and planning of the supply rather than the market side. The reader should understand that in case of negative evaluation, it is likely that further research would demonstrate better policy and planning in other areas of the nation.

A review of these examples reveals that much is being done in policy and planning formulation, a relatively recent event. The chapters that follow offer depth in how the tourism system functions, the several components, new concepts of tourism planning, some emphasis on the community, and some planning conclusions and principles. More complete descriptions of policies and plans can be found in the Appendix.

# Recent Decades Of Planning

After reorganizing the International Union of Tourist Organizations (IUOTO) into the World Tourism Organization (WTO) and moving its venue from Geneva to Madrid, the organization took on a strong advisory role. In 1978 two documents resulted from staff study of the tourism planning issue.

*Integrated Planning* was prepared as a handbook guide for developing countries in their quest for expanded tourism. Local planning examples included Sappora in Japan, Ayers Rock in Australia, Carthage, Tunisia, and a few other specific projects. Regional planning studies included those for Greece, Central Australia, Yugoslavia, and some South American areas. National planning had begun for Pakistan, Benin, Venezuela, Ireland, Sweden, Haiti, Madagascar, Syria, Cameroon, Poland, Iran, and Iraq.

A second document, *Inventory of Tourist Development Plans* (1978, iii) summarized results from a survey of some 118 national tourism administrations, mostly developing countries. Out of a total response of 37 nations, 26 stated that they had a plan underway, eight replied that a plan was in preparation, and three said that there was both a plan underway and in preparation. A sampling of these plans shows their diversity.

Colombia earmarked (1977–1986) one-third of its total 10-year economic development budget of $1,500 million for tourism. Priorities included: Atlantic Coast and San Andres Island, frontier areas, areas of urban expansion, San Augustin archeological route, Boyaca tourist route, and Amazonian forest area.

Israel set aside $700 million for its 10-year tourism plan (1974–1983) to develop accommodations, attractions, beaches, parks, forests, convention centers, and marinas.

Thailand's plan was for a five-year period (1977–1981) to be spent on marketing and promotional programs, conservation and development projects, and upgrading facilities.

Another survey, *Report on Physical Planning and Area Development for Tourism in the Six WTO Regions* (1980), revealed that the 1,619 plans identified were of several levels: 184 local, 348 regional, 180 domestic, 266 intraregional, 42 sectoral, and 599 at the site scale.

Most plans called for utilizing governmental funds for direct intervention into physical development and marketing. Baud-Bovy (1982, 308) drew several conclusions regarding why so few plans were implemented, stating that several factors had been overlooked:

The difficulties of land control were underestimated and the plans generally induced land speculation.

The plans were not sufficiently adaptable to rapidly changing conditions.

The mechanisms of the tourism sector, the structures of the tourism industry, and its integration into the nation's policies and priorities were not given sufficient attention.

Essential qualitative aspects were not quantified or otherwise taken into consideration in the evaluation of the socio-economic impacts.

The possible competition of alternative destinations was disregarded.

These observations together with the application of planning techniques led to the elaboration of Product's Analysis Sequence for Outdoor Leisure Planning (PASOLP) (Lawson and Baud-Bovy: 1977).

## Developing Countries

During this period, the World Bank and UNESCO sponsored a seminar on tourism development with participants from 18 countries, including 11 developing nations in Asia, Africa, and Latin American (DeKadt: 1979). This conference examined tourism planning and effectiveness to date, evidence that many countries were engaged in some phases of planning. Plans varied from those that were remedial, attempting to cover earlier mistakes, to others, such as in Bermuda, that were broad enough to encompass social and environmental as well as economic issues (DeKadt: 1979, 15). These early examples of tourism planning identified several concerns that had not yet been fully addressed: lack of local involvement; concern over only gross returns, not *net* returns; exaggerated promises of employment; poor selection of markets; and failure to develop tourism slowly.

Tourism planning for authoritarian countries involves strong governmental inputs all the way from overall policies to funding, building, and managing of transportation, attractions, services, and promotion/information. For example, in the seashore and mountain resorts of Bulgaria, "During the high season, a large number of skilled workers are sent from the interior to the resort areas to provide a variety of services according to a previously established plan" (Planning and Development: 1976, 102). But not all agree that centralized authoritarian tourism planning is effective. Seekings (1975, 1) stated that:

> . . . first, it is difficult to reconcile the volatile, unpredictable nature of the tourism market with the rigid realities of centralized, bureaucratic planning process; second, the level of personal motivation in socialist societies often results in a standard of personal service which many tourists consider inadequate; and third, for one reason or another, the human relationship between visitors and their hosts is—at least as far as eastern Europe is concerned—curiously lacking in warmth. . . .the

planned economy and tourism—which go so nicely together in theory—perhaps do not go together in practice.

## Canada

In Canada, a form of tourism planning—collaboration between the provinces and federal government—officially began in 1971 as a functional part of the recognized Canadian Government Office of Tourism and Commerce (CGOT). The Policy Planning and Industry Relations branch became committed to the planned development of tourism in Canada. CGOT asked each province for a "tourism development plan overview" as the first step in planning. Financial support for planning came from the Department of Regional Economic Expansion and provincial sources. Two planning conferences, in 1973 and 1974, provided a forum for idea exchange and demonstrated that provinces were just beginning to plan (Federal: 1973; Travel Industry: 1974).

Blythell (1974) identified problems and constraints to tourism industry development and the need for planning in Canada:

Provincial draft plans (TDPs) are not yet well thought through.
Project development is not well identified or prioritized.
Staff backup is inadequate.
TDPs have poor market appraisals.
Projects disproportionately are skewed to government, not private sector.
There is lack of recognition of time needed.
"Heavy fog" persists over which provincial departments are responsible.
Financial community remains skeptical over tourism.

One example of response was a plan for tourism development in Ontario (Tourism Development: 1976) that studied the various components of tourism and set out guidelines for expansion of travel markets and zones that had the greatest potential for development.

Two basic forms of regional planning have emerged in Canada. Statutory regional planning, primarily land use, is authorized and administered by provincial governments and carried out by voluntary associations of communities. Regional development planning concerned with socio-economic planning is done both by federal and provincial governments, independently and jointly. Robinson and Webster (1985, 30) argue that these mechanisms are poorly integrated, that the Ministry of State for Economic and Regional Development will stimulate greater participatory planning, that societal subgroups should not be overlooked as economic policies are emphasized, and that leadership at both the federal and provincial levels needs to be strengthened, particularly in the resource sector. (Descriptions of recent tourism planning in Canada, particularly for Alberta and Saskatchewan, are presented in the Appendix.)

## United States of America

In the U.S.A. where tourism is viewed mainly as a private sector responsibility and where planning is mostly delegated to the local levels, there has been a gradual increase in concern over tourism planning at the federal level (Gunn: 1983). The first expression of need for federal involvement in tourism policy was the 1940 Domestic Travel Act, which gave the Department of Interior, through the National Park Service, the charge to "encourage, promote, and develop" travel within the United States. The emphasis was on domestic travel. The program was dropped during World War II, revived slightly from 1968 to 1972, and was then absorbed by the Office of International Travel, which in 1961 became the United States Travel Service (USTS). But by this time the emphasis politically had shifted from domestic to foreign travelers to the United States to help offset a balance of payment deficit. In 1970, Congress created the ad hoc National Tourism Resources Review Commission, charged to study "tourism needs and the resources to meet these needs at present and by 1980."

This was the first comprehensive analysis of tourism in the United States, and it employed five task forces to study: (1) tourist needs and demands for facilities, (2) tourist resources and institutions, (3) tourism and transportation, (4) tourism and information, and (5) determining the needs and resources of foreign tourists. In the process, several other task forces were added: economic impact, environment, national parks, airlines, minorities, and handicapped. This effort resulted in a six-volume report, *Destination USA*. It recommended the creation of four bureaus, including the Bureau of Tourism Research and Planning, which would be "established to determine future tourism needs and prepare plans that ensure adequate development of resources and to coordinate these plans with other agencies and private industry" (Destination USA: 1973, 1/17).

Implications for national policy are found in the objectives of this proposed bureau:

Ensure that tourism facilities be adequate to meet future needs.
Ensure that the quality of the tourist's experience is maintained.
Provide leadership and a basis for cooperation with private industry.
Increase efforts to reduce the travel portion of the balance of payments deficit.
Ensure optimum utilization of public lands by tourists consistent with environmental protection policies.
Provide assistance to states for tourism research and development.

Take the lead in dispersing tourism activity both to relieve congested areas
and to aid economically depressed regions (Destination USA: 1973, 1/17).

Six of these seven objectives emphasized other than the balance of payment
issue. After several drafts of legislation, a National Tourism Policy Act was
passed in 1981. Title I revised the USTS into a new U.S. Travel and Tourism
Administration (USTTA) with the following charges: to optimize economic
development of tourism, assure universal travel, encourage educational values
of travel, stimulate foreign travel to the United States, stimulate historic restoration, relate tourism policy to national energy and conservation policy, harmonize public-private development, stimulate competition, upgrade quality of
tourism service, and assist in research and planning.

This was the first time that tourism was federally declared as important, and it
has social as well as economic implications. Thus far, budget and political
constraints have not allowed the USTTA to be active beyond promoting
incoming travelers and a very minimum research role, primarily in flight
analysis of travelers to the United States.

Meanwhile, in order to avoid the confusion of many different methodologies
and scopes of economic travel studies within the individual states, a private
nonprofit U.S. Travel Data Center was established. It is now sponsored by the
Travel Industry Association of America.

In the United States, tourism planning was not centralized but was being
influenced by public opinion and local planning legislation. By the late 1960s
there were at least 14,000 separate jurisdictions (out of 3,000 county governments, 18,000 municipalities, 17,000 townships) that exercised some form of
land use control (Healy: 1976, 7). The federal legislation and agencies relating to
air quality, water quality, and even quality of life have dramatically changed the
setting for tourism land use development.

Probably the first regional plan for tourism development in North America
was *Guidelines for Tourism-Recreation in Michigan's Upper Peninsula* (Blank et
al.: 1966). The study utilized regional tourism planning concepts completed in
1965 (Gunn: 1965). The strategy utilized 15 local county voluntary committees
in addition to tourism specialists from Michigan State University and professional landscape architects. The steps taken over three years included analysis
and conceptualizing potential areas for development.

The major recommendation identified 10 viable destination zones and the
kinds of projects needed within each. Emphasis was placed on development of
new attractions on the assumption that greater development of attractiveness
was needed if normal market forces for businesses were to follow. Secondary
recommendations included project sketch concepts for entrance highway
gateways, travelways, waterways, wildlands, Great Lakes shoreline, historic
redevelopment, and community development. Through the project a prevailing

planning policy was that of protecting the unusual resource assets of the region, at the same time making them available for tourism. Within 10 years following, it was found that tourism had been stimulated, resulting in a great amount of public and private investment (Gunn: 1973).

A cursory survey of state tourist offices for this book revealed some dominant and special tourism policies under which they operate. Nearly every state and territory use legislated state funds for promotion of tourism. Many provide matching funds for promotion. A few support research, such as marketing and visitor surveys, advertising conversion studies, and image and economic impact studies. Only a few identify supply side assistance as a role of government. In these cases, the support is for historic protection and redevelopment, advisory services (hospitality training, marketing, new business assistance), and direct grants for civic improvements and new attractions. Several states indicated cooperation between the tourism office and other state agencies, such as parks, recreation, highways, fisheries, and forests. Although this survey showed limited involvement of state tourist offices in planning for the supply side, this does not mean that governments are not involved in tourism. Many other agencies at the federal, state, and local levels are responsible for many of the physical developments important to tourism. It does mean, however, that states have not yet accepted tourism supply side planning as a governmental function.

State legislatures have reacted to both local demand and federal incentives with new land use guidelines that relate closely to tourism planning. Two states, Hawaii and Colorado, have established strong land use planning foundations that increasingly affect tourism development.

Throughout the 1950s and 1960s, Colorado welcomed the boom in winter sports because it meant increased economic development, job opportunities, and increased taxes for public revenues. In the early 1970s, local constituencies, fanned by the environmental movement and federal legislation, began to question this development. As a result, Governor Lamm, on May 20, 1975, directed the preparation of the Colorado Winter Resource Management Plan, which has more far-reaching implications than only winter tourism. This plan was to formulate, with assistance of both citizen and expert inputs:

State guidelines for impact analysis for winter recreation development.
Alternative policies for the development and conservation of the state's winter resources.
Alternative management policies for the state's natural resources economic growth and human settlement patterns (Ohi: 1975, 128).

As Hawaii entered statehood, tourism was favorably viewed as the answer to a dwindling agricultural economy. Tourist arrivals rose from 243,000 in 1962 to 2.6 million in 1973, and their spending leaped from $109 million to $890 million in the same period (Meyers: 1976, 71). In the 1950s, however, there was increasing awareness of Hawaii's finite resources, and an innovative land use act became law in 1961.

Although implementation has had its ups and downs in Hawaii, there has been less suspicion and greater general acceptance than is usual in the continental United States (Meyers: 1976, 103). The impact has been more effective in the outer islands than on Oahu, which contains most of the population, but with recent revisions the environmental resources have been protected to a greater degree at the same time that tourism has become the number one economic producer.

More specifically, in response to legislative resolution, the Department of Planning and Economic Development entered into tourism planning studies in 1970. However, emphasis was placed on the integration of tourism and not the preparation of a specific government tourism plan. This approach is justified on the basis that the tourist product consists of much that is involved in overall community and state planning. Concern over unbridled tourism growth has fostered tourism planning even to the extent of controlling capacities of destination areas—a maximum of 375 people per acre in the pleasure resort of Waikiki (Jonish: 1973, 5).

A more recent indication of Hawaii's sincerity about planning tourism was the legislative act in 1976, whose purpose was "to establish an Interim State Policy on Tourism for the orderly planned growth of tourism so as to result in the maximum benefit to the people of Hawaii" (Act 133: 1976, 1). The objectives of this act are to:

1. Provide an optimum of satisfaction and high quality service to visitors.
2. Protect the natural beauty of Hawaii.
3. Preserve and enrich the understanding, by visitors and residents, of our native Hawaiian heritage as well as the cultural and social contributions to Hawaii of all its ethnic groups and people.
4. Sustain the economic health of the visitor industry to the extent that such economic health is compatible with the aforesaid objectives (Act 133: 1976, 3).

## South Asia

Although not uniform in policies or their application, planning for tourism has been related to the special resources and markets for South Asia (India, Pakistan, Bangladesh, Sri Lanka, Nepal, Bhutan, and the Maldive Republic) (Richter and Richter: 1985, 201-217).

In all these countries, there are mixed public and private roles. Whereas airlines, railways, and some hotels are government-owned, most tourist services are privately owned and managed. In India, the public role sets national policy and coordinates programs of the states. Their functions are primarily publicity, training, liaison, and statistical research. But public corporations, such as the India Tourist Development Corporation (ITDC), have been established to develop commercial facilities in areas that have not yet been responsive but seem to have potential. These are also demonstrations intended to set standards for the rest of the industry. South Asian countries have in several cases followed India's policy of planning and development, although varying greatly in degree.

Of interest is the extent of centralizing rather than decentralizing power and policy. India maintains coordinated planning and foreign visitor promotion at the federal level, but delegates much to the state, regional, or local authorities. As a consequence, much domestic tourism has been stimulated by the states. For example, the state of Haryana has been very successful in developing "highway tourism." This has created over two dozen attraction complexes on well-traveled roads leading from Delhi to Agra, Jaipur, Chandigarh, and outside destinations. Generally, the other countries in South Asia with less experience and less-established plants have centralized tourism planning and development.

The smaller countries, especially islands, have carried out stronger "enclave" rather than decentralized development. The intent is to avoid the impact problems experienced in the 1970s by transporting tourists directly to destination zones planned to absorb tourist impact. Although this may tend to prevent tourist-host contact, one of the objectives of social travel, it helps to control the management of tourism. Of course, enclave development is less functional for markets that seek closer cultural contact. For isolated beach resorts, the enclave policy seems quite suitable.

According to the World Tourism Organization (World Tourism 1983-1984: VIII, 2) significant changes in tourism policy are taking place in this region. Originally directed only toward commercial development, policy has shifted more towards social and humanitarian values. Relationship to living standard is a new concern. The desire to shift the emphasis of tourism from international to domestic visitors has not been well received by the commercial sector or the population at large. Pleasure and recreational travel go against traditional values. Greater emphasis in tourism in this region has not been able to fare well against the many other pressing demands upon public funds and policies. Several positive steps have been taken to simplify frontier formalities and new borders are being opened. The shift from an agricultural to an industrial economy is creating a larger middle class with new ability to travel. A major challenge is to stimulate greater travel among the young—40 percent of the population is under 15. But political tension within the region continues to be a problem. Finally, policymakers and planners are seeking more and better data on tourism in order to understand it and particularly the linkage between the tourism sectors and other economic activity.

## People's Republic of China

Relatively new is the policy of developing tourism within the Republic of China; less than a dozen cities were available to tourists in the 1970s compared to more than 100 by 1982 (Richter: 1983, 400). With the first spurt of tourism, accommodations and other services were overtaxed and traveler complaints were abundant. A slower paced development in recent years has improved the quality of travel there.

Dramatic changes in tourism development have resulted from sharp changes in governments and policy. During the Cultural Revolution, tourism was paralyzed—483 tourists in 1969 (Uysal et al.: 1986, 114). Some growth took place following President Richard Nixon's visit in 1972, but tourism was limited by the influence of the "Gang of Four." With more open political trends and a growth tourism policy, visitors have increased greatly—180,000 in the Shanghai branch area in 1983.

Tourism is managed at the federal level by the China Travel and Administrative Bureau (CTAB); the issuing of entrance permission is delegated to the China Travel Service and Overseas Travel Service. Within the country, the China International Travel Service makes all arrangements for travel to specifically approved destinations. Because information is controlled at the federal level, the ideology of open cultural exchange through travel is yet far away.

In China also enclave tourism planning is exercised to its fullest. The tourist is taken to specifically planned destinations and given higher quality services than residents. The increased numbers of tourists are more demanding of international standards of travel and less susceptible to regimentation. Tourism has been a strong stimulant to the arts and cultural preservation.

Rapid increases of travelers to China have exposed some problems that are to be met by new federal plans by the State Administration for Travel and Tourism (SATT) (Guangrui: 1985, 14). Some of the key issues cited were: underestimation of rapid growth in domestic tourism, preoccupation with maintenance instead of new development, lack of mobilizing local developers, inadequate long-term market studies, and lack of improved services. As a result two phases of development have been planned—increase the attractiveness of China as a world-class destination and improve services and facilities. For reaching these objectives, specific measures have been undertaken:

Restructure tourism administration, decentralize management.

Improve transportation and communications.
Strengthen control of prices.
Develop new scenic sites and attractions.
Increase profitability of enterprises.
Build more modestly priced hotels for domestic markets.
Improve personnel.
Strengthen control of domestic travel.

## Yugoslavia

An evaluation of tourism development in Yugoslavia shows that both long-and short-range planning have been effective (Sallnow: 1985, 113). A nation complicated by 12 major nationalities, five alphabets, and six republics, Yugoslavia has long demonstrated a strong interest in tourism, both foreign and domestic.

Although major communities such as Zagreb and Belgrade and mountain regions such as Sarajevo have become destinations for both business and personal travel, the Adriatic Coast has experienced greater growth as a travel destination. This has resulted from market segment demand for and governmentally focused planning for the resources of this area.

Tourism development was strongly affirmed in the overall Act of the National Economy Plan 1966–70 (Kobasic: 1981, 233). This act stimulated banking and financial support for tourism development, primarily accommodations. Concentration of planning toward resort centers has resulted in highly seasonal operation and revenues.

The social-economic plan for 1981–1985 set out eight main priorities for tourism development: (1) dynamic expansion for both domestic and foreign travel, (2) increase in foreign currency through tourism, (3) improvement of social and economic effects of tourism, (4) better use of tourist resources, (5) evaluation of better development possibilities, (6) improved service quality, (7) better international marketing, and (8) presentation of the national environment and heritage (Sallnow: 1985, 115). This demonstrates the national policy interest in tourism.

Tourism has been the major development to revitalize many coastal communities that were experiencing economic decline. Two forces are cited as important for this growth. First is the development of the international package combined with automobile camping. Second is a rising domestic market. Although much emphasis has been placed on promoting foreign visitors, domestic travel has increased to a greater degree. Much of governmental monetary policy (credits, exchange of dinars for dollars) has been altered to stimulate tourism. In mid-September 1984, of the 470,000 tourists in Yugoslavia, 365,000 stayed on the Adriatic Coast, and this total included 290,000 foreigners (Sallnow: 1985,

123). This suggests that the national plan to foster international travel has been effective.

Yugoslav tourism planning has fostered enclaves of services near but not threatening to prime historic or natural resources. For example, the historic core of Dubrovnik and the natural resources of nearby islands such as Lokrum have been managed under strict conservation policies while new hotels and other tourist services were planned and developed in adjacent Lapad and Babinkuk.

Planning potential for Yugoslavia's future tourism seems to rest primarily in extending the summer season, developing winter sports, optimizing the use of existing resources, increasing conventions, and adding new self-catering resorts. Tourism development, especially in Croatia, is in line with the long- term ethical heritage of its people toward love for their native land, tightly knit family bonds, honor, fidelity, and especially hospitality.

## Sri Lanka

It may be that islands have received the greatest tourism planning attention because the limits to physical space and resources are more apparent. In 1967, a Ten-Year Ceylon Tourism development plan was launched and later evaluated for its success. Key elements of the plan included:

Delineate five destination regions.
Each destination to capitalize on its unique resources.
Destinations to be marketed for separate segments.
Design controls to emphasize protection of local economic base, use of indigenous materials, theming of dominant styles.
Training of manpower: management, plant services, visitor services.
Community relations program.
Phased review and plan adjustment.
Development that would preuse and enhance cultural, social, and historical aspects of Ceylon.
Provide opportunities for resident travel and recreation and social exchange with visitors (Case Study: 1983, 6).

Evaluation over a decade later showed a high degree of success. Almost all hotels do have a local character, but a few are esthetically jarring. Water and waste controls prevent pollution of beaches and protect health. Some erosion of coasts, coral, and fish resources has occurred. Solid waste remains a problem. The federal financial aid program was matched with environmental and design approvals. By 1983 the contribution of tourism to the overall economy had risen from nearly nothing to 2.4 percent. Direct employment in 1982 was 26,776.

Considering costs of infrastructure, maintenance of national parks, museums, and federal tourism operations, the economic cost-benefit ratio shows an 18 to 20 percent return on investment (Case Study: 1983, 18).

Despite an overall positive impact of tourism based on this plan, the actual development has revealed some impacts still deserving of attention. The development of resort centers has not upset local villages and has dramatically increased local economies. Bentota Resort Centre, formerly obtaining revenues of $33,300 primarily from coconut production, has now risen to $2.6 million and employment of 500 persons (Case Study: 1983, 24). But in several instances, the local level private sector and government have not played a strong role because of lack of understanding of tourism. Culturally, there has been great expansion of the production and sale of crafts, although accompanied by some erosion of quality. Tourist behavior that lacks respect for local religions and customs is sometimes abrasive to local residents. The expected dramatic influx of prostitution, drugs, and crime as a result of tourism growth has not occurred. There are some complaints about the changes brought by tourism, such as outside investment, cultural shock for young people, and coping with new demands. But other residents suggest that opportunities have increased, that the serenity of village life remains, and that aculturation was inevitable.

These issues raise the question of the need for greater research into social, cultural, environmental, and economic impacts—carried out regularly rather than sporadically.

## Great Britain

Tourism planning in Great Britain has been a mixture of advisory plans and federal and local regulations such as for land use. From time to time, due to heavy influxes of tourists creating housing shortages, there have been political moves deliberately not to promote some areas, particularly London (Davies: 1979, 11).

The British Tourist Authority (BTA) is primarily a promotional and coordinating body as are those of Scotland, England, and Wales. Selective financial assistance is provided. The English Tourist Board has set up a network of 12 independent regional tourist boards to "develop, encourage, service and coordinate tourism resources; to advise local authorities on tourism matters generally, including the tourism aspects of structure and local plans; and to provide an input to national tourism policies and assist in their implementation when agreed" (Local Government: 1979, 3). In certain cases, BTA may assist local authorities in joint ventures.

The *Strategic Plan* of the British Tourist Authority (1981) stated several policies and programs directed toward better management of tourism. Whereas

London remains a primary destination, the government policy should spread visitors wherever feasible to new outlying destinations. Marketing programs between BTA and local areas should be on a joint cooperative basis. New attractions and activities need to be stimulated. Better product development is needed to meet new market demands. Information and traveler guidance must be improved. Destinations are overlooked because of poor communication. Rather than statutes requiring them, better standards dictated by markets are needed. Local authorities need better liaison with tourist organizations to make sure local places meet the needs of tourism alongside those of residential sectors. New financial programs, such as tax abatement, may be productive.

Although not entering into physical planning, the BTA and the several other authorities maintain policies that plan for growth and improvement of tourism.

England has experimented, with limited success, with federal assistance to economically depressed regions to stimulate employment (Elias: 1982, 341). But these have been successful only when integrated with overall economic policy.

The experience of "enterprise zones" in Great Britain may have political planning merit if applied to tourism. Launched in 1980, by 1984 there were 25 established to stimulate economic activity primarily in small communities that needed a boost (Hall: 1984, 296). The key incentives for new business included 10-year exemptions from taxation, reduced loan rates, and shortcuts through planning red tape. These are not necessarily suitable for widespread use, but a method of intervening under special circumstances. Of course, they were developed at considerable cost: £16.8 million in tax relief, £38.0 million in capital allowances, £39.8 million in public sector development, and £38.3 million in other public costs. This means on average a public cost of £16,600 per job created. Some critics wonder if these jobs would have been created anyway. The greatest lesson for planning is that the greatest advantage cited by developers is the reduction of red tape, suggesting that the planning process generally has become bogged down in permitting and approvals.

Duffield and Long (1981, 403) found in their study of the "highlands and islands" region of Scotland that the introduction of tourism was not necessarily solving outmigration and employment problems in the area as had been planned. The seasonality of markets and their preference for touring rather than resorting have produced less economic enhancement than was planned. The social change of tourism—different mores, religions, and ethics—has fostered greater polarization than adaptation. These changes of new tourism threatened the existing order of political power. Because most tourists were housed in homes rather than in hotels, the local people literally had to live with tourists. Entrepreneurs who came in from the outside were better accepted than second-home owners. Much resentment came from inflated land prices. Tourism has brought new infrastructure, equally usable by residents. Tourism, as a tool for regional development, has difficulty in not eroding the social and environmental structure that make it appealing for tourism.

## Bali

The island of Bali has been described as having the best planned tourism in Southeast Asia. With generally good regional and site planning, Bali resorts have expanded tourism greatly between 1982 (152,364 visitors) and 1985 (202,421) (Mugbil: 1986, 40). Catering to outsiders with distinctly different socio-economic status, the major enclave was planned for Nusa Dua, located at the southernmost tip of the island, some distance from villages. The cultural heritage and environmental assets have been protected at the same time that the resort hotels are thriving on trade primarily (81 percent) from Australia, Japan, the United Kingdom, and the U.S.A. Although the hotels are concentrated near the beaches, other advantages are short trips to many cultural attractions such as the art center of Ubud and the villages of Celuk and Mas, famous for jewelry and woodcarving.

Rodenburg (1980, 177) has raised the issue of scale of enterprise in tourism development, using Bali as an example. The tourism plans for Bali in the 1970s stressed large-scale development on the premise of economies of scale and transformation of a "backward" into a modern society. This required major hotel development of an "international standard"—air conditioning, swimming pools, restaurants, shops. Through greater economic muscle, greater promotion has attracted investors to these large-scale tourist facilities. However, large-scale development in Bali has increased cultural clash, nonresident financial control, and massive infrastructure costs by government.

Instead, the author recommends that opportunities for small-scale development may have been overlooked. Craft and small industrial tourism can better adapt to local attractions, can offer a higher rate of return to local economy, can reduce the need for huge governmental infrastructure, can reduce costs of employment, can increase production yield, can stimulate entrepreneurship, and can minimize the social and cultural impact of tourism.

This point of view is challenged by Jenkins (1982, 229) who contends that the developers of large-scale tourism are almost necessary and are not this detrimental. He suggests that they may function best when accompanied by small-scale development. When local areas are mindful of the dangers of large-scale projects, any potential impacts can be avoided or at least mitigated through good planning.

## Turkey

In 1982, Turkey enacted its Law for the Encouragement of Tourism No. 2634

(Turkey Tourism: 1986). The stated purpose is "to ensure that necessary arrangements are made and necessary measures are taken in order to regulate, develop and provide for a dynamic structure and operation of the tourism sector." This act contains six chapters governing roles of the private and public sectors, establishment of investments, financial arrangements, and regulations.

Chapter 1 covers scope and definitions in addition to purpose. Chapter 2 includes procedures for certification of investments, protection of natural resources, preparation of plans, rights of control, and tariffs. Chapter 3 includes several fundamentals pertaining to investment incentives, such as tourism credits, employment, sale of alcoholic beverages, establishment of the tourism development funds, and other incentives. Chapter 4 concerns the operation of yacht harbors and yachting activities. Chapter 5 involves inspection procedures and penalties. Chapter 6 authorizes the formation of study groups to determine tourist regions, tourist areas, and tourist centers, communication with the several ministries, and authorizes the Supreme Board of Coordination of Tourism Affairs to implement the law.

Other pertinent documents include "Foreign Investment Law," "Regulation for State Owned Land Relocation for Tourist Investments," "Investment Procedures," and "Encouragement Conditions and Incentives."

Turkey utilizes a mixed economy principle with the government providing infrastructure and the private sector the superstructure of facilities. At the Turkish Grand National Assembly, December 19, 1983, the government established two policies: (1) the adoption of an export-led development strategy, and (2) the determination of economic and financial activity through the operation of free-market forces. The Ministry of Culture and Tourism (MCT) is responsible for tourism promotional activities, domestically and abroad. It has 16 offices in major cities of the world.

In its publication, *Turkey Tourism—Opportunities for Investors,* the relationship of the government to private enterprise is described. The MCT has identified 14 areas that have been planned based on studies within the Five-Year Development Plan. These represent an estimated 600,000 bed capacity when completed. South Antalya, Side, and Koycegiz were completed by 1985. Plans for other areas have been prepared by the MCT and are available to prospective investors. The types of planned areas include: seaside resorts (primarily on the Mediterranean Sea), yachting and skiing centers, thermal springs resorts, and major cities such as Istanbul, Izmir, Antalya, and Ankara.

Public bodies have surveyed the potential of areas within Turkey for tourism development. General criteria for selection includes resource assets, such as favorable weather, attractive flora, water and beach resources, and ancient settlements. Ease of flight access from European markets is also cited as an asset. One region favored for development is southwestern Turkey (Southwest Turkey: 1986?). Illustrated in Figure 2-1 are 13 areas that have been identified by the

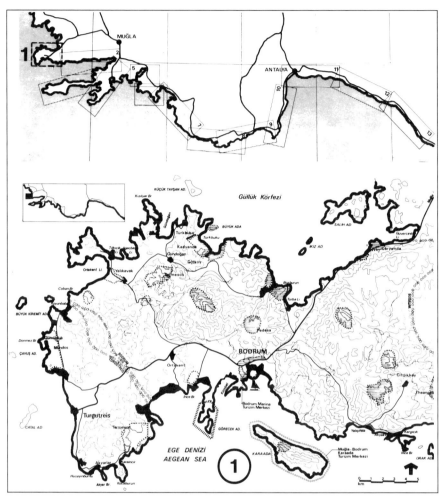

**Figure 2-1.** Areas planned for tourism in Turkey. The top map shows 13 areas along the Aegean Sea designated by the government for tourism planning and development. The bottom map illustrates (in black) specific zones planned by the government for tourism investment and development of resorts in Area 1 along the coast of the Bodrum peninsula (Southwest Turkey: 1986, 14, 15).

government. A map of one of the areas is also illustrated showing the several designated Touristic Installation Zones.

Policies for development allow private investors to lease lands from the MCT in designated areas. Private lands within the tourist regions may be expropriated by the government and, in turn, may be allocated for tourism development. Investors are required to submit development plans for approval. If the lands include sites declared as historical or natural, approval must be obtained from the Supreme Council of Immovable Cultural and Natural Assets. The government is making improvements in infrastructure to service potential sites. Land development within municipal jurisdictions is subject to the zoning and building regulations of the municipality.

## Ireland

The Irish Tourist Board has its objective:

> To optimize the economic and social benefits to Ireland gained by the promotion and development of tourism both to and within the country consistent with ensuring an acceptable economic rate of return on the resources employed and taking account of:
> —tourism's potential for job creation,
> —the quality of life and the development of the community,
> —the enhancement and preservation of the nation's cultural heritage,
> —the conservation of the physical resources of the country,
> —and tourism's contribution to regional development (Report: 1985, 2).

Tourism interests in Ireland have prepared a "white paper" outlining new policies for tourism planning and development. Included is the establishment of a Tourism Development Team that will act as a catalyst identifying tourism opportunities and matching suitable developers and investors to appropriate projects. Criteria for selection include:

1. Principally be directed at generating out-of-state tourism.
2. Have a realistic prospect of providing a real return on investment.
3. Have the potential to create sustainable jobs (Link: 1986, 1).

## Finland

Tourism promotion and development in Finland are the responsibility of the Finish Tourist Board, whose goals are:

> Improving the operating prospects of the tourist industry.
> Developing services and the content of holidays.
> Improving the availability of tourist services.
> Working on attitudes with a view to making them more receptive toward tourism (*Tourism Statistics:* 1986).

This official body grew from earlier stages of development of the Finnish Travel Association, replaced by the Finnish Tourist Board in 1972. Its policies strive to increase domestic tourism and tourism from abroad, to promote holiday recreation opportunities for Finns, and to improve operating conditions for the domestic travel industry.

The Finnish Tourist Board consists of a chairman and five other members appointed by the Ministry of Trade and Industry. These include representatives from the Ministries of the Interior, Foreign Affairs, State Finance, and Communications.

Advisory to the board is the Tourist Affairs Council, appointed by the Council of State. The chairman is a minister dealing with tourism and 20 members representing the tourist industry, including representative experts from certain regions and representatives of labor unions. The operation of the Finnish Tourist Board staff is under a director with three departments: general affairs, marketing, and development.

Attractions and services have been developed for domestic and foreign markets, consisting of 45 percent leisure and 55 percent business (including 14 percent conference guests). In recent years, coach tour tourism has shown the largest increase, coming primarily from Sweden and the Federal Republic of Germany. Domestic tourism was most popular at campsites, cottages, and farmhouse accommodations. The largest numbers of foreign tourists come from Sweden, Norway, the Soviet Union, and the Federal Republic of Germany. Much foreign tourism comes to Finland on the way to the Soviet Union.

In terms of destinations, metropolitan areas such as Helsinki are the most popular for foreign tourists. Swedish tourists gravitate to Swedish settlements; Germans to lake districts and Lapland; and Soviets prefer the southern urban areas of Finland. Nature–related activities have increased in development in recent years—hiking, fishing, hunting, cruises, canoeing, and camping. Because of the climate, tourism fluctuates greatly.

Tourist services in Finland are provided mainly by the private sector, but government aid is provided in several ways. Special aid is given for joint marketing and public relations projects. Large-scale developments, such as complexes for family holidays, water activities, and skiing, may be given financial aid up to one-third the cost. Other service businesses may qualify for grants especially at the start-up stage (Marketing Strategy 1985: 1986, 210).

## France

Clarke (1981, 447) has provided a critical review of tourism development in one area of France. This review supports the principle that planning should incorporate careful market analysis, particularly seasonality.

Based on studies of a national planning commission, the government in 1963 intervened to develop the Languedoc coastal area of south-central France because of increased evidence of tourist demand. The region had not developed because of flooding, mosquitoes, and lack of access and drinking water. But the dry climate, nearness to airports and main highways to Spain and Western Europe, plus demand suggested that it be developed. A special directorate for this area was created—the Mission Interministerielle Pour L Amenagement Touristique du Littoral Languedoc-Roussillon. This body acquired the land and developed roads, ports, and forestry and mosquito control. Then local input to the site was made by Societes Departementale d'Economie Mixte. Then private developers were to invest in building. The plan (1972 Schema Plan) included several large and some small resorts separated by conservation areas.

By 1979 public and private investment had reached six billion francs, accounting for 65 percent of the total industrial investment. An assessment of its development reveals moderate success. It has not attracted many foreign visitors, contributing only small input to the balance of payments. The pervasive French tradition of vacation travel concentration in July and August creates severe seasonality problems at the resorts—14 percent occupancy in January and 75 percent in August. The spurt of employment during construction has not resulted in a significant increase in permanent jobs in the area. Clarke indicates that most of the multiplier effects have been spread outside the area and many of the social objectives have not been reached. Without balance with agricultural and industrial development, tourism (of this kind) does not appear to have justified the heavy investment.

## Austria

Tourism in Austria represents 10 percent of the GNP (Annual Report: 1986?) giving it a prominent place in national policy. The constitution allocates responsibility for tourism to the nine provinces, but overall coordination is maintained by the Austrian National Tourist Office (ANTO) under the Austrian Federal Ministry for Trade, Commerce, and Industry. ANTO is responsible for general matters pertaining to tourist policy, tourist advertising, representation abroad, relevant aspects of development aid, statistics, area planning policy, handling of complaints, screening of tourist projects, and liaison with the European Recovery Program. In recent years, as a result of the development and promotion of winter sports, the winter season now accounts for 40 percent of all overnight stays in Austria (the Tyrol province accounts for 34.4 percent of all-night stays annually).

By policy, the government of Austria introduced tourism in the 1950s to the villages of Hermagor and Passriach, located on the shores of Lake Pressegg in the Gail Valley of southern Austria (Gamper: 1981, 433). Initial plans did not include new hotels but the use of existing homes and farms. The local populations were directly involved. Within a decade the local people have transformed a depressed area into a prosperous one. Stores have been remodeled, new buildings have been erected, new dairy farms have been established, manure piles (from former farming) have been removed, streets have been paved, and campgrounds have been established. Traditional costuming, once frowned upon as being low-class peasant dress, has been revived with new pride. Whereas the two communities had little communication for generations (one even spoke a separate language—Wimbish), they have cooperated on tourism development, forming the Hermagor-Pressegersee Tourist Association. Tourism, combined with improved highway and rail access and labor from outside, became the catalyst for adaptive change by local people for local objectives.

## Australia

In Australia, promotion of incoming travel is the dominant function of the federal agency, the Australian Tourism Commission; planning and development are carried on by state, regional, and local levels. An example is Melbourne's Western Region tourism development plan. Located in the state of Victoria, the area includes nine municipalities, an area of 1,958 kilometers (1,213 miles), and a population of 436,000 (Tourism Potential (1): 1984, 2).

Research has documented the existing tourism situation, such as tourism features, use volumes, regional infrastructure, travel and expenditure patterns, economic significance, image perceptions, existing policies and strategies, and general tourism outlook for the region. Conclusions included:

Need for upgrading existing and developing new attractions.
More, better, and more widely distributed information.
Upgrading of facilities.
Economic expansion opportunities from tourism.
Need for regional and local strategies for development.

The concept for development (Tourism Potential (2): 1986) addresses these issues and is directed toward dynamic interaction between the key components of *people* with the desire and ability to participate in tourism/recreation activities; *attractions* based on natural and cultural features of tourism interest; *services* for users and to support the activities; *transportation* to move people to and from attractions and facilities; and *information/direction* to assist users in knowing, finding, and enjoying the attractions.

The overall conceptual plan for Melbourne's Western Region is illustrated in Figure 2-2. An example of one of the several precinct plans is illustrated in Figure 2-3—the Werribee South Tourism Precinct. Physical development recommendations for this precinct include: Western Gardens Theme Park, Na-

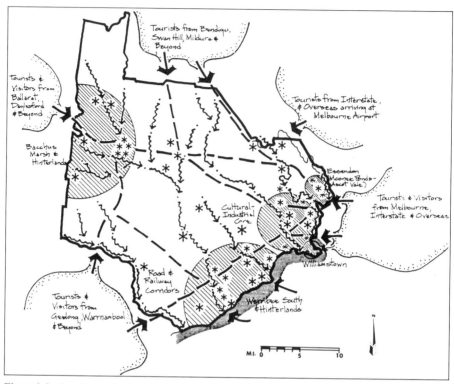

**Figure 2-2.** Tourism potential, Melbourne's Western Region. A conceptual plan showing tourism precincts (shaded), attraction clusters (stars), and linkage corridors with market access points (Tourism Potential: 1986, 20).

tional Aviation Museum, upgraded Werribee Park and Point Cook Park, Fishing Resort Village, a coastal discovery trail, new birdwatching facilities, and addition of new events. The development of the features of Williamstown, such as revitalizing Nelson Place and the Strand, historic restoration, new restaurant and entertainment facilities, new landscape vistas, better train and water transport linkage with Melbourne, and a major historic theme attraction in Williamstown are recommended. Special emphasis is placed on waterbased tourist transport, especially for the Maribyrnong and Werribee rivers and along the coastline.

For organization and administration, it is recommended that tourism development planning be added to the present functions of the Western Region Commission (WRC). It would provide information, analysis, coordination, and integration. Local governmental roles would be through appropriate controls, identification of potential tourism sites, and promotion and liaison with other agencies. The private sector role would be investment in and development of attractions, facilities, and services.

Further recommendations are made for funding, marketing, and implementation. Australia's approach to tourism planning and development could be

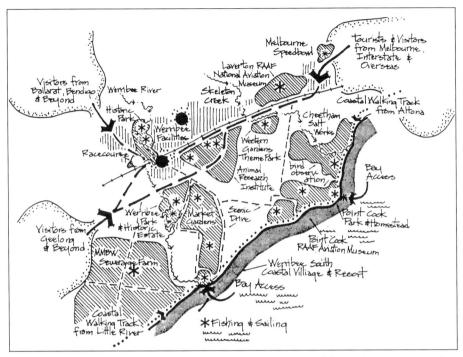

**Figure 2-3.** Concept plan, Werribee South Precinct. Conceptual detail for market access, attractions (stars), development zones (shaded), local communities (dots), and transportation routes (Tourism Potential: 1986, 46).

described as a democratized process whereby the several governmental levels integrate their separate roles, where local publics are involved, and in which private enterprise maintains its entrepreneurial role.

## Valencia, Spain

The democratic constitution of Spain in 1978 decentralized much of the government to the several provinces, called "autonomous communities," such as Communidad Valencia, an area of 23,305 square kilometers (9,322 square miles) centrally located in eastern Spain (Aventura: 1986). Overall promotion of Spain and general policies are set at the federal level in the office of the Secretary General of Tourism within the Ministry of Transport, Tourism, and Communication. The Tourist Board of the Generalitat de Valencia, under the auspices of the Department of Industry, Trade, and Tourism of Valencia, oversees tourist supply and is responsible for inspecting tourist firms, authorizations, systems of financial aid, and promotion. It is the main organization involved in planning and fostering development of tourism for Valencia. The cities of Valencia, Castellon, and Alicante provide administrative services for tourism under the auspices of the tourist board.

In order to expedite promotion as well as development, the Valencian government has established the Institut Turistic Valencia (ITVA). Its objectives are:

1. Survey of the tourist trade for drawing up marketing plans.
2. Promotion of existing resources through publicity, direct promotion, and
   public relations.
3. Stimulate new supply facilities, such as better opportunities for playing
   sport, activities focused on Valencian culture and history, congresses,
   conventions, association activities, and federations.
4. Coordination of tourist promotional drives.
5. Foster positive image, optimize markets, and tap new markets.

## Israel

Based on its tourism master plan of 1976. Israel provides several governmental
incentives for development (Israel Tourism Project: 1982). In the decade prior
to the report, the government invested over $513 million in tourism projects of
which $485 million was in the form of incentives for the private sector.
    Policy favors projects that will:

Contribute to extending the tourist's stay.
Increase the number of low-priced accommodations.
Develop new destinations and improve existing areas.

The principal incentives are grants, loans, and tax exemptions. Grants up to
12 percent of the value of the investment are given to hotels, apartments,
camping sites, youth hostels, and field schools. Other projects may be granted
moneys up to 5 percent of the value of the project. The extent of loans varies
with the development area, some up to 60 percent. An approved enterprise is
eligible for reduced income tax during a period of seven years from its first year
of reported profits, up to a maximum of 14 years from the date of approval.
    The governmental bodies involved in the approval process are:

Ministry of Tourism
Investment Center Board
Tourism Industry Development Corporation
Israel Lands Administration
Government Tourism Corporation Ltd.

## Jordan

Jordan's tourism development plan of 1981–1985 resulted in definitive accomplishments and yet fell short of meeting all objectives (Five Year Plan: 1986, 357). Tourism activities were expanded, a hotel training college was established, private sector training was provided by the private sector, information offices were established, and the Department of Tourism participated in many conferences and exhibitions. Yet, due to lack of funds, several objectives were not met. It was hoped that domestic skills in large hotels could be improved. The plan failed to develop the market for Arab visitors and Jordanian expatriates. Trades and handicrafts were not upgraded. Marketing did not keep pace. And other items did not materialize—new North Shouneh recreation park, King Talal Dam Club, a playland park for children, the lagoon at Aquaba, a camp at Wadi Rum, and the new phase of the tourist site at Azraq.

The plan for 1986–1990 identified several characteristics of tourism important to its planning and development:

Vulnerability of tourism to political events and economic fluctuation.
Contributions to tourism by Jordanians working abroad.
High proportion of Arab tourists.
Year-round availability of attractions.
High contribution of foreign currency to GMP.
Contribution to cultural cooperation and international understanding.
Attention to archeological digs and restoration of antiquities.
Good geographical distribution of such sites.

However, the report also identifies some key problems facing the nation for further development of tourism:

Poor balance of payments from tourism.
Relatively low expenditures on gifts and antiques and programs.
Dearth of services for tourists—Arab and foreign.
Overconcentration of tourism in Amman and Agaba regions.
Poor coordination of control agencies.
Insufficient emphasis on feasibility studies.
Inadequate legislation for projects.
Shortage of trained personnel—marketing, planning, administrative.
Excessive foreign investment.
Paucity of data—volume, expenditures, revenues.
Insufficient allocations for archeological sites.
Visitor erosion of archeological sites.

In light of these problems, the new plan set goals and objectives toward their solution. By means of a fund obtained through license fees to be administered by the Tourism Authority, federal expenditures are to be increased to cover many projects and programs:

Expanded marketing and promotion.

42 site projects (visitor centers, attractions, rest houses, museums, training center for handicrafts, tourist village, amusement park, summer-winter resorts, curative seaside resorts, hotels, parks, diving center, and yacht club.

11 antiquities projects, including renovation of ancient cities, archeological excavations, purchase and protection of antiquities, new archeological museums, and collection of ancient Arabic scripts.

These tourism planning and development efforts are to come from the following sources (amount scheduled through 1990):

| | | |
|---|---:|---|
| Tourism authority (general budget) | 4,395 | (JD'000) |
| Aqaba regional authority (general budget) | 7,350 | |
| Social security | 3,500 | |
| Private sector | 38,300 | |
| National fund | 5,350 | |
| Department of Antiquities | 4,983 | |
| Total | 63,833 | (JD'000) |

## Korea

In recent years, Korea has become a significant tourism competitor using a mixed private-public sector approach (Korean Tourism: 1985?). The Government Tourism Organization sets policy. Under the supervision of the Minister of Transportation are the Korean National Tourism Corporation, the Korean Tourist Association, and the Bureau of Tourism. The Bureau has five divisions concerned with governmental policy: planning, facility, promotion, domestic tourism, and guidance.

The Korean National Tourism Corporation (KNTC), a quasi-governmental body, is charged with implementation. It can make direct investment in facilities as well as provide aid and incentives for the private sector. As in most governments, it installs the basic infrastructure for tourism development. Established in 1974, the KNTC is supervised by a president and a board of directors drawn from government, education, and trade. Its functions are "to perform tourism promotion, tourism resources development, education and training of employees engaged in travel business and domestic tourism guidance" (Korean Tourism: 1985?, 120).

The KNTC is made up of four divisions: the Planning Division has three parts—department of planning and research, department of emergency planning and department of general affairs. The Promotion Division is divided into two parts—departments of overseas promotion and promotional publication. The Development Division contains a department of development, department of domestic tourism, tourism training institute, and the Kyongju Hotel School. The Business Division handles all business affairs including procurement, office supervision, several airport business offices, and several information offices.

The Law of Tourism Promotional Development Fund identifies specific purposes:

1. Construction and repairs of tourism facilities including hotels.
2. Repairs and development of the means of transportation.
3. Construction and repairs of infrastructure needed for the development of tourism business.
4. Other necessary items concerning tourism business set by presidential order.

In addition, certain implementation measures are provided by this law:

1. The establishment of overseas offices operated by travel agents.
2. Public relations activities for the attraction of foreign tourists.
3. Rationalization of management in the travel industry.
4. Construction and expansion for accommodation and shopping facilities undertaken by the state or local government-run enterprises and KNTC.
5. Lodging and shopping facilities ordered by the Minister of Transportation.

Korean tourism policy recognizes the need for planned balance between resource protection and development. Development purpose is to "increase income of the residents, growth of employment and the conservation of natural resources and cultural assets." For example, the Kyongju tourist destination plan objectives are:

1. To enhance cultural heritage through protection, conservation, and development of traditional resources.
2. To build an international resort city famous for ancient Korean culture through the comprehensive development of tourism.
3. To increase resident income by developing and promoting tourism-related industries in the area.

This destination includes the focal city of Kyongju, Pomun Lake, and relics and ruins in 14 historic sites. Funding comes from several federal agencies, the Kyongju Tourism Agency, and the private sector.

## Conclusions

A review of governmental reports and tourism development literature shows not only that many nations are involved in tourism policymaking and planning, but also that the interest is rather recent. Many new policies and programs have appeared in only the last few decades.

Whereas the dominant interest among nations continues to be promotion, concerns over improving the supply side are conspicuous. Many nations provide subsidies and the more socialist countries intervene directly with physical development of resources and services. In market economy countries, private and public roles are rather well defined. Federal governments provide the bulk of nationwide promotion and information services. Development of infrastructure and physical planning policies and regulations are often delegated to regional and community governments. The private sector is most frequently involved in hotels, food services, entertainment, tours and transportation engineering, and production. An emerging sector—volunteerism and nonprofit organizations and cooperatives—appears to hold promise in many nations, especially in small-scale and indigenous development.

Even though countries vary in how tourism planning and development are organized and carried out, attention seems to be drawn to common issues. Most all are driven by economic goals. But, increasingly, concerns over social, environmental, and economic costs are appearing. In many instances, a desire for greater planning and slower development is expressed.

## Bibliography

*Act 133,* Relating to the Interim Tourism Policy Act. Honolulu: The Legislature, Hawaii, 1976.

*Annual Report, 1985.* Vienna: Austrian National Tourist Office, 1986?.

*Aventura Valenciana.* Valencia, Spain: Institute Turistic Valencia, 1986.

Baud-Bovy, Manuel. "New Concepts in Planning for Tourism and Recreation," *Tourism Management,* 3 (4), 1982, 308–313.

Blank, Uel, Clare A. Gunn, and Johnson, Johnson and Roy, Inc. *Guidelines for Tourism-Recreation in Michigan's Upper Peninsula.* East Lansing: Cooperative Extension Service, Michigan State Univ., 1966.

Blythell, D.C. "As to Problems and Constraints to Industry Development," Travel Industry Development Seminar, Ottawa: Canadian Government Office of Tourism, 1974.

Car, Kresa. Presentation at Tourism Seminar—Techniques of Planning and Development, September 14, 1972, Zagreb, Yugoslavia.

"Case Study on Tourism Environment—the Sri Lanka Experience," WTO/UNEP Workshop on Environmental Aspects of Tourism, Madrid, July 5–8, 1983. Madrid: World Tourism Organization.

Clarke, Alan. "Coastal Development in France," *Annals of Tourism Research.* 8 (3), 1981, 447–461.

Davies, Ednyfed Hudson. "Tourism in British Politics," Tourism Society Discussion Meetings Series on British Tourism at the Crossroads, March 12, 1979. London: The Tourism Society.

DeKadt, Emanuel. *Tourism, Passport to Development?* New York: Oxford Univ. Press, 1979.

*Destination USA.* Report of the National Tourism Resources Review Commission, Vols. 1–6. Washington, DC: U.S. GPO, 1973.

Duffield, Brian S. and Jonathan Long. "Tourism in the Highlands and Islands of Scotland—Rewards and Conflicts," *Annals of Tourism Research,* 8 (3), 1981, 403–431.

Elias, Peter. "The Regional Impact of National Economic Policies: A Multiregional Simulation Approach for the UK," *Regional Sciences,* 16 (5), October 1982, 335–343.

*Federal Government Programs Relevant to Tourism Development.* Travel Industry Development Planning Papers, No. 3. Ottawa: Canadian Government Office of Tourism, 1973.

*Five Year Plan for Economic and Social Development.* Hashemite Kingdom of Jordan. Amman: Ministry of Planning, 1986.

Gamper, Josef A. "Tourism in Austria—A Case Study of the Influence of Tourism on Ethnic Relations," *Annals of Tourism Research,* 7 (3), 1981, 432–446.

Guangrui, Zhang. "China Ready for New Prospects for Tourism Development," *Tourism Management,* 6 (2), June 1985, 141.

*Guide for the Investor in Israel's Tourism Industry.* Jerusalem: Ministry of Tourism, 1986.

Gunn, Clare A. *A Concept for the Design of a Tourism-Recreation Region.* Okemos, MI: BJ Press, 1965.

Gunn, Clare A. "U.S. Tourism Policy Development" (special issue), "Leisure Today." *Journal of Physical Education, Recreation & Dance,* 54 (4) April 1983, 32–35.

Gunn, Clare A., Uel Blank, and Ray Gummerson. "Vacationscape: A Case Study—Government, University and Landscape Architects," *ASLA-NICLA Interface '73.* Edward Fife, ed. Proceedings of the National Instructors in Landscape Architecture, Mackinac Island, MI, 1973.

Hall, Peter. "Enterprises of Great Pith and Moment," *Town & Country Planning,* 53 (1), November 1984, 296–297.

Healy, Robert G. *Land Use and the States.* Baltimore: Johns Hopkins Univ. Press, 1976.

*Integrated Planning.* Madrid: World Tourism Organization, 1978.

*Inventory of Tourist Development Plans.* Madrid: World Tourism Organization, 1978.

*Israel Tourism Projects Development Information for the Investor.* Jerusalem: Ministry of Tourism, 1982.

Jenkins, C.L. "The Effects of Scale in Tourism Projects in Developing Countries," *Annals of Tourism Research,* 9 (2), 1982, 229–249.

Jonish, James E. and Richard E. Peterson. "Impact of Tourism: Hawaii," *Cornell H.R.A. Quarterly,* August 1973.

Kobasic, Anton. "Lessons from Planning in Yugoslavia's Tourist Industry," *Tourism Management,* 2 (4), November 1981, 223–239.

*Korean Tourism Annual Report, 1984.* Seoul: Ministry of Transportation and Korca National Tourism Corporation, 1985(?).

Lawson, F. and Manuel Baud-Bovy. *Tourism and Recreation Development.* London: Architectural Press, 1977.

*Link,* December 1986, The Irish Tourist Board, Dublin, Ireland.

*Local Government and the Development of Tourism.* Circular 13/79, March 27, 1979. London: Department of the Environment.

"Marketing Strategy 1985" (from Marketing Plan—Finnish Tourist Board). In: *Perspectives: Leisure Travel and Tourism,* Edward M. Kelley, ed. Wellesley, MA: Institute of Certified Travel Agents, 1986, 206–212.

Meyers, Phyllis. *Zoning Hawaii.* Washington, DC: Conservation Foundation, 1976.

Mugbil, Imtiaz. "Beach Resorts in South East Asia," *Travel & Tourism Analyst,* November 1986, 37–48.

Ohi, James M. "Colorado's Winter Resource Management Plan: The State's Responsibility for Comprehensive Planning." In: *Man, Leisure and Wildlands: A Complex Interaction,* Eisenhower Consortium. Springfield, VA: National Technical Reformation Service, 1975.

*Planning and Development of the Tourist Industry in the ECE Region.* Proceedings, Symposium on the Planning and Development. New York: United Nations, 1976.

*Report and Accounts for the Year Ended 31st December, 1985.* Dublin: Irish Tourist Board, 1986.

*Report on Physical Planning and Area Development for Tourism in the Six WTO Regions.* Madrid: World Tourism Organization, 1980.

Richter, Linda K. "Political Implications of Chinese Tourism Policy," *Annals of Tourism Research,* 10, 1983, 395–413.

Richter, Linda K. and William L. Richter. "Policy Choices in South Asian Tourist Development," *Annals of Tourism Research,* 12, 1985, 201–217.

Robinson, Ira M. and Douglas R. Webster. "Regional Planning in Canada," *APA Journal,* 51, Winter 1985, 23–33.

Rodenburg, Eric E. "The Effects of Scale in Economic Development—Tourism in Bali," *Annals of Tourism Research,* 7 (2), 1980, 177–196.

Sallnow, John. "Yugoslavia: Tourism in a Socialist Federal State," *Tourism Management,* 6 (2), June 1985, 113–124.

Seekings, John. "Comment," *Tourism International Air-Letter,* mid-October 1975.

*Southwest Turkey Touristic Investment Areas.* Ankara: Department of Physical Planning, 1986(?).

*Strategic Plan, 1981 to 1985.* London: British Tourist Authority, 1981.

*Tourism Development in Ontario: A Framework for Opportunity.* Prepared by Balmer, Crapo & Assoc. for Ministry of Industry and Tourism. Toronto: Tourism Development Branch, 1976.

*Tourism Potential in Melbourne's Western Region, Report No. 1: Existing Tourism Situation in the Region,* prepared by John Henshall & Assoc. and Scenic Spectrums, Melbourne, Australia, Feb. 1986. *No. 2: Framework for a Regional Tourism.*

*Tourism Statistics.* Helsinki: Finnish Tourist Board, 1986. Other publications: "The Development Program of Tourism in 1982–1990," "Administration of Tourism in Finland," "Development of Tourism in Finland."

*Travel Industry Planning and Development Seminar.* Travel Industry Development Planning Papers, No. 4, Ottawa: Canadian Government Office of Tourism, 1973.

*Turkey Tourism—Opportunities for Investors.* Ankara: Ministry of Culture and Tourism, 1986. Supplementary publications, January 1986: "Turkey Tourism Encouragement Law," "Turkey Foreign Investment Law," "Turkey Regulation for State Owned Land Allocation for Tourism Investments," "Turkey Encouragement Conditions and Incentives."

Uysal, M., Lu Wei, and L.M. Reid. "Development of International Tourism in PR China," *Tourism Management,* 7 (2), June 1986, 113–119.

*World Tourism 1983-1984.* Madrid: World Tourism Organization, 1985.

# Chapter 3

# Planning Approaches

The review of tourism policies and plans in Chapter 2 makes it clear that the influence of politics on tourism is considerable. The private sector bias of service businesses tends to overshadow the reality of political influence over tourism, local to federal. Individual project feasibilities that fail to take into account political exigencies are very incomplete estimates indeed. Political influence does not remain solely with those areas developing tourism. "For whatever else may be said of tourism, it now involves a substantial number of voters in the generating countries as well as in the host countries" (Burkart: 1980, 4).

It is at the level of politics as well as organization, management, and promotion that tourism must be viewed in its overall context—as a functioning system. When viewed in this manner, the compelling need for integrated planning becomes crystal clear. Whereas much of tourism development is driven by markets, the supply side that includes transportation, attractions, services, and promotion/information cannot function effectively without each other to satisfy markets. In addition, the basic functioning system does not operate in isolation; it is influenced by many external factors.

The political realm and the functioning tourism system are the focus of this chapter.

## Tourism And Politics

Richter (1983, 313) laments over the general neglect of tourism by political scientists as well as the body politic. Most political activities are not optional; they are brought on by constant pressure from many sources for political action. Tourism does not generally fit this stereotype, especially in developing countries where governments seek tourism.

Richter identifies many political science challenges that have not yet received much attention. Needed are studies of how tourism infrastructure is interrelated to that of other sectors of the economy. Political concern over how much to centralize or diffuse tourism has not received much attention. Comparisons between policies of nations provide a fertile field for the political scientist. Political implications of some identified spinoffs of tourism, such as inflation, pollution, housing shortages, crimes, and political corruption, need study. Should scarce resources be devoted to tourism development as has happened in many developing countries? How nations should posture themselves regarding tourist flows is not necessarily founded in what is best for either sending or receiving countries. Soviet statistics equate the average "profit" from one tourist to the export of nine tons of coal (Richter: 1983, 325).

Yet, not only economic motives set policy. Witness the involvement in the Olympics and in world's fairs, which seldom pay for themselves in financial terms. "China gains something when Americans and others line up for hours in Knoxville to see their exhibit" (Richter: 1983, 325). For many years the United States showed its opposition to the political ideologies of China and Cuba by prohibiting travel to them. Now these bans have been lifted but placed on Libya instead. Further political complexities are exhibited by international organizations that foster tourism development, such as the World Tourism Organization, the Pacific Area Travel Association, and the Caribbean Tourism Association. And more recently, the political problem of terrorism becomes intimately connected with policies and activities of tourism and on both sending as well as receiving countries. "Unfortunately, political factors seem to supercede the mostly technical factors that stand in the way of travel and tourism" (Ronkainen: 1983, 425).

Politics and foreign policy have much to do with international tourism. Increasingly in recent years, countries are reaching beyond their boundaries to establish better tourism relations. The People's Republic of China, for example, after being closed to foreign visitors for many years, opened its door to travelers in response to the Shanghai Communique, signed by China and the U.S.A., in 1972. The Soviet Union and the U.S.A. on September 15, 1985, signed an Accord of Mutual Understanding and Cooperation in Sports to foster better interchange between athletic events. These two international powers further promoted greater familiarization between their peoples through tourism by means of a cultural agreement in 1985 on contacts and exchanges. Memorandums or more formal agreements currently are in effect between the U.S.A. and several countries—Mexico, Egypt, Philippines, Yugoslavia, Morocco, Canada, Israel, and the People's Republic of China (Edgell: 1987, 18).

A review of tourism politics generally suggests that the dimensions and results of tourism planning will be influenced most by the nation's ideology and its interpretation for overall social, political, and economic goals (DeKadt, 1979, 33). Policies and practices for tourism will follow the overall policies and practices of the nation as a whole. This is reflected in the relative roles of government and private enterprise, how profits are divided, the sectors most likely to benefit, domestic versus foreign travel influence, and relative dependence on tourism.

Politics as well as tradition have had much to do with tourism planning and development in Spain, for example (Morris: 1985, 77). Because the force of industrialization was felt much less in Spain than in the northern European countries , the need for planning was less visible when tourism began. As in all market economy countries, private property rights are sacrosanct. And Francoist policies of a strong centralized government did not prepare local governments for accepting planning responsibilities after his regime.

As a consequence, there has tended to develop a polarization of conservation and planner interests as opposed to hoteliers and tourism interests in Spain. Unfortunately, there is no voice for tourist constituents who probably would endorse neither extreme but rather a stand toward resource conservation as necessary to tourism growth. Plans by outside consultants are often attacked as irrelevant.

In recent years, tourism has become a major agenda item of the European Community, a coalition of western European nations concerned about unified economic development. Guidelines for tourism, published in 1985 (Tourism: 1985), replace the original draft of 1982. Several perspectives and achievements are outlined. In order to improve social tourism, rural tourism, and cultural tourism, the Community encourages members to expand policies and financial assistance for specific projects. Already, many transport problems between member nations have been alleviated and new measures are being initiated for rail, air, and sea transport , as well as better systems of border crossing control. Measures are being taken to improve security for travelers and to modify insurance programs. The Community is active in supporting education and training in tourism with emphasis on technology exchange.

One of the most unclear and controversial of the tourism political issues is that of public-private balance. Neither command nor market economy countries have completely solved this issue. Governmental intervention, especially through subsidies, is practiced in many countries. Even in the United States, labeled a capitalistic nation, many aspects of tourism are publicly supported— parks, recreation areas, roads, airports, infrastructure, promotion, and education. The capital improvements and operational budgets for many publicly owned attractions, such as national parks and national forests, are nonmarket priced and depend more on governmental appropriations than on fees. Hughes (1984, 15) identifies three areas that justify governmental intervention even in market-driven tourism:

Social costs and benefits (externalities).
Public goods (special cases of externalities).
Merit wants.

Within the personal recreational side of tourism, there may be enough mental and physical benefit to society as a whole for government to subsidize park and recreation areas. Restored historic sites, festivals, and arts programs related to tourism may be considered of sufficient value to society to need public support beyond that available from a market price system. A community may believe that the added social value of tourism is worth the incremental investment in expanded streets, sewers, water supply, and other publicly supported services.

For tourism, governments may feel justified to assume some public good costs, such as for pollution or erosion control related to greater volumes of visitors. Additional litter control and policing are not exclusively for tourists but also for local residents and total populations who seek a quality landscape for the entire region or nation.

There is a point of view, especially in the countries of eastern Europe, that the social and personal values obtained from travel should not be limited only to those able to pay; there are "merit wants." Extra subsidy to low income and handicapped populations has not received great favor in the West. "The free enterprise philosophy would suggest that consumers should be sovereign in the choice of goods and services and 'distortion' of choice by way of low prices is to be avoided" (Hughes: 1984, 17). There is danger that excessive governmental intervention, because there is no relationship to market demand, can promote activities and investments that nobody wants. Instances of this can be found throughout the world.

In recent years the term "social tourism" has become a part of tourism lexicon. In part it is reactionary to the economic, business, and commercial emphasis so frequently placed on tourism. But, in a larger part, social tourism has come to describe the individual achievement through travel, "not as one more piece of merchandise to acquire" (Haulot: 1980, 2). Social tourism as a movement originated in Europe and is promulgated primarily by the International Bureau of Social Tourism. Much of the focus has been on stimulation of travel by lower income populations, increased and better scheduled nonwork time, and greater involvement by nonprofit organizations. Whereas these objectives would appear more tightly aligned with socialist than market economy countries, it was a surprise to learn of the long-time practice of "social tourism" in North America. Canada and the U.S.A. for many years have provided low-cost public use of many travel and recreational opportunities, such as children's camps, low-cost accommodations in colleges and universities, employee travel associations, service club park and recreation facilities, and a host of governmentally supported recreation and park opportunities used by all classes and income groups at a fraction of the cost of operation (Ruest: 1980, 53; Reid: 1980, 57). For example, it is not well known that in the U.S.A., nominally a capitalist country, there are more public lands available to travelers than in any other nation of the world. Federal agencies alone (exclusive of state, county, and city lands for recreation) administer more than 755,683,000 acres of lands available for public use at very little cost.

Political accountability for tourism, as with other governmental programs, is less efficient than that of enterprise. Profits provide the commercial sector with rather clear measures of efficiency, productivity, and accomplishment of objectives. In politics, this is more ambiguous. First, governmental objectives and goals are often unclear and imprecise. Second, tests for accomplishment rely on the ultimate response by the public, which can become bogged down in polemic rhetoric that makes measurement impossible (Brown: 1982, 4). This realism of politics makes it even more imperative that all governmental agencies and their political actors become more fully informed on tourism issues. There is great need for sharper policies, not only for tourism but for the many political implications of tourism throughout all publics.

## Planning And Political Ideologies

How planning is accomplished varies with political ideologies of nations. In a market economy, planning is diversified and usually accomplished at the lowest level possible; in a command economy, it is centralized.

In the U.S.A, for example, planning is highly decentralized and done primarily at the local level. Municipalities exercise the greatest amount of planning through the development of plans and support through legal tools of zoning, subdivision regulations, and building codes. Planning deals mostly with physical space rather than ideologies as in other countries (Vasu: 1979, 11).

Localities derive their planning powers from the states. State governments authorize localities to engage in certain planning and related activities and directly operate substantive programs in planning and urban development. It is at this level that land use is regulated.

Planning at the federal level in the U.S.A. is primarily through financial grants and incentives, such as for highways, sewers, water facilities, and mass transit. Planning is also accomplished on federal lands and waters through the several federal agencies, such as the National Park Service, Fish and Wildlife Service, Bureau of Land Management, Army Corps of Engineers, and U.S. Forest Service.

Planning in the Soviet Union is not only centralized but also directed toward socialist ideologies. In theory, this means that the planning of cities and rural areas is designed to be classless and controlled for optimum density. However, great growth of U.S.S.R. cities has occurred in recent decades, just as everywhere else. And, although there is much uniformity in housing in some cities, consider-able differentiation exists between high officials, economic managers, top Com-munist party officials, and those of lower status (Brunn and Williams: 1983, 148, 159). Soviet planning that reflects the socialist ideology is more evident in cities such as Moscow than in those whose past history was strongly influenced by western Europe, such as Prague, Warsaw, and Budapest.

Great Britain and many other countries have established planning policies and practices between the extremes of the United States and the Soviet Union. For example, both social and physical objectives stimulated Great Britain to experiment with concepts of "new towns" (Brunn and Williams: 1983, 8). This concept began in the early 1900s and refers primarily to a comprehensively planned new community to incorporate all the ideal facets of city economic and social life. In both Great Britain and Canada, the market economy and private land philosophy is dominant, but much social and economic planning takes place by government. Even though the ideal community of the new town concept did not meet expectations because of high costs of construction, failure to cope with the complexity of social structure, and reluctance of people to be told where to work and live, there remain many land use controls throughout the region.

Decentralized planning has been the trend in recent years. Denmark's "plan-ning jungle" (Ostergard: 1981) in the 1960s prompted planning reform of 1970. This reform identified three planning thrusts—physical, economic, and sector. Regional planning responds to guidelines from the municipalities. The country has been divided into rural and urban zones with purposes and uses defined. The new emphasis on local planning and planning that is more comprehensive to include social and economic as well as physical planning is commendable. But, Ostergard expresses concern that politicians who make policy decisions may not be able to solve the needed diversity of problems all at once. "If the living conditions shall be improved, this comprehensive aspect of future social plan-ning must be safeguarded, as otherwise experts will fight experts and by means of sector arguments try to promote sector aspects."

Planning, as a concept, is accepted in virtually all political ideologies. But how it is carried out is influenced by the tradition, organization, and political bias of a country. For example, an important element of Zionist ideology was anticity, favoring the creation of a prosperous peasantry (Gradus: 1980, 411). A study comparing Israeli and American planners (Kaufman: 1985, 361) revealed that both held similar attitudes toward citizen participation and somewhat similar beliefs toward the issues of environment, mass transit, and low-income people. However, American planners were less inclined to let personal values influence their work and more inclined to use political solutions. "Because fewer of them [Americans] than of the Israelis have their ethical compasses whirling around, in the sense that more of them see ethical grays, professional life may be less problematical" (Kaufman: 1985, 362).

A superficial view of Canada and the United States would reveal few economic, social, or political differences—same language, industrial development, and apparent urban structure. But beneath this surface there are some fundamental differences pertinent to planning. In Canada, entrepreneurship is less aggressive and planning is much more bureaucratic. Public involvement and processes of planning are not the issues among planners in Canada as in the U.S.A. (Hodge: 1985). A key difference is the concept of land ownership. The U.S. Constitution spells out personal property rights and the limits of state intervention; Canada's does not. Private citizens are not land owners in the same sense. Since the 1950s, the permitting system of zoning invokes controls at the time of a building proposal and dominates the workload of planners. The official planner is removed from goal-setting, plan-making, and public input. He is more dependent on lawyers and ministry planners.

From this brief discussion, it may be concluded that no matter the political ideology, planning takes place. However, the separate political aspirations and organizational mechanisms vary greatly. Any effort toward programs of tourism planning must be adapted to and compatible with the political structure and ideologies of the nation. Regardless of ideology, all the political, managerial, and promotional aspects of tourism must center on the way in which tourism functions.

## Foundations For Planning

Reviews of tourism planning and development, even in a brief sampling of nations with varied political ideologies and traditions, show some common dimensions. It is these commonalities that can provide the foundations for planning. It would appear that no matter the nation, the greatest differences lie in *how* tourism is managed, not *what* constitutes tourism development. If it is true that there are some common elements everywhere, it would seem that these make up the foundation upon which planning for the future must be built.

Perhaps no segment of the human ecosystem is any more sensitive to maladjustment and misuse than the segment utilized by tourism. The human ecosystem, since the Industrial Revolution, has expanded greatly. Now, instead of resources determining man's activity, as was true in early man and continues to be true in other animal life, human decision-making determines resources. Sargent (1974, 16) indicates that this is true throughout man's activities, including travel. Projections suggest that the demand for this will continue to mount, increasing the pressure on man's ecosystem. The answer is neither blind conservation in the sense of blanket resource protection nor continued blind resource

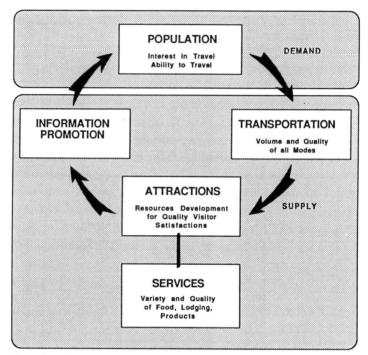

**Figure 3-1.** The Functioning Tourism System. A model of the key functional components that make up the dynamic and interrelated tourism system. Important is the relationship between demand and supply.

exploitation for tourism, but integrative, collaborative, and cooperative planning action by those who make decisions on tourism land use.

How does tourism really function? Is it merely a matter of setting up promotion? What really drives tourism? What keeps it running smoothly? Questions such as these must be addressed by planners of tourism. Just as an automobile consists of thousands of parts that must operate in harmony, so it is with tourism. The misfiring of the ignition can disrupt overall automobile function—jerky driving performance, inefficient gas consumption, and probably excessive carbon deposits. Any one small part of tourism, such as a change in market preference or international currency exchange, can upset the rest of the system. Within tourism there are several important components whose functions greatly affect one another.

But, worldwide, preoccupation with promotion has overshadowed the other functional components. Promotion is essential, but even more important are several other components that must be in reasonable balance if the system is to run smoothly. By running smoothly is meant proper satisfactions of travelers, rewards to business and other sectors, protection of basic environmental assets, both social and physical, and enhancement of local quality of life.

One way of illustrating how tourism works is shown in Figure 3-1; where the major components of the functioning tourism system are diagrammed.

## The Functioning Tourism System

A beginning point of tourism planning is that of the functional tourism system. As discussed earlier concerning the problems of tourism development today, the fragmented approach to tourism tends to overemphasize the separateness of the structural elements, such as hotels, airlines, and advertising. Important as these elements are, they tend to confuse rather than help in identifying components important to tourism planning. Without attractiveness to lure travelers, the hotels, airlines, and advertising would not be needed. From a regional planning point of view, it is difficult to organize planning on the basis of the structural elements alone.

Generally the functioning tourism system, the heart of all tourism development and operation, consists primarily of a *demand* (market) side and a *supply* side (Figure 3-1). The four components of the supply or "plant" side could be described as: transportation, attractions, services, and information/promotion (Gunn: 1972, 21). Whereas other researchers and writers arrange the titles and groupings in slightly different ways, the functions included are similar. Jafari (1982, 2) refers to these as the "market basket of goods and services, including accommodations, food service, transportation, travel agencies, recreation and entertainment, and other travel trade services." To these he adds resident-oriented products and market differentiation. Murphy (1985, 10) also has a "demand" and "supply" side with similar components. Mill and Morrison (1985, 2) have described the components in a similar manner, combining attractions and services into a "destination" component.

For tourism, the demand and supply sides are in a slightly different relationship than for manufactured goods that are distributed to markets. Instead, markets for tourism are distributed to the "goods" or destinations that make up the product. A great amount of planning attention to the product (the supply side) must be given because it is fixed geographically.

All components of the supply side are essential to a properly functioning tourism system. All must function in a delicate but tightly integrated balance. If tastes change, if transportation cost or mode changes, if new attractions are built, if new generations of service are developed, or as new information and promotion are created, the balance is upset and adjustments in all other components must be made to compensate. This dynamism is not yet fully understood or widely applied, and for several reasons.

Many hoteliers still believe that their product is rooms, whereas their service is greatly dependent on promotion/information, access, and on attractions within reach. Highway departments tend to believe that their product is graded concrete ribbons, whereas the true product is personal traveler satisfaction with safe, convenient, interesting, and dependable access from home to destinations and return. *Misunderstanding of the tourism product* is often a constraint on a smoothly functioning tourism system.

*Jurisdictional boundaries* often obscure the relationships between components. Many public agencies, created for other than tourism purposes, are often heavily involved in tourism nonetheless. Federal, provincial, state, and city departments of highways, parks, wildlife, forests, waters, police, health, taxation, airways, and planning—all impinge upon tourism. Yet, these generally have no policies or staff functions that overtly include tourism. They become entrenched in narrowly prescribed bureaucracies that make integration for tourism functioning very difficult.

For tourism the *private organizations* tend to be fragmented, causing further disintegration of the tourism system. Hotel, restaurant, attraction, airline, bus, rental car, and many other organizations have become increasingly proliferated. This scatter tends to increase the difficulty of integrating their functions into a whole for tourism.

Another complication is the *resident-traveler mix* of markets for most businesses serving tourism. Not one of the key businesses that provide the greatest economic impact for tourism is exclusively directed to traveler markets. Even hotels often provide many functions for residents—banquets, meetings, dining, shopping. This duality of markets means that location, design, construction, and management must be directed to both market segments—local and traveler. These businesses cannot set policy exclusively for travelers.

In many regions, *ideological stress* sometimes reduces the efficient functioning of the tourism system. Polarized and antagonistic approaches to development versus conservation can slow and even deny tourism development. Even when planning, design, and management practices can ameliorate conflict, they are not given a chance because of heated ideological arguments.

Although better planning cannot solve all these issues that constrain better tourism functioning, it has much better predictability than merely muddling through. Chapters 4 through 8 are devoted to discussion of the several components and conclusions that are important for planning. Briefly, the key components are described as follows.

## Markets

To say that the demand side of tourism is a major one is certainly an understatement. Without volumes of people-markets who have both the desire and the ability to travel, tourism cannot develop and thrive. Important planning consid-erations center around where they live, what they prefer, what they regularly participate in, and their expenditures. Planners must be aware of how markets relate to the following four components of the supply side of tourism.

## Attractions

A wide variety of physical settings and establishments provide a pull for travelers to visit destination regions. Whereas almost anything at one time or another may become an attraction, the functional tourism system requires identification, planning, and management of physical developments and pro-grams that provide visitor satisfactions. Attractions, no matter who owns and provides them, not only attract visitors but provide for their participation. Attractions, as considered here, may be owned publicly (as Yosemite Falls, California), commercially (Astroworld, Texas), or by nonprofit organizations (Greenfield Village, Michigan).

## Services Facilities

The most important functional component from an economic input sense is that of the facilities and services, such as hotels, motels, restaurants, bars, retail sales, and other services. Tourist spending on facilities and services provides the major economic input in most destination regions.

## Transportation

Linkage between place of residence and place of destination is a very important component. Although automobile and air travel tend to dominate most tourist travel, other modes of transportation, such as boat, ferry, train, horseback, aerial tram, and hiking, frequently are critical links of the transportation system.

## Information Promotion

People accumulate information and develop values relating to travel experi-ences that influence their decisions to travel and where to travel. All those functions that relate to learning about travel objectives are important. Whereas advertisers and promoters provide heavy input, people are informed and guided by a great variety of sources, not the least of which is their own past experience and recommendations from friends and relations.

## Interdependence

Certainly, this model of a tourism system is a very generalized approach, but it is useful as a tool for planning. In addition to the functions within each component, there are strong interdependencies between components. The simple model in Figure 3-1 has many cross-linkages as well as the dominant flow as shown. The cross-interpendencies are very strong, forming an extremely sensitive and dynamic whole.

Often, physical development includes multiple functions. For example, a restaurant can serve both as an attraction (if it becomes the target of travel purpose) and as a service. But not all restaurants serve both functions. A bus is a means of transportation, but a bus tour is an attraction by virtue of its interpretive program about the scenic and historic sites along the way. Even so, this does not argue against understanding the separate function of each component.

The strong tie between people and attractions is very important in tourism planning. This linkage has made it extremely difficult for scholars to separate "demand" from "supply." It is the qualities of the destination area—the way people value them—that open them to travel. Therefore, it is extremely difficult to make a precise measurement of demand because it can change rapidly with changes in supply. For example, before Disney World was established in Orlando, the demand for northern Florida was no greater than that for the interiors of other southern states.

The provision of services has much to do with attractiveness of a region. Until hotels and airline service were provided to the outer islands of Hawaii, their tourism popularity was relatively low as compared to Waikiki or Oahu. A similar importance of service was demonstrated in the development of other Pacific Islands such as Fiji and Samoa. Of course, the great importance of the indigenous scenic and cultural attractiveness of these islands should not be overlooked.

The interlacings of tourists, attractions, services, facilities, and transportation are expressively illustrated by the following excerpt by travel writer Sutton (1977, 19):

> And, so we were, in short order, and more easily than you think, in a safari bus with a Kikuku driver, rolling into the private estates where animals live.
>
> Before lunch, the first day, we were trying to stare down the elephants who were staring into our wagon trying to decide, perhaps, whether we were animal or vegetable. . . . We lodged one night in some style at Salt Lick Lodge, and had drinks on the terrace while we watched the elephants bathe and then powder themselves with dust. There were honey badgers with the fruit cup and hyenas with the soup . . . . Ah, Africa, Africa.

The component of "information/promotion" is intimately linked with all the others. Advertising and promotion must always be about something—transportation, services, facilities, and attractions. And the prime reason for advertising is to influence the component "people."

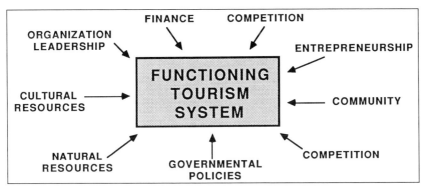

**Figure 3-2.** Influences on the Functioning Tourism System. A diagram illustrating the several factors external to the functioning tourism system that influence its development and success.

Many other interrelationships could be identified supporting the principle that the tourism system is extremely dynamic. Change in any one component can dramatically influence the others. This principle of dynamic, not static, interaction is fundamental to planning—which of necessity also must be dynamic.

This concept of the tourism system defines the scope of tourism planning and the work of all actors involved. Tourism planning must include but cannot stop with consideration of the location and development of hotels, motels, and restaurants. Tourism planning must include but cannot stop with the development of economically positive inputs to areas and regions. Tourism planning must also include the development of transportation systems for travelers, the needs and interests of tourists, the information/promotion system for travelers, the location and many other factors important to the development of attractions, and, finally, how all are integrated into a functioning whole.

## External Factors

Such a core of functioning components are greatly influenced by many external factors (Figure 3-2). Planning cannot be concerned solely with the core only because all sectors may be subject to outside influences as those inside their own control. Several factors can have great influence on how tourism is developed. A brief examination of these may help in understanding this complicated reality of tourism, critical to planning.

### Natural Resources

Because so much of the travel objective for many markets is related to natural resource assets, the quantity and quality of these factors are very important to tourism development. Key natural resource assets for tourism include: climate and atmosphere, water and waterlife, vegetative cover, wildlife, topographic conditions, and surface geology. Travel destinations having poor quality or being void of these assets will not be able to support development to meet certain market needs. Some of these factors are subject to manipulation. Air conditioning, heating, wildlife management, forest management, improvement of streams, landscape plantings, and altering land forms can modify some natural resource characteristics, but most are related to specific pieces of geography.

## Cultural Resources

Many travel markets today seek destinations with interesting man-made factors. Increased interest in heritage and roots has stimulated the fields of history, anthropology, geography, and archeology. Places that have especially important cultural characteristics are being favored for tourism development over bland and lackluster areas. Historic sites, historic buildings, archeological digs and artifacts, pilgrimage shrines, locales of lore, unusual technology, ethnic concentrations, engineering feats, crafts, industrial plants and processes, and theme parks make up the pulling power of many destinations. Some can be artificially created almost anywhere (museums, theme parks), whereas others (historic sites) are anchored to place.

## Entrepreneurship

Because tourism is dynamic, needed are entrepreneurs who visualize opportunities for new developments and creative ways of managing existing developments. The ability to see an opportunity, to obtain the proper location and sites, to engage designers to create physical settings, to gather the human resources needed for operation, and to manage the physical plant and services is important for travel development. For industrialized nations, entrepreneurship is a part of the culture. It is known that this factor alone is a major one that increases the difficulty of tourism development in many undeveloped countries.

## Finance

Certainly, for the development of tourism, capital is required. But the ease of obtaining the financial backing for tourism varies greatly. Public and private lenders are often skeptical and have a negative image of the financial stability of tourism. Because so much of the tourism physical plant is small business and has attracted inexperienced developers, some of this reputation is justified. However, recent trends have demanded much greater business sophistication and higher capital investment. Tourism does take considerably more capital than is popularly believed. Financial backing is an important factor for both public and private tourism development.

## Labor

The availability of adequately trained workers in an area can have considerable influence on tourism development. As markets demand higher levels of service, well-trained and competent people are in greater need. The popular view that the untrained can perform all tasks needed in the diversity of tourism development is false. When the economic base of any area shifts, those taken out of employment may be retrainable, but they are not truly available for tourism jobs unless such training is provided. The labor capacity of an area has much to do with tourism development.

## Competition

The freedom to compete is a postulate of the free enterprise system. If a business can develop and offer a better product, it should be allowed to do so in order to satisfy market demand. However, before an area begins tourism expansion, it must research the competition—what other areas can provide the same opportunities with less cost and with greater ease? Is there evidence that the tourism plant has already saturated a market segment? Certainly, competition is an important factor.

## Community

A much more important factor influencing tourism development than has been considered in the past is the attitude toward tourism by the several community sectors (see Chapter 11). Whereas the business sector may favor greater growth of tourism, other groups of the local citizenery may oppose it on the grounds of increased social, physical, and economic competition for resources and other negative impacts. Political, environmental, religious, cultural, ethnic, and other groups in an area can make or break the development of tourism.

## Governmental Policies

From federal to local governing levels, statutory requirements may foster or hinder tourism development. How the laws and regulations are administered— loosely or rigidly—can influence the degree of tourism development. Policies on development of infrastructure by public agencies may favor one area over another. The policies of the many departments and bureaus can have a great bearing on how human, physical, and cultural resources are utilized.

## Organization, Leadership

Only recently being recognized is the great need for leadership and organization for tourism development. All planning is subject to implementation by many sectors. Many areas have hired consultants to identify tourism opportunities, but frequently such plans for development have not materialized for lack of organization and leadership. Both private and public organization and leadership are essential if much tourism development is anticipated.

Without doubt, as tourism development research and experience broaden, more influential factors will be found. Any planning for tourism in the future must take into account the core of the tourism functional system and the many factors influencing it.

# Conclusions

This chapter centers on three major topics of importance as foundations for planning. First, all regions and nations must develop planning of tourism within their political and ideological frameworks. Second, no matter the political ideology, tourism functions are common throughout the world. Several components are functionally interdependent. Third, this basic functional system is influenced greatly by several external factors. It is from these three principles that the following conclusions have been derived.

*Political implications of tourism have received little study.* Whereas a few scholars have begun to show concern over the politics of tourism, generally little is known. Most nations have accepted promotion as a federal or regional role in tourism, but little about other tourism-oriented political decisions has been documented. It appears that the field of political science of tourism is wide open to investigation. Certainly the planning for future tourism will depend greatly upon more and better information on how federal-to-local political entities see their roles and make decisions.

*Some governmental roles appear to be universal.* In spite of great political diversity, some tourism roles have been commonly accepted by government. From market- to command-economy countries, the provision of basic infrastructure has been accepted by government. Throughout the world, governments have dominated expenditures on promotion of tourism. Governments, rather than the private sector, dominate the provision of highways, airports, harbors, and parks and the protection of rare cultural and natural resources. In many countries, development of hotels, food services, information centers, bus lines, travel agencies, and airlines has been an accepted governmental responsibility. The planning of tourism is increasingly seen as a cooperative governmental-private role. It is very clear that government, even in market economy countries, is much involved with tourism.

*The functional tourism system is in a dynamic but delicate balance.* The past focus on sites—individual businesses and public establishments—has prevented an overview of tourism as a system. When viewed as a system, it can be seen readily that every tourism site is in some way related to and dependent on every other site and program. This fact shows that no site can be planned independently, that only planning on a larger scale can reflect the true dynamics of tourism as a system. This statement does not imply command planning by a super agency but rather that each individual enterprise must consider the larger context in its planning and decision-making. Each enterprise is delicately balanced with all others at any one moment.

*Demand and supply are major forces in the tourism system.* Planning for tourism development must be driven by two forces at the same time. Demand, the tourism markets, determine what the traveling public wants, needs, and is willing to pay for. Markets can be segmented but change over time. The supply side must be developed, not only in response to markets but also in light of many geographic and management factors related to the destination region. These factors of the supply side can be placed in four categories—attractions, transportation, services, and information/promotion. Planning for tourism must take all demand and supply factors into account.

*Constraints limit proper functioning of the tourism system.* Whereas a smoothly and efficiently running tourism system is ideal, several constraints tend to hamper its functioning at its best. A narrow and nonintegrated view is fostered by misunderstanding the tourism product, by protecting turf, by lacking understanding of dual markets (local and visitor), and by conflicting ideological views. Planning must be directed toward the elimination of these barriers if tourism is to function.

*The tourism system is influenced externally.* The market side and the four key components of the supply side that make up the functioning tourism system are greatly influenced by a number of external factors. The extent and quality of natural and cultural resources; the availability of entrepreneurs, finance, and labor; the extent of competition; the social economic standing and attitude of the communities; and the policies of governments and organizations can influence greatly the development of supply and satisfaction of market demand. Planning is hollow if it does not consider externalities as well as the internal functioning of the tourism system.

# Bibliography

Brown, Claire. "Resource Planning for Recreation and Tourism," *Essays in National Resource Planning,* Centre for Urban and Regional Studies. Birmingham, England: University of Birmingham, 1982.

Brunn, Stanley D. and Jack F. Williams. *Cities of the World—World Regional Urban Development.* New York: Harper & Row, 1983.

Burkart, A.J. "The Impact of Politics," *Tourism Management,* 1 (1), March 1980, 3–4.

DeKadt, Emanuel. *Tourism, Passport to Development?* New York: Oxford Univ. Press, 1979.

Edgell, David L. *International Tourism Prospects 1987-2000.* Washington, D.C.: U.S. Travel and Tourism Administration, 1987.

Gradus, Yehuda and Eliahu Stern. "Changing Strategies of Development: Toward a Regiopolis in the Negev Desert," *APA Journal,* October 1980. 410–423.

Gunn, Clare A. *Vacationscape: Designing Tourist Regions.* Austin: Bureau of Business Research, Univ. of Texas, 1972.

Haulot, Arthur, "Social Tourism as a Universal Phenomenon: Sociological and Cultural Characteristics" In: *Condensed Report of the 1980 Convention of the International Bureau of Social Tourism.* Montreal, Canada: IBST Commission for North America, 1980,1–8.

Hodge, Gerald. "The Roots of Canadian Planning." *APA Journal,* Winter 1985, 8–22.

Hughes, Howard. "Government Support for Tourism In the UK," *Tourism Management,* 5 (1), March 1984, 13–19.

Jafari, Jafar. "The Tourism Market Basket of Goods and Services" In: *Studies in Tourism, Wildlife, Parks, Conservation,* Tej Vir Singh et al, eds. New Delhi: Metropolitan, 1982.

Kaufman, Jerome L. "American and Israeli Planners," *APA Journal,* Summer 1985, 352–364.

Mill, Robert Christie and Alastair Morrison. *The Tourism System.* Englewood Cliffs, N.J.: Prentice Hall, 1985.

Morris, Arthur. "Tourism and Town Planning in Catalonia," *Planning Outlook,* 28 (2), 1985, 77–82.

Murphy, Peter. *Tourism: A Community Approach.* New York: Methuen, 1985.

Ostergard, Niels. "The Overall Danish Planning System and Planning for the Sectors of Land Use and Natural Resources," International Conference, "The University and Rural Development—The Road Between Theory and Practice," Backaskog, Sweden, June 30, 1981.

Reid, Leslie M. "Social Tourism in the United States" In: *Condensed Report of the 1980 Convention of the International Bureau of Social Tourism.* Montreal, Canada: IBST Commission for North America, 1980, 57–60.

Richter, Linda K. "Tourism Politics and Political Science—A Case of Not So Benign Neglect," *Annals of Tourism Research,* 10 (3), 1983, 313–335.

Ronkainen, Ilkka. "The Conference on Security and Cooperation in Europe—Its Impact on Tourism," *Annals of Tourism Research,* 10 (3), 1983, 415–425.

Ruest, Gilles. "Social Tourism in Canada" In: *Condensed Report of the 1980 Convention of the International Bureau of Social Tourism.* Montreal, Canada: IBST Commission for North America. 1980, 53–56.

Sargent, Frederick, II, ed. *Human Ecology.* New York: American Elsevier, 1974.

Sutton, Horace. "Vacation Quandary," *Houston Post* (Spotlight Section), June 26, 1977.

*Tourism and the European Community.* European File 11/85. Commission of the European Communities. Brussels: Directorate-General for Information, 1985.

Vasu, Michael Lee. *Politics and Planning—A National Study of American Planners.* Chapel Hill: Univ. of North Carolina Press, 1979.

# Chapter 4

# Markets

A major building block for tourism planning and development is the demand side. Until one understands the "push" side of tourism, coming from people who have the *interest* and *ability* to travel, planning is impossible. It is *for* tourists (here encompassing all kinds of travelers) that planning and development take place. Unless the interests and objectives of travelers are met, the many desired derivatives from tourism, such as economic, will be vacuous indeed.

Some readers may object to the term "markets" for those people with interests in and ability to travel. This objection would be on the basis that much of travel, such as for mental refreshment through recreation, does not fit a market price system; it is a social good. However, it is difficult to conceive of a costless travel experience even when prices of certain services, such as parks, may be very low.

This chapter provides brief insight into the personal characteristics of travelers. It introduces readers to the characteristics of markets that require deeper study for planning. Some historical background is offered to set the stage. How tourists are defined and some of the popular market research approaches are described briefly. This introduction is followed by discussion of people's decision-making process, characteristics of travelers, and some useful classifications. The chapter ends with some important planning implications and conclusions regarding people as travelers.

## Historical Background

As has been pointed out by many authors (such as Bridges: 1959; Murphy: 1985), the histories of travel and civilization are parallel. Early man kept on the move from one pasture to another. Trade and pleasure travel have been intertwined for centuries. Early travelers were explorer-traders, such as Marco Polo, who sought new functions and societies out of curiosity and yet had increased trade motivations. Pilgrimages to religious meccas, resorting around the Mediterranean, and participation in the spas were popular in Greek and Roman times.

Many early travelers were doctors seeking new experiences while researching for plants and herbs with medicinal properties. For many periods of history, travelers from several socioeconomic classes sought and experienced travel in similar ways—visited historic sites, preferred the better modes of travel, encountered both good and bad inns and restaurants, listened to guides with rote speeches, left graffiti, and shopped for souvenirs.

The lowest point in tourist travel came with the early Middle Ages because of feudalism and the stringent necessity of work for survival. Religious values were coupled with prohibitions against recreational travel.

Before modern science and technology identified many causes of disease, the curative powers of sea water and spring waters containing many minerals became popular health objectives. These attractions sparked fashionable resorts the world over.

In the seventeenth century, in the heyday of British history, no young man was thoroughly educated unless he had experienced the arts, architecture, women and foods on the Grand Tour. The trip was met with many risks from bandits and bad food, but it usually included visits to the seats of government of France, Italy, Switzerland, and Germany.

The priority given travel throughout history is evident no matter the mode of transportation. First, hiking and horseback riding, followed by stagecoach travel, proved that the desire was there. Major technological advances, such as the train and steamboat, gave new impetus to travel, especially by the wealthier classes. Travel exploded with the automobile, not because of roads but because it opened up new markets. For the first time, a personalized form of transportation became available to the middle and lower classes.

> Working men and working women—factory operatives, plumbers, waitresses, bank clerks, farm hands, stenographers, storekeepers, subway guards, mill hands, garment workers, office boys, truck drivers—found countless pleasures and amusement readily available that had once been restricted to the privileged few (Dulles: 1965, 397).

Whereas the recreational tourist trip represented the first surge of automobile use, sales and other business people began to popularize the flexibility of automobile business travel.

But it took jet travel to shrink the world and expose a new abundance and mosaic of travel markets. Records have demonstrated that regardless of economic shifts over many decades, the desire to seek out travel experiences has continued to grow worldwide. In spite of great strides in the communications, Boorstin's (1961) prediction that travel would become atavistic has not been borne out in fact. Travel, for many reasons, now has become almost a right of life, evidenced by the resistance to restrictions arbitrarily imposed upon it.

# Tourists Defined

Historically, the term "tourist" meant a pleasure traveler influenced by leisure time and the desire for adventure and vacation experience. Residents of destinations began to perceive of tourists also in a pejorative sense—boors, insensitive to local mores, and noninhibited.

In recent years, the search for a more precise measurement of tourists and their activities has stimulated more elaborate definitions. Statisticians and market analysts sought better descriptors for more accurate accounting. The World Tourism Organization in 1978 developed a working definition, further updated in 1983 to include:

1. *Domestic visitors*—residents who travel within a country for not more than one year; purpose not remuneration.
   a. *Domestic tourists:* at least a 24-hour stay and not more than one year for pleasure, recreation, holiday, sport, business, visiting friends and relatives, mission, meeting, conference, health, studies, religion.
   b. *Domestic excursionists:* staying less than 24 hours.
2. *International visitors*—residents of one country visiting another (Chadwick: 1987, 48).

Other definitions state a minimum travel distance from home as 25, 50, or perhaps 100 miles. Leiper (1979, 396) and others prefer to include one night's stay and a "discretionary" temporary tour in their definitions and separate travelers from tourists. Canada and the United States have an agreement to include resident travelers, nonresident travelers, and other travelers in their statistics (Chadwick: 1987, 49). Business and pleasure travelers are included.

More and more there is a move to include *all* travelers in statistics. This was the decision when the United States Travel Data Center was organized. Throughout this text, tourists are defined as all travelers (except commuters) in order to provide a comprehensive scope of planning concerns and solutions.

# Research Approaches

Today many types of research approaches are taken in order to provide greater insight into the phenomenon of the tourist. One of the most comprehensive sources of the many approaches is *Travel, Tourism and Hospitality Research* (1987). This handbook contains current information on techniques and methodologies of research on several aspects of tourism as well as markets and marketing.

Chadwick (1987, 52) classifies tourist analysis studies into three groups. *Household surveys* are made at places of travel origin and often cover non-travelers as well as travelers. A statistically random sampling process can reveal information about the entire population within reasonable limits of accuracy. Data on frequency of travel volume, such as person trips, party trips, and vacation trips are popularly obtained. Travel expenditures, on or before trips, are important economic data obtained by household surveys.

*Location surveys* are made at sites on trips, such as inflight surveys, exit surveys, entry surveys, and highway counts. These surveys cover one visit and may relate to the entire trip or only to the site experience. Data may be obtained on expenditures, activity participation, opinions and attitudes, as well as social economic status of travelers.

*Business surveys* approach travel from the other side—the supply side. Surveys of travelers in hotels and at theme parks can reveal many important facts about such visitors. Sources of travelers, extent of visits, size of parties, place of residence, social economic characteristics, and modes of travel are often measured.

One of the most popular forms of traveler research has been measures of economics (Frechtling: 1987, 325). Nations, states, and communities often wish to distinguish between expenditures of foreign and domestic travelers. This is based on the concept of tourism as an export, creating economic impact only from "new" dollars coming from outside. *Direct observation* of expenditures is often used as a method but is cumbersome and costly. Secondary effects are difficult to measure in this way. Estimation by *simulation,* a model of key relationships, is set up in equations and data are collected for basis impact. A complex equation has been established by the U.S. Travel Data Center for measuring the economic importance of tourism in all states of the country. Frechtling identifies the following criteria for evaluating economic studies: relevance, coverage, efficiency, accuracy, and applicability.

As yet, economists have not agreed upon a standardized methodology for tourism research. Therefore, a reader of reports must be alert to definitions and scope, especially when comparing study results.

*Forecasting* of travel demand is desired by the planner, but is most difficult to accomplish. Forecasting is defined as the art of predicting the occurrence of events before they actually take place (Archer: 1980, 5). As the uncertainties of travel increase—taste, terrorism, international currency exchange, diversity of destinations—projections become less reliable. Because planners, developers, and promoters are in constant need of forecasting, the concept continues to occupy an important place in market evaluation. Although scientific research methods are used increasingly, forecasting, as defined, remains an art based on experience and judgment.

Uysal and Crompton (1985, 7) have provided helpful descriptions of qualitative and quantitive approaches to tourism forecasting of demand. Under qualitative approaches, three methods used by experts are described. *Traditional approaches* include review of survey reports to observe consistent trends and changes. Sometimes surveys within originating market sources are made to obtain the past history of travel as well as opinions of future trends. The *Delphi method* is an iterative type of research inquiry using opinion of knowledgeable experts. It consists of several iterations by a panel, which responds to specific questions about trends. Each panel member is anonymous to one another. Of course, this method relies heavily on the extent of expertise of the panel members and the influence of the director. But it is a useful tool, especially when used alongside other measures of prediction. A *judgment-aided model* (JAM) uses a panel in face-to-face contact and debate to gain consensus on several scenarios of the future. Each scenario is based on a different set of assumptions, such as political factors, economic tourism development, promotion, and transportation.

Among quantitative approaches, Uysal and Crompton describe three kinds. *Time series* studies are often statistical measures repeated year after year. Here it is assumed that all variables are working equally over time. In order to reflect changes in influential variables, transfer function models have been developed but involve complex mathematical and statistical techniques. *Gravity and trip generation models* assume that the number of visits from each origin is influenced by factors impinging upon those origins. The primary factors are distance and population. Some researchers criticize gravity models on the basis of not reflecting price, not accounting for shrinking of distance perception by new modes of transportation, and other difficult variables. *Multivariate regression models* allow the use of many variables in predicting travel. Income, population, travel cost, international context, and other variables can be introduced.

This brief discussion is offered only to suggest that much experimentation of methods for forecasting demand is taking place. Some quantitative and statistical approaches can provide clues to future tourist flows. However, as Archer (1980, 12) points out, "Despite the introduction of scientific methods of analysis and structural approaches to formulate opinions, forecasting still involves an element of crystal-ball gazing." A combination of rigorous quantitative analyses and expert opinion for forecasting should always be an implicit part of tourism planning.

Whereas most forecasting has been applied to developed countries, an application of demand analysis was applied by Uysal and Crompton (1984, 288) to Turkey, a developing country. The variables selected as likely to be important in predicting demand were: income, relative prices (between sending and receiving countries), exchange rate, transportation costs, promotional expenditure, and other factors such as special events and political unrest. Using secondary data sources from both inside and outside Turkey, the following model was derived:

$T_{RG}$ = (PCI,RPI,EXR,PE,DUMV)
where:
$T_{RG}$ = a measure of the demand for travel services by country G to country R (number of tourists, expenditures),
PCI = a measure of per capita income in country G,
RPI = relative prices—the ratio of prices in R to prices in tourist-generating countries, GS,
EXR = relative exchange rate between unit of R's currency and unit of G's currency,
PE = promotional expenditure,
DUMV = social/economic instability in country R,
R = country exporting services (Turkey),
G = tourism-generating countries.

This model was applied using estimates from available data in Turkey and the generating countries of West Germany, U.S.A, France, Italy, the United Kingdom, Greece, Austria, Yugoslavia, Spain, Canada, and Switzerland. This study revealed that "the variables of income, price, and exchange rate were consistently significant factors in the determination of international tourist flows to Turkey for all the tourist generating countries" (Uysal and Crompton: 1984, 296). The demand for tourism in Turkey is highly price elastic. The exchange rate was important, except for sources of Canada and Italy. Countries closer to Turkey showed greater sensitivity to exchange rates. The impact of promotional expenditure was found to be minimal.

In the opinion of those specializing in forecasting techniques, continuing efforts toward forecasting demand should be part of tourism planning. Both short-term and long-term forecasts should be made and measured against these criteria:

Provide the information requested by tourism managers and planners.
Cover the specified time periods for which the forecast is needed.
Be of sufficient quality for its purpose (Archer: 1987, 77).

## Traveler Image

One aspect of tourist behavior that has been puzzling to suppliers has been the great variation in response to the same product by a range of visitors. It would appear that preimage of a destination, and anticipated rewards from a trip there, has much to do with visitor satisfaction. It has been suggested (Gunn: 1972, 114) that a cyclical sequence takes place, that each trip is experienced in three steps: (1) hypothesis, or preimage of the experience, (2) input, the actual participation, and (3) check, an evaluation and revised image for the next experience. The planning of tourism development, strongly oriented to physical resources, must be broadened to include understanding of this sequence of visitor perception.

## Preimage (Hypothesis)

No tourist comes to a tourism situation without some mental preimage. Over a lifetime everyone accumulates generalized and specific knowledge, habits, and attitudes toward tourist situations. Individual personality impinges upon tourist decisions. The difficulty with attempts to describe the average tourist is that the range of predisposition of tourists is very wide and little understood. One might speculate that the degree of "success" of tourist attractions depends greatly upon the psychological "set to respond" of each tourist. For example, an historic site might be evaluated quite differently by a high school history teacher familiar with many facts regarding the site and a traveler who has an hour to kill before his motel room is made up. Each is approaching the site with quite different sets to respond.

Before participating in travel and attractions, a tourist—as compared to a nontourist—is already predisposed to travel again. In other words, the stored information he or she holds, based on a lifetime of information-gathering, possibly including previous participation, provides a foundation for making the next travel decision. A Canadian study of the United States showed that 37 percent of the population were "almost certain" of their chance of taking a vacation trip within the following 12 months (Hocklin: 1976, 3). An additional 39 percent indicated that they "probably will" and 21 percent said there was an even chance they "may" or "may not."

Another study was made of flyers versus nonflyers (Plog: 1974, 57). Those predisposed not to fly when they travel were labeled "psychocentrics"—those who seemed to center their thoughts on the small problem areas of life. They tend to prefer:

The familiar in travel destinations.
Commonplace activities at travel destinations.
Sun'n'fun sports, including considerable relaxation.
Low activity level.
Destinations they can drive to.
Heavy tourist accommodations, such as heavy hotel development, family-type restaurants, and tourist shops.

Familiar atmosphere (hamburger stands, familiar type entertainment, absence of foreign atmosphere)

Complete tour packaging.

On the other hand, those who were plane travelers were called "allocentrics" and showed preference for:

Nontouristy areas.

Sense of discovery and delight in new experiences, before others have visited the area.

Novel and different destinations.

High activity level.

Flying to destinations.

Tour accomodations that include adequate-to-good hotels and food, not necessarily modern or chain-type hotels, and few "tourist"-type attractions.

Tour arrangements that include basics (transportation and hotels) and allow considerable freedom and flexibility.

Studies of preimages of destinations have been revealing. Hunt's study (1975) of images in four states in five geographical subdivisions of the United States showed the importance of proximity. Those areas closest to destinations seemed to have a more accurate image of the destinations than those farther away. But "whether perceptions are right or wrong is not as important as what potential visitors believe" (Hunt: 1975, 6).

A study in 1981 (McLellan and Foushee: 1983) examined preimages of the United States as held by potential travelers from nine countries: Canada, Mexico, Japan, the United Kingdom, West Germany, France, Brazil, Argentina, and Venezuela. The study was performed with the cooperation of 20 tour operators who had led many tours from these countries and had observed problems. They administered questionnaires to travelers. Personal safety was perceived to be the major problem (35.7 percent for French, 91.2 percent for Japanese). French travelers were concerned about cost (71.4 percent), compared to only 34.5 percent for West Germans. Over 30 percent of all respondents were "very concerned" over information available.

Kale and Weir (1986, 4) found that a sampling of American potential travelers provided clues to the relatively low selection of India as a travel destination. Although the respondents held positive images (culture, things to "see and do," exotic, low cost, scenery), they indicated that they were reluctant to go because of many negative images. Among these were: poverty/beggars, political instability, lack of cleanliness, hostile climate, and fear of the unknown.

It is clearly evident that potential travelers hold varied images of travel destinations and that these images have strong bearings on selection of trips and satisfaction from the trip after arrival. Research and analysis of preimages are necessary for effective planning.

## Participation and Evaluation (Input and Check)

The actual visit to a destination has many implications for tourism planning. The direct exposure to the business or pleasure event has the possibility of exceeding, equaling, or not coming up to expectations. If the experience provides better satisfactions than anticipated, the traveler has upgraded his image. This is certainly an ideal scenario with several implications. It suggests that the destination development be of the highest quality and market match. It also suggests that promotion should not exaggerate the destination's merits. If a destination at least meets expectations, the experience has not lowered the traveler's impressions. Of course, the worst scenario is that in which actual experience falls below expectations. Whereas factors of the immediate experience outside the realm of planning such as social group or weather could influence this dissatisfaction, the quality of experience depends greatly on what has been planned. Certainly it endorses the need for improved destination development and care in promotion. Modern media technology and expertise has reached levels of sophistication whereby destinations can be pictured far more ideal than they can be expected to perform.

Throughout each tourist's travel activity, the experience is being evaluated, consciously or not. Although it may not be done in the studied manner of a drama critic, nevertheless it *is* done. Measurement of this phase in tourist behavior is not yet a popular area of research, but some insight may be obtained by study of consumer behavior generally.

One measure is a price-quality relationship. If price is considered as a total budgeted vacation item and experience, there may be a correlation with quality (as perceived by the tourist) of the attraction. For example, within one vacation day, a tourist may rank sightseeing side trips not only by the satisfaction the attraction produced, but also by the time and effort to drive to it. A less complete attraction may be rated higher than a "better" one only because it could be accomplished more readily. Those attractions that are features of package tours may be more rewarding than others of higher quality only because they were more readily available as part of the tour. In general, and within limits, the tourist, as other consumers, views price as an indicant of quality.

This presents public agencies with a fee dilemma. If they set prices at a level to cover investment and operational costs, the fees would be so high that tourists would question their social welfare status as government agencies. However, if they lowered fees to offer only a token control or to cover only operating costs, comparable commercial enterprise service (camping, for example) would object to unfair competition. For most tourist attraction activities, the product is not subject to standardization, and therefore comparison is difficult. Pricing does not always correlate with quality.

Another important consideration is that of meeting expectancies. The nation-wide survey of public tourism officials and private sector representatives conducted by a special Senate Committee (National Tourism Policy Study: 1977, 9) showed that "difficulties in satisfying the public's expanding and increasingly sophisticated demand for tourism and travel resources, opportunities, facilities, and services. . . ." was a basic need not being met. If the set to respond or predisposition of the tourist toward attractions, transportation, and other services are met in real participation, tourist rewards are high. If they are not met, rewards are low. Several years ago the developers of Hawaii were admonished not to install dog fighting, bullfighting, or replicas of attractions elsewhere. Rather, the tourist's anticipation of Hawaii as a paradise with brown-skin natives, surf, and swaying palms should not be denied in reality. Because the tourist to Hawaii "is in an auto-suggested trance of romance" (Barnet: 1966, 18), the spell should not be broken.

Pearce (1982, 146) described image studies as generally using three measures: mental mapping, multidimensional scaling, and repertory grid analysis. He applied the latter technique to British subjects to determine changes of their images of Morocco and Greece as travel destinations after actually going there. Travelers to Greece found that the holiday environment was more adventurous, better for cheap shopping, and freer from other tourists than anticipated. Travelers to Morocco saw that country as more adventurous, better for cheap shopping, having more interesting sights, and bringing them into contact with more peasant people than they had expected. This suggests that images can change through actual experience.

Studies of image change can be very constructive for planners and planning bodies to assess new programs of development.

## Consumer Behavior

Tourism planning and development strategies are increasingly using principles of consumer behavior to obtain better information on traveler decision-making. Consumer behavior has been defined as the "acts of individuals directly involved in obtaining and using economic goods and services, including the decision processes that precede and determine these acts" (Engel et al.: 1973, 5). Whereas the professionals in this field recognize that evaluations of behavior face all the difficulties inherent in psychological research, much progress has been made in identifying influential factors . Assael (1984, 10) lists three factors of greatest influence in decision-making by consumers: characteristics of the consumer (needs, perceptions, demographics, life-style), environmental influences (culture, social class, face-to-face groups, situations), and marketing strategies (product, price, advertising, and distribution).

Many consumer behaviorists emphasize that decision-making is extremely dynamic and changeable over time. This behavioral function has a much more dramatic impact on tourism than on manufactured products. Motivations vary and at any one time more than one may be operative. A single project development, for example, a resort complex within a given destination, may be viewed quite differently as a travel objective even by the same person but at different times.

Crompton (1979, 408) sought to understand motives for pleasure vacation travel with a special study of 39 respondents using in-depth unstructured interviews. The data revealed four main components of disequilibrium—tension in the motivational system that occurs when a need (perhaps to travel) arises. First, the study supported the descriptor of disequilibrium among the respondents—all felt the need for change. Second, the respondents felt a desire for a break in routine. Third, three alternatives were expressed regarding the nature of the break from routine—stay at home, take a pleasure vacation, or travel for other purposes such as visiting friends and relatives or going on a business trip. Fourth, motives for pleasure vacation travel were revealed. Two groups of motivations were found: *sociopsychological* and *cultural.*

Seven psychological motives were identified for taking a pleasure vacation trip. (1) Temporary escape from a mundane environment was frequently expressed. This motive was not necessarily related to specific kinds of environments, but more to preimage of what a change might do. (2) Some people viewed the vacation as an opportunity for exploration and evaluation of self. By becoming immersed in a new setting, it was believed self-discovery would emerge. (3) Many believed the trip would offer relaxation although the trip might be physically exhausting. (4) Prestige was not a strong motive, although this may be difficult for the respondent to articulate. Perhaps travel has become so commonplace that prestige is no longer important. (5) The motive of regression appeared to be important. The opportunity for puerile, irrational, and uninhibited behavior was important to several respondents. (6) Many respondents saw pleasure vacations as opportunities for family gatherings when relationships could be enhanced or enriched. (7) Interaction with residents at destinations motivated several respondents. The preferred medium for this was the organized tour because other forms of travel were less conducive to meeting local people in a meaningful way. Tours also provided desirable interaction with others on the tour.

| HOWARD/SHETH | MCDANIEL/CRISSY | MCDANIEL | AIDA | IUOTO |
|---|---|---|---|---|
| Attention Comprehension Attitudes | Awareness | Awareness Knowledge Liking | Attention | Unawareness Awareness Comprehesion |
| Intention Purchase | Interest Evaluation | Preference Conviction Purchase | Interest Desire | Conviction |
| | Trial Adoption | | Action | Action |

**Figure 4-1.** Tourist Buying Stages Compared. A comparison between several authors' perceptions of the sequence of stages that tourists engage in during their process of purchasing travel (Mill and Morrison: 1985, 71).

In addition, several cultural motives were cited. Novelty, such as curiosity and adventure even if returning to a previously visited destination, was mentioned. Even though adventure was a motivation, it was often tempered by realities such as language and custom barriers. A second cultural motivation was that of educational value of a destination, especially for children. There seemed to be a moral compulsion to see and experience the attractions of an area.

Understanding the array of motivations, revealed by such research probes, is of value to overall planning, particularly in the search for potential destination development. In other words, the consumer behavior side of destinations is as critical to planning and development as are the resource qualities of supply.

Consumer behavior modeling by many researchers shows a similar sequence of steps from initial awareness through action. Mill and Morrison (1985, 71) have summarized these as shown in Figure 4-1.

In recent years, market researchers are finding behavioral (decision-to-purchase) differences between products and services. Zeithaml (1984, 192) has found that the greatest difference, certainly of importance if applied to tourism, occurs within three characteristics. The *intangibility* of services refers to the fact that they cannot be sensed, as can products. They cannot be displayed, felt, tasted, or touched. For this reason, services cannot be evaluated during a search process prior to purchase. *Nonstandardization* of services makes it difficult for both providers and consumers to maintain standards of quality. During periods of high demand, quality of service may be reduced. As new workers replace former ones, the amount and quality of service may vary because they are personal. Much depends upon the actual experience customers have with the service rather than upon prepurchase tangible qualities. And finally, the *inseparability of production and consumption* separates services from goods. Services are sold, produced, and consumed simultaneously, whereas tangible goods are produced, sold, and then consumed, most often at entirely different times and locations.

These differences between services and goods suggest that services are less adaptable to traditional promotion such as advertising. Instead, consumers turn to others who have experienced these services for advice prior to purchase. If masses of tourists are influenced more by word-of-mouth information than advertising, promoters may wish to readjust budgets and efforts toward tourism promotion.

Planning for tourism development becomes even more complicated when the product—part physical development and part service—is fully understood. Designation of areas that appear to have potential for tourism may or may not be successful because of the dependency upon the service variable.

## Activities

Planning that seeks a demand-supply match requires data on the array of activities that the originating populations not only prefer but actually engage in when they travel. Because of the almost limitless extent of activities associated with travel, cataloging and classifying become difficult.

Chadwick (1987, 50) has outlined a typology that divides activities into four groups, as shown in Figure 4-2. These groups are business, other personal business, visiting friends, and relatives and pleasure.

Perhaps most frequent is the division of all travel activities into simply *business* and *personal*. Mill and Morrison (1985, 98) use this broad grouping and divide business travel activities into three parts:

Regular business travel activities
Participating in meetings, conventions, congresses
Engaging in incentive travel activities (business-pleasure mixed)

The pleasure/personal traveler activities are divided into groups:

Resorting                          Beach pleasure activities
Family travel activities           Handicapped activities
Elderly travel activities          Casino gambling
Single and couple activities

Another classification that may be more helpful to planners is based on whether a tourist seeks to gather experiences along the touring circuit or at longer-stay destinations. Whereas there is much overlap in this grouping, the activities sought, and therefore the attractions, services, and facilities, can be quite different.

Behavior at a longer-stay destination demands a variety of activities that can be repeated, such as fishing and boating. Some of the categories of destination activities, reflecting destination behavior of tourists are:

Organization camping                Casino gambling
Conventioneering, business meetings  Using vacation homes
Visiting friends, relatives          Attending sports events
Resorting

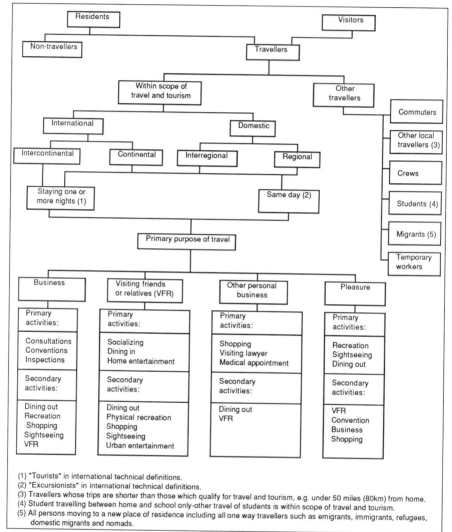

(1) "Tourists" in international technical definitions.
(2) "Excursionists" in international technical definitions.
(3) Travellers whose trips are shorter than those which qualify for travel and tourism, e.g. under 50 miles (80km) from home.
(4) Student travelling between home and school only-other travel of students is within scope of travel and tourism.
(5) All persons moving to a new place of residence including all one way travellers such as emigrants, immigrants, refugees, domestic migrants and nomads.

**Figure 4-2.** A Classification of Travelers. All planning for tourism must encompass the full range of travelers. This classification provides the basic definition of travelers used by statisticians in Canada (Chadwick: 1987, 50).

Behavior on touring circuits requires a new location for each activity. Place-ness becomes as important as the kind of activity. It is part of the human curiosity to seek new places in order to learn more, to experience new settings, and to prove to peers that one is not parochial. Tourists require attraction settings spaced along the travelway that emulate the special place characteristics of the area and reflect the special activities that can be enjoyed en route. Some of the categories of touring circuit types of activities are:

Viewing roadside scenery
Camping
Visiting friends, relatives
Attending festivals, pageants
Shopping
Visiting natural areas
Visiting historic sites
Outdoor sports
Visiting unusual institutions
Engaging in usual recreation activities

Because outdoor recreation has been identified as a special category, dominantly requiring travel, it was surveyed for the United States President's Commission on Americans Outdoors (Participation, 1986). Thirty-five of the most popular outdoor recreation activities were grouped into six categories as listed in Table 4-1.

**TABLE 4-1**
OUTDOOR RECREATION ACTIVITY GROUPINGS

Results of a U.S.A. survey to determine recreation activities for the President's Commission on Americans Outdoors.

1. BALL GAMES AND RUNNING (41% participate often)
   - Basketball
   - Softball or baseball
   - Football
   - Running or jogging
   - Attend sports events (those who participate in ball games are more apt to attend them)
   - Soccer

2. SPECTATOR OUTINGS (76% participate often)
   - Sightseeing
   - Driving for pleasure
   - Picnicking
   - Visit historic sites
   - Attend zoos, fairs, or amusement parks
   - Walking for pleasure (this semiactive recreation combines with spectator activities)

3. FISHING, HUNTING, AND HORSEPOWER (37% participate often; activities using motorized vehicles are more apt to be done by the same people who fish and hunt.)
   - Fishing
   - Hunting
   - Motorboating
   - Recreational vehicle camping
   - Driving off-road on motorized vehicles such as motorcycles, snowmobiles, trail bikes, 3-wheelers

4. OBSERVING NATURE (31% participate often)
   - Backpacking
   - Day hiking (note that running/jogging, walking for pleasure, and day hiking fall in different activity groups)
   - Tent camping (note that tent camping and recreational vehicle camping fall in different activity groups)
   - Other camping
   - Canoeing/kayaking/rafting
   - Bird watching or nature study

5. WATER AND GOLF (48% participate often)
   - Sailing/windsurfing (note that three types of boating—motor boating, canoeing and sailing—fall in three different activity groups)
   - Swimming in an ocean, lake, or river
   - Swimming in an outdoor pool
   - Golf

6. WINTER SPORTS (11% participate often)
   - Ice skating
   - Downhill skiing
   - Cross country skiing
   - Sledding

(Source: Participation in Outdoors: 1986, 2)

## Segmentation

Until recently any tourist was considered like all other tourists, and all planning and management strategies treated tourists as a homogenous whole. As has been found in marketing other products, there is much merit to dividing the totality of tourists into groups with similarities.

Market segmentation has been defined by Kotler (1972, 166) as "the subdividing of a market into homogeneous subsets of customers, where any subset may conceivably be selected as a market target." He offers three basic conditions that should be met for segmentation. First, there must be great enough numbers in each segment to warrant special attention. Second, there must be sufficient similarity of characteristics within each group to give them distinction. Third, the subsets must be viable—worthy of attention. When planning for physical development, as well as assessing social, economic, and environmental impact, it should be very helpful to have segmented refinements of potential tourist groups who might travel to the area.

Earlier segmentation was directed toward grouping tourists by demographic characteristics—age, sex, income, ethnicity, stage in life cycle, and occupation. Generally, it has been found that grouping according to these characteristics has not been as useful as anticipated. Whereas some extensive foreign vacations are relatively costly and require higher income markets, income is more of a limitation than a determinant. Many people with a wide diversity of incomes are found at tourism destinations. Even though ethnicity has not been widely researched, there seem to be similar traveler characteristics across several racial and national groups.

Ages of travelers do have a bearing on what is developed. But it has been found that age groupings are of less importance than believed. One bracket that has increased in importance in the United States is the 50-plus traveler. Norvell (1985, 126) found that convention travel is just as popular with 50-plus travelers as with others. The 50-plus travelers are more likely to travel for entertainment, sightseeing, theater, historical sites, and shopping than for outdoor recreation. Regarding regional destination preference, there was little difference from other travelers. Older travelers tend to spend more time on trips but stay less frequently with friends and relatives than do younger travelers. Whereas the use of recreational vehicles (RVs) is greater among the 50-plus group, this use declines in favor of package tours over the age of 65. In 1984, the 50-plus traveler accounted for 30 percent of all domestic travel, 30 percent of all air trips, 32 percent of all hotel/motel nights, and 72 percent of all RV trips. Continuing research on age segmentation may be of value in planning destination development.

Crask (1981, 29) surveyed two regions of the United States to segment the vacationer market—the Great Lakes area and the southeastern United States. The study resulted in five segments, each having different age distribution, income, and interests. The *rest and recreation vacationer* (R & R) was dominantly middle-aged with no children at home, upscale income, and interested in no specific destination but primarily to recuperate from career stress. The *sightseer* group planned their vacations, were not concerned over travel distance and cost, were older, and with higher education levels. *Cost-conscious/attraction-oriented* vacationers seek man-made attractions, such as theme parks and shops, and were generally younger, less well educated, and more family oriented than the first two groups. *Sports enthusiasts* were unconcerned over sightseeing opportunities, liked sports activities, and tended to be single, young adults with lower to middle income status. The last segment, *campers,* was much interested in facilities as well as sightseeing, was composed of young adults with children, less education, and lower incomes. Studies such as this provide insight into demand for purposes of overall regional planning.

A study by Solomon and George (1977, 14) demonstrates the limitations of demographic analysis for planning guidance. A special market segment, historian travelers, was examined and compared to other travelers. This study was related to the bicentennial celebration of 1976 and sampled out-of-state pleasure travelers going through a southern state. The results showed remarkable similarity between most demographic factors between historian and nonhistorian travelers. The vast majority of both were married and had similar educational backgrounds. Historians were slightly older with somewhat higher incomes. Data such as these are not as helpful for planning as those that identify different interests at destinations.

For example, Mayo's study of visitors to national parks (1975) identified seven segments having different interests and motivations at national parks. Visitors to three different types of destinations—educational-historical, scenic, commercial—were asked about national park destinations. Although fewer than one in five felt they had good knowledge about the parks, all had definite images of these parks. Whereas the study showed similarities in demographic characteristics, there were significant differences in psychographic variables. The researchers identified seven clusters: the adventurer, the nonplanner, the impulse decision-maker, the action-oriented person, the outdoorsman, the escapist, and the self-designated opinion leader.

## Lifestyle

Tourist lifestyle is becoming a more important basis for market segmentation than demographics alone. Lifestyle is a concept that includes factors of great importance in an individual's values and choices of preferences, including demographics.

Of particular importance is the VALS program, a copyrighted lifestyle analysis, developed by Arnold Mitchell, founder and director of SRI International Values and Lifestyles Program from 1978 to 1981 (Mitchell, 1983). Although it may not be equally applicable to other populations, it is based on analysis of American lifestyles, identifying nine lifestyles grouped into four categories:

> Need-driven groups:
>> survivor lifestyle
>> sustainer lifestyle
> Outer-directed groups:
>> belonger lifestyle

emulator lifestyle
achiever lifestyle
Inner-directed groups:
I-am-me lifestyle
experiential lifestyle
societally conscious lifestyle
Combined outer-and inner-directed group:
integrated lifestyle

The state of Pennsylvania applied this system to its tourism markets in its primary source areas: Pennsylvania, New Jersey, New York, Ohio, Maryland, Delaware, Virginia, and West Virginia (Shih, 186, 2). By means of scientific survey research whereby the respondents did not know that the purpose was for Pennsylvania, some valuable results were obtained.

The largest group coming to Pennsylvania was identified as "achievers" (36.6 percent) characterized as successful, happy, hardworking, self-reliant, and reasonably affluent. Again, safety, good accommodations, and friendly hosts were cited as most important travel factors. This group was more interested in business contacts and longer distance travel than shopping, camping, or public transportation.

The next largest group making trips to Pennsylvania were "belongers" (35.6 percent). This group is the solid, comfortable, and middle-class group characterized by uncomplicated, conservative, conventional lives centered around the home. They are conforming and expressively patriotic. They place higher than average importance on safety, reasonable prices, good accommodations, and friendly hosts.

The "societally conscious" group was third largest (17.8 percent). These people are generally like achievers except that they are more mission-oriented and stress qualities such as frugality, simplicity, and conservation. They ranked reasonable prices, relaxing vacations, and safety higher than good accommodations. They expressed interest in destinations with a wide variety of activities including hiking and cultural attractions.

Studies such as this can be of great value to planning for tourism development. In this case, even though the three groups dominated the market profiles, the others should not be ignored by Pennsylvania. For certain destinations there may be potential for development of tourism attractions and services for "survivors" (2.1 percent), "sustainers" (1.7 percent), emulators," (0.9 percent), "I-am-mes" (0.6 percent) and "experientials" (2.0 percent).

Although the objective of the Pennsylvania study was primarily for evaluation of their advertising, the results are equally valuable in assessing the adequacy and opportunity for the supply side.

Regional and destination planning must take into account the seasonality differences among market segments. Capital investments at destinations are often directed to single season markets and have difficulty in obtaining adequate financial returns for success.

A study by Calatone and Johar (1984, 14) demonstrated great variation among market segments of automobile tourists traveling the Massachusetts turnpike. They found that not only do the same segments seek different benefits for each of the four seasons, but that different segments seek similar benefits at several seasons. Analysis suggested that the total travelers could be divided into five segments: frequent visitors, sightseers, sports and relaxation seekers, young nature buffs, and representative subgroups.

Frequent visitors were characterized as higher income, larger family, and longer-distance travelers, and came at all seasons. Sightseers were generally older and less frequent travelers seeking high-quality attractions, highways, and accommodations especially in summer and spring. Sports and relaxation travelers sought summer and winter sports, were less interested in high-quality accommodations, and tended to be younger. Young nature buffs were active outdoors people, young, artistic, and as much interested in spring and fall as the summer. The representative subgroup was less well defined, did not like to take chances, had lower incomes, and were interested in one-day visits with a great variety of activities.

Travelers in spring looked for good roads, sporting activities, historical and cultural activities. Summer travelers were not only interested in sporting and cultural attractions, but also sought a clean and scenic environment at relatively low cost. The travelers in fall were interested also in modest accommodations and cultural attractions. Winter travelers looked for winter sports, good commercial facilities, and shopping.

One of the best summaries of tourist market segmentation is that prepared for use in Canada. The following seven categories, travel market segments, are described and brief comments are offered regarding their effectiveness in tourism planning and promotion (Table 4-2).

**TABLE 4-2**
TRAVEL MARKET SEGMENTS

1. Purpose of Trip/Use Segmentation
   - Pleasure travel
   - Personal business
   - Other business
   - Convention/meetings
   - Tournaments/sports groups

   • This is usually the most effective segmentation approach because the target market is actively seeking a specific kind of product.

2. Channel of Distribution Segmentation
   - Direct customer sales
   - Travel agents
   - Tour operators
   - Tour wholesalers
   - Airlines
   - Government marketing
   - Regional/local tourist ass'ns

- This approach is effective in further afield markets that cannot be reached directly at reasonable cost or where travel trade companies have a market that is closely matched.

3. Socioeconomic or Demographic Segmentation

| | |
|---|---|
| Age | Family life cycle |
| Sex | Social class |
| Education | Home ownership |
| Income | Second-home ownership |
| Family size | Race or ethnic group |
| Occupation | |

- This is commonly used segmentation approach, since these segments are often easy to reach and information on them is usually available.

4. Product-related Segmentation
   - Recreation activity
   - Equipment
   - Brand loyalty
   - Benefit expectations
   - Length of stay
   - Transportation mode
   - Experience preference
   - Participation patterns

- These are difficult segments to reach, but they are well matched to the use of specific products.

5. Psychographic Segmentation
   - Personality traits
   - Lifestyle
   - Attitudes, interests, and opinions
   - Motivations

- In tourism, this can be an effective segmentation approach, since tourism product use is extensive among certain psychographic groups. Also, many advertising media are segmented this way.

6. Geographic Segmentation
   - Country
   - State, province, and county
   - Region
   - Urban, suburban, and rural
   - City size
   - Population density

- This is the most common segmentation approach because these markets are clearly defined and accessible. It is often not an efficient approach, however, unless it is used in combination with other approaches.

7. Use Frequency/Seasonality Segmentation
   - Heavy users
   - Moderate users
   - Infrequent users

- Data should be readily available on these customers, so this method is likely to be cost-effective.

(Source: Marketing: 1986, 60)

# Conclusions

*Travel is a constant phenomenon.* If there is any lesson from history, it is the constancy of travel. Travel has its roots in ancient history and continues to grow as a fact of civilization. Once considered frivolous and capricious, travel has become a firm and continuing reality throughout the world. The "push" side, demand, is expressed in markets that regularly have the desire and the ability to travel. It is because of this continuing market pressure that the full scope of economic, social, and environmental planning is involved. This involvement is at all levels of government—federal to local as well as by the private sector.

*Tourists are now redefined.* To the old definition of tourists as only vacationist (holiday) and recreation travelers must be added the full scope of all travelers, no matter the purpose. Increased research and monitoring of tourism has revealed the need for this broader definition. The "industry" side—service businesses— finds business and duty travel as important as pleasure travel for success. The public side—parks, museums, convention centers—finds that their visitors come from a variety of markets. In today's travel, much of business travel is linked with pleasure travel requiring the scope of planning to include both.

*Planning starts with visitor activities.* Markets, within places of origin, seek activity opportunities at travel destinations. Planning for tourism must include information on activities sought by travelers. Over history, many of these activities are constant. For generations, travelers have engaged in activities such as visiting friends and relatives, conventioneering, sightseeing, shopping, and visiting historic sites. But there is continuing change in the variety of activities and particularly how these are to take place. Boating and attending conferences today require far different destination development—higher quality, increased technology, greater investments—than even a few decades ago.

*Consumer behavior research is growing.* More frequent, more varied, and more scientific research of travel consumer behavior is providing new insights of value to tourism planning. Three types—household surveys, location surveys, business surveys—increasingly provide qualitative and quantitative data on what travelers do, what they prefer, when they travel, where they travel, and how much money they spend. New principles of consumer behavior generally are providing new understandings applicable to travel behavior, such as the sequence of decision-making, factors influencing decisions, and value of travel to participants.

*Planning requires market forecasting.* Once largely guesswork, forecasting of travel behavior is utilizing better and more reliable techniques. Whereas not all factors of behavior are predictable, trends can be measured for clues to future travel. Because market attitudes and behavior are dynamic, it is essential to continuously monitor traveler characteristics. Planning for modifications in present plants and opportunities for new development for tourism require regular updates on market shifts. As destination development becomes more costly and more complicated to manage, the need for current market trends becomes ever more critical.

*Images are more important than realities.* Studies of traveler psychology show that what is physically true at destinations may not be market images of those destinations. Because of many external influences, people at their places of origin have mental images of travel destinations. Awareness of these images is important for tourism planning. For tourism development, it is important that the actual participation at a destination equal or surpass the preimage perception of the destination in order for the experience to be satisfying. Because images are slow to change, regular monitoring of traveler images becomes important.

*Market segmentation, an important planning tool.* The generalization that travelers are all alike is being disproven by segmentation research. Clusters of factors are useful in dividing the overall markets into subsets, each of which has homogeneous characteristics. Recently, lifestyles, combinations of sociodemographics and attitudes, are being used to segment travelers for planning and management purposes. Segmenting by season is as important as by interest. When better information is available on market segments, destination resources can be evaluated for potential development that is more likely to succeed.

# Bibliography

Archer, Brian H. "Forecasting Demand: Quantitative and Intuitive Techniques," *Tourism Management.* 1 (1), 1980, 5–12.

Archer, Brian H. "Demand Forecasting and Estimation," Chap. 7. In: *Travel, Tourism and Hospitality Research,* J.R. Brent Ritchie and Charles R. Goeldner, eds. New York: John Wiley & Sons, 1987, 77–85.

Assael, Henry, *Consumer Behavior and Marketing Action,* 2nd. ed. Boston: Kent Publishing, 1984.

Barnet, Edward M. "Please Don't Wake the Traveler," *Honolulu,* August 1966.

Boorstin, Daniel J. *The Image.* New York: Harper Colophon Books, 1961.

Bridges, J.G. "A Short History of Tourism." In: *Travel and Tourism Encyclopaedia.* London: Travel World, 1959), 28–42.

Calatone, Roger J. and Jotinder S. Johar. "Seasonal Segmentation of the Tourism Market Using a Benefit Segmentation Framework," *Journal of Travel Research.* 23 (2), Fall 1984, 14–24.

Chadwick, Robin A. "Concepts, Definitions and Measures Used in Travel and Tourism Research," Chap. 5, In: *Travel, Tourism and Hospitality Research.* New York: John Wiley & Sons, 1987, 47–61.

Crask, Melvin. "Segmenting the Vacationer Market: Identifying the Vacation Preferences, Demographics and Magazine Readership of Each Group," *Journal of Travel Research,* 20 (2), Fall 1981, 29–34.

Crompton, John L. "Motivations for Pleasure Vacation," *Annals of Tourism Research,* 6 (1), October/December 1979, 408–424.

Dulles, Foster Rhea. *A History of Recreation.* New York: Appleton-Century-Crofts, 1965.

Engel, James F. et al. *Consumer Behavior,* 2nd ed. Hinsdale, IL: Dryden Press, 1973.

Frechtling, Douglas C. "Assessing the Impacts of Travel and Tourism—Introduction to Travel Impact Estimation," Chap. 27. In: *Travel, Tourism and Hospitality Research,* Ritchie and Goeldner, eds. New York: John Wiley & Sons, 1987, 325–332.

Gunn, Clare A. *Vacationscape: Designing Tourist Regions.* Austin: Bureau of Business Research, Univ. of Texas, 1972.

Hockin, Pamela. *A Report of Travel Intentions of Americans in 1976.* Ottawa, Canada: Government Office of Tourism, 1976.

Hunt, John D. "Image as a Factor in Tourism Development," *Journal of Travel Research,* 13 (3), 1975, 1–7.

Johnson, Walter H. "Tomorrow's Traveler—A Marketing Analysis," *Tourism Management,* 4 (2), 1983, 129–132.

Kale, Sudhir H. and Katherine M. Weir. "Marketing Third World Countries to the Western Traveler: The Case of India."*Journal of Travel Research,* 25 (2), Fall 1986, 2–7.

Kotler, Philip. *Marketing Management: Analysis, Planning and Control,* 2nd. ed. Englewood Cliffs, NJ: Prentice-Hall, 1972.

Leiper, Neil. "The Framework of Tourism: Towards a Definition of Tourism, Tourist, and the Tourist Industry," *Annals of Tourism Research,* 6 (4), October/December 1979, 390–407.

McLellan, Robert W. and Kathryn Dodd Foushee. "Negative Images of the United States as Expressed by Tour Operators from Other Countries," *Journal of Travel Research,*22 (1), Summer 1983, 2–5.

*Marketing, Management* (Marketing management program, Maclean Hunter Ltd). Toronto; Canadian Hotel and Restaurant, 1986.

Mayo, Edward. "Tourism and the National Parks: A Psychographic and Attitudinal Study," *Journal of Travel Research.* 14 (1) 1975, 14–17.

Mill, Robert Christie and Alastair M. Morrison. *The Tourism System.* Englewood Cliffs, NJ: Prentice-Hall, 1985.

Mitchell, Arnold. *The Nine American Lifestyles.* New York: MacMillan, 1983.

Murphy, Peter E. "Evolution of Tourism," *Tourism. A Community Approach.* New York: Methuen, 1985, 17–29.

*National Tourism Policy Study: Ascertainment Phase.* Senate Committee on Commerce, Science, and Transportation. Washington, DC: U.S. GPO, 1977.

Norvell, Hal. "Outlook for Retired/Older Traveler Market Segments," *1985-86 Outlook for Travel & Tourism,* Proceedings of Eleventh Annual Travel Outlook Forum. Washington, DC: U.S. Travel Data Center.

"Participation in Outdoor Recreation Among American Adults and the Motivations Which Drive Participation," report prepared for the President's Commission on Americans Outdoors, by Market Opinion Research, 1986.

Pearce, Philip L. "Perceived Changes in Holiday Destinations," *Annals of Tourism Research.* 9 (2), 1982, 145–164.

Plog, Stanley C. "Why Destination Areas Rise and Fall in Popularity, "*Cornell H.R.A. Quarterly,* February 1974.

Ritchie, J.R. Brent and Charles R. Goeldner, eds. *Travel, Tourism and Hospitality Research.* New York: John Wiley & Sons, 1987.

Shih, David. "VALS a Tool of Tourism Market Research: The Pennsylvania Experience," *Journal of Travel Research.* 24 (4), Spring 1986, 2–11.

Solomon, Paul J. and William R. George. "The Bicentennial Traveler: A Life Style Analysis of the Historian Segment," *Journal of Travel Research.* 15 (3), 1977, 14–17.

Uysal, Muzaffer and John L. Crompton. "Determinants of Demand for International Tourist Flows to Turkey." *Tourism Management,* 5 (4), December 1984, 288–297.

Uysal, Muzaffer and John L. Crompton. "An Overview of Approaches Used to Forecast Tourism Demand," *Journal of Travel Research,* 23 (4), Spring 1985, 7–14.

Zeithaml, Valarie A. "How Consumer Evaluation Processes Differ Between Goods and Services" In: *Services Marketing: Text, Cases & Readings* by Christopher H. Lovelock. Englewood Cliffs, NJ: Prentice-Hall, 1984.

# Chapter 5

# Attractions

Although all components of the tourism system are important to its functioning, attractions provide the energizing power. Attractions, however, are not well understood and are often interpreted as only "commercial." In fact, attractions are the on-location places in regions that not only provide the things for tourists to see and do but also offer the magnetism for travel. Attractions are the mirror side of market interests, the places where personal and social expectations from travel are realized. Because markets change over time, so do attractions even though they often have physical roots. Furthermore, economic impact, even though directly derived largely from commercial services and facilities, is dependent upon the attracting power of a region to draw travelers. Therefore, those regions contemplating new development or expansion of tourism need to place high priority on the planning, establishment, and quality operation of attractions.

## Historical Background

Although no one knows about the very first attractions of the world, we do know that the early Greeks and Romans frequented resorts and traveled to distant sites for pleasure. The main attractions for domestic travelers were the delightful climate, remarkable temples, sculptures, and paintings of Athens as well as the scenic appeal of the Greek islands. But "no country had more to show the tourist than had Egypt" (Balsdon: 1969, 299). Up the Nile, tourists were fascinated by the statue of Amenophis III, believed to utter miraculous exhortations at sunrise. For the Romans, a 12-day trip to Alexandria brought them the exotic pleasures of "the Pharos (a great lighthouse that was numbered among the Seven Wonders of the World), the tomb of Alexander the Great, and the massive temple complex known as the Serapeum" (Turner and Ash: 1975, 27). Roman tourists also frequented resort villas along waterfronts outside the cities to fish, boat, and bathe.

Tourist attractions were not of much concern during the Middle Ages because eking out a living consumed most of one's energies. Religious shrines, however, did provide the motive for many to travel great distances from home. More than a religious exercise, the shrines provided an escape, an opportunity to socialize and a chance to see new surroundings.

Renaissance travel, stimulated by conquest, gave the social and political elite the opportunity for pleasure travel to exotic lands. There they visited health spas, churches, shrines, and "pleasure grounds."

In spite of earlier evidence of the functions of attractions, the *Oxford English Dictionary* states that it was not until 1869 that the word began to take on its present meaning: ". . . a thing or feature which draws people by appealing to their desires, tastes, etc., especially an interesting or amusing exhibition which 'draws' crowds" (Murray: 1933, 1/555).

Places of business and commerce as well as destinations for pleasure appeared early in travel history. Thomas Cook's package travel innovation was first stimulated by his zeal for trips to temperance meetings and business objectives (Burkart and Medlik: 1981, 15). Wells Fargo, and later American Express, were founded on the transport of goods and bullion.

Beaches and mountains brought many travelers to pleasure destinations in the 1800s. Seacoasts and lakefronts, long before emancipation of women's dress allowed bathing, offered beautiful vistas, boating, socializing, and cooling relief from inland cities. Mountain climbing and winter sports stimulated resort development at sites with appropriate physical amenities. Because travel markets were affluent, elaborate castlelike structures and ornate landscapes with gardens, lawns, statuary, and fountains developed during the period. Travel attractions for resorting were quite different from those for touring and sightseeing.

The concept of resorts—complexes providing a variety of recreations and social settings at one location—came early, even in Roman times. Over many years, several factors have contributed to the success of resort attractions. Scientific medical treatment was not yet generally available and certain locations and waters were believed to hold curative powers. Thousands of visitors gave testimony to the magical healing value of "mineral spring waters" when consumed or used for bathing. An added array of sports, recreations, food services, and lodging seemed a logical outgrowth. Attractions at transportation nodes, especially in the railroad-steamboat era, stimulated local resorts rather than enroute sightseeing. Probably the socializing of gambling, dancing, and other people attractions had much to do with increased popularity of spas such as at Bath and Buxton in England, St. Moritz in Switzerland, and Baden-Baden in West Germany (Gee: 1981, 18).

The United States, as well as Europe and Asia, was caught up in the development of attractions that became resort spas. It began on the eastern seaboard with places such as Yellow Springs, Bath Springs, Warm Springs, Healing Springs, Saratoga Springs, White Sulphur Springs, and many others. But as settlement of the central and western parts of the United States grew, so did the mineral spring resorts, especially along rail lines. Hot Springs, Arkansas, established as a preserve in 1832 and a national park in 1921, and the Broadmoor in Colorado Springs remain as evidence of this resort era. The Grand Hotel on Mackinac Island, Michigan, although not having mineral springs, was justified as a health resort. A New York doctor in 1857 described the setting:

> The three great reservoirs of clear and cold water, Lakes Huron, Michigan and Superior, with the Islands of Mackinaw in the hydrographical center, offer a delightful hot weather asylum to all invalids who need an escape from crowded cities, polluted exhalations, sultry climates, and officious medication (Mansfield: 1857, 15).

Sightseeing and its attractions had quite different origins. For many years a tourist was not defined as anyone who traveled for any purpose that included business and resorting but only as a vacation pleasure traveler. Even before Marco Polo brought back tales of the exotic attractions he had experienced in the Orient, travelers had toured for no other reason than to see the sights. The attracting forces of the Grand Tour—spanning most of the seventeenth and eighteenth centuries—were primarily historic and cultural sites in France and Italy (Clarke: 1986, 135). For example, the young Englishman was attracted to the

> courts of princes and courts of justice, churches and monasteries, walls and fortifications, havens and harbours, antiquities and rivers, libraries, colleges, disputations and lectures, shipping and navies, armouries, arsenals, magazines, exchanges, bourses, warehouses, exercise of horsemanship . . . (Turner and Ash: 1975, 34).

Sightseeing had a pleasurable but also a purposeful cultural mission. "The houses of the English eighteenth-century nobility soon began to look like carbon copies of Italian sixteenth-century originals, their interiors suitably adorned with Italian statues, busts and painting" (Turner and Ash: 1975, 36).

But at first the landscape, especially forests, were not viewed by travelers as attractions. Quite the contrary, trees, craggy cliffs, and rocks were ugly and frightening until acculturation of landscape painters romanticized them. Even cleared agricultural lands became desirable travel vistas because of bucolic farm paintings and romantic poetry emulating rural living as the epitome of life-styles. Neatly tilled land was congruent with growing industrialization that demonstrated man's orderly control over the earth. Man-made attractions—

bridges, mills, railroads, and waterworks—were soon added to natural phenomena as travel attractions.

Interest in scenic places, as tourist attractions, was stimulated by political support of large dedicated tracts of the public domain for scenic protection and use. The U.S.A. was the first to establish huge natural resource reserves open to the public as national parks. Perhaps the most appealing types of national parks were those based on spectacular scenery. An English adventurer, James Mason Hutchings, assisted by two Indian guides, took the first tourist party into Yosemite Valley in 1855. When he returned home, he stimulated great interest in this magnificent attraction through his writing, illustrated with sketches by Thomas Ayres, an artist who had accompanied him on the trip. By 1916, when the National Park Service was established, there were 16 U.S. national parks and 21 national monuments with a total area of 7,426 square miles (19,233 square kilometers). These firmly established the concept of national parks as major tourist attractions.

A sketch of the history of attractions would not be complete without mention of the rise of sports. The establishment of major arenas and stadia throughout several countries has stimulated considerable travel. Although there were indications of the growth of sports in the U.S.A. before the Civil War, the major development in this area came later. "No transformation in the recreational scene has been more startling than this sudden burgeoning of an interest in sports which almost overnight introduced millions of Americans to a phase of life shortly destined to become a major preoccupation among all classes" (Dulles: 1965, 183).

But, the real explosion in the numbers and diversity of attractions has come in the twentieth century. Driven by faster and affordable automobile and plane travel as well as a burgeoning middle class with money to spend, nearly every location of the world became reachable by millions more people. Natural and cultural resource assets worldwide were now accessible, fostering an entirely new and massive host-guest relationship.

The beginning of the twentieth century, stimulated by industrial growth and prosperity, was also the beginning of varied faddish amusement and recreation activities. The attractions of nightclubs, dance halls, golf, and card-playing were added to the outdoor recreational appeals of sailing, hunting, and fishing.

Although only slightly interrupted by World War I, attraction development and growth continued to expand into new dimensions. No longer was travel affordable to only the rich. As new ships were built, European travel to castles, art galleries, cathedrals, and museums greatly expanded. This period also introduced the attraction of the package tour—stringing several attractions, travel, and other services together into a prearranged agreement between suppliers and tourists.

As domestic travel to the many natural resource attractions of North America expanded, attention began to broaden into man-made attractions as well. Science and technology, such as at world's fairs, demonstrated that many more things could be created to attract travelers. There followed a rash of roller coasters, merry-go-rounds, circle swings, shoot-the-shoots, and ferris wheels, not only popular with nearby residents but with long-distance travelers as well. Coney Island, on the Long Island shore, for example, not only drew throngs of New Yorkers to the beach, boardwalk, carousels, shooting galleries, and dance halls, but it also became a national attraction.

This interest in other than natural attractions represented a major turning point in U.S. tourism development. Heretofore, the natural environmental assets had dominated attractions. Unless an area had special natural resources, such as geysers, lakes, ocean beaches, warm winter climate, or mountains, it had little potential for tourism. Although cities had offered amusement parks for many years, it took the innovative features, the variety, and the impeccable maintenance and operational standards of Disneyland to prove the validity of such tourist attractions.

Not until the 1960s was there more than sporadic interest in the development of cultural attractions such as historic sites. Although the U.S. National Park Service had provided museums and interpretive talks on history for many years, most Americans did not become aware of their heritage until the last few decades. The movement gained considerable momentum during preparations for the nation's bicentennial celebration. Actually the first efforts were stimulated not by tourism but by those groups interested in historic restoration and preservation for their own sake. Colonial Williamsburg, Mt. Vernon, Charles Towne Landing, the Alamo, Mystic Seaport, for example, although becoming major tourist attractions, were first conceived as necessary to capture a vanishing culture in America.

Because attractions represent the core of the tourism "product" and because such products are oriented to geographic places, there is a need to understand the strong linkage between physical placeness and tourism experience throughout history. Even though some activities have been added and styles of development have changed, many attractions such as developed beaches, mountains, scenic areas, historic sites, outdoor sports areas, resorts, amusement parks, nature centers, shopping areas, convention centers, and unusual constructed developments have provided remarkably consistent tourist appeal throughout the last century.

Today there is hardly a nation that has not identified and extolled its attractions. Jordan speaks of its unspoiled beaches and ancient ruins of the Lejjun fortress; Finland of its cottaged lake districts and campgrounds; Austria of its musical heritage, sports, and festivals; Ireland of its verdant country villages and coastal resorts; and Japan of its many natural and cultural attractions, but especially its "meet-the-Japanese-at-home movement, called the 'Home Visit System' " (Tourism in Japan: 1986. 19). These and hundreds of other attractions appear regularly in tourist promotional literature throughout the world.

## Scope

Important to planning is the scope—the great breadth—of attractions available to tourists. The scope includes great diversity and widespread distribution providing attraction development that meets a variety of traveler interest. This almost incalculable scope complicates the task of the planner and the delineation of planning policy. The variables are so great that the rationale of choice for development becomes a joint responsibility of policymakers, planners, designers, developers, and managers with heavy input of data concerning users.

For example, Austria promotes a wide variety of attractions, from places where one can engage in health-related activities to sites for music festivals. This variety was reflected in their literature emphasis on "Get into Shape in Austria," "Nature the Healer," "Cycling in Austria," "Wonderful Austria," "Music Land Austria," "Children's Wonderland" (Annual Report: 1986). In spite of the many cultural and historic attractions of Vienna—Ringstrassen, Opera House, Parliament Building, St. Stephen's Cathedral, Spanish Riding School, Burggarten, Schonbrunn Palace, Prater park, Stadtpark—the city accounted for only 4.9 percent of total bednights in 1985. Nearly all other federal provinces exceeded the 5,535,803 bednights for Vienna. The most attractive province, in terms of bednights of visitors, was the Tyrol with 38,793,982 bednights. This westernmost tip of Austria, between Italy and Germany, contains the spectacular Northern Limestone Alps, affording all-season natural resource assets for visitor enjoyment and enrichment. Innsbruck has become a year-round resort city.

The Philippines, containing over 7,000 islands, has many cultural and natural resource attractions for visitors. Within Manila, trade travel has been fostered by the Philippine Center for International Trade Exhibitions, the Philippine Trade Center, the Tourism Pavilion, the Philippine Plaza, and the Philippine International Convention Center, all built on reclaimed land from Manila Bay. Intramuros, Chinatown, Makati, and Rizal Park are significant attractions within the city. In outlying areas of Luzon are the waterfalls of the Pagsanjan River, 2,000-year-old rice terraces, and beautiful vistas across volcanic lakes some 2,200 feet below a lodge near Tagaytay. Within short flight range are attractions of nearby islands, such as the church towers and mosques of Christian and Islamic origin in western Mindanao.

A travel guide for New York State lists attractions for 10 regions outside New York City (Liberty Centennial: 1986). Literature for the Finger Lakes Region,

located in the west-central portion of the state, lists 42 museums (ı
art, science, transportation, photography, Indian, doll, carriage, coın......
tion), 29 historic sites (homes, battlefields, shrines), 23 wineries, 17 centers for
arts, crafts, science, nature, 15 parks, 5 plant and farm tours, and other attractions in smaller numbers such as racetracks, sports arenas, farmer's markets,
commercial theme parks, zoos, universities, trails, and fish hatcheries.

## Classification of Attractions

One way of classifying attractions is on the basis of two very important types of
tourism: *touring circuit* and *longer-stay (focused)* (Gunn: 1972, 74). Touring
circuit attractions must satisfy touring markets—those traveling for business or
pleasure on tours that include many separate locations. They need not be of
qualities that bear repetition by the same users. On the other hand, longer-stay
attractions are at or clustered about destination areas. These attractions are used
by the same users repeatedly over entire vacation or business trip periods. Some
examples of tourism attractions classified by touring circuits or longer-stay
categories are listed in Table 5-1.

TABLE 5-1
A CLASSIFICATION OF ATTRACTIONS

| TOURING CIRCUIT ATTRACTIONS | LONGER-STAY (FOCUSED) ATTRACTIONS |
|---|---|
| Roadside scenic areas | Resorts |
| Outstanding natural areas | Camping areas |
| Camping areas | Hunting/water sports areas |
| Water touring areas | Organization camp areas |
| Homes: friends/relatives | Vacation home complexes |
| Unusual institutions | Festival, event places |
| Shrines, cultural places | Convention, meeting places |
| Foods, entertainment places | Gaming centers |
| Historic buildings, sites | Sports arenas, complexes |
| Ethnic areas | Trade centers |
| Shopping areas | Science/technology centers |
| Crafts, lore places | Theme parks |

Of course, there are many kinds of attractions within these categories and
many combinations of attractions, providing a rich storehouse of possibilities
for the planner as he or she considers the potential of a region. However, the
distinction between attraction defined as experience, and attraction defined as
physical entity must be clear. The ultimate personal and social good derived

from tourism is the individual experience. The perception of obtaining interesting, worthwhile, exciting (and many other kinds of) experiences is the goal of the tourist. But no planner can guarantee this. The best he or she (and designers of attractions) can do is to plan for the physical development and programs that in turn have high probabilities of providing the experiences sought by the tourists.

Examples of touring circuit travel attractions are abundant. An unusual touring circuit is the Cherry Blossom Front tour that follows the beginning of cherry blossoming from the southern islands of Japan on to their fading in northern Hokkaido. Included is participation in local *hanami,* or flower-viewing parties (Japan Travel: 1987, 1). A motorcoach tour of the eastern U.S.A. (America's Best: 1987, 16) over a period of nine days includes visits to many urban attractions of New York, Philadelphia, and Washington, D. C., as well as Niagara Falls, the Corning Glass Center, Appalachian Mountains, the Hershey factory, and the Amish farm areas. A helicopter tour of the Canadian Rockies (Tauck: 1987, 7) includes the attractions of Banff, Lake Louise, Yoho and Kootenay national parks, and Kicking Horse Pass.

Longer-stay and focused attractions are equally abundant. The Orlando, Florida, area illustrates a huge attraction complex within close proximity (Anthony: 1987, 66). In 1986 there were 1,282 major meetings and conventions in Orange County with 680,611 participants. Orlando has surpassed all but two American cities (Los Angeles and New York City) with 53,344 hotel rooms. Walt Disney World, EPCOT center, convention centers, Sea World, Spaceport USA, and Church Street Station, a Dixieland entertainment-shopping complex, are among the many attractions.

The Caribbean Islands abound with longer-stay attractions (Caribbean: 1987). Puerto Rico, Virgin Islands, Cayman Islands, Barbados, Antigua, Jamaica, and St. Lucia contain many resort complexes. In addition to fine beaches, attractions include health and fitness centers, water sports, tennis, casinos, nightclubs, convention facilities, and short drives or walks to historic buildings and sites.

A contemporary hybrid, the cruise ship, has become popular worldwide. It offers attractions in a touring circuit context by putting in at many attraction complexes in port cities. It also offers on board ship all the attraction amenities of many resorts—swimming, dancing, gourmet foods, entertainment, games, and socializing. Being self-contained, the cruise ship often carries a mix of business and pleasure travelers. Incentive groups represented 20 percent of the business of the Holland America Line in 1986 (Milligan: 1987, 16). Many observers attribute the great growth of cruise tour markets to the "Love Boat" TV show. However, the featured line, Princess Cruises, has about 40 percent group business, much of which is incentive.

## Resource Foundations

Developers are always seeking better information on what makes a place attractive to visitors. Because this phenomenon of attractions is based on physical as well as market variables, the task becomes very complicated.

Based on observation and general experience, Ferrario (1979, 24) selected six criteria for attractions. A long *season* of use is considered an asset, whereas a short season is a limitation. The factor of *access* is related to both time and space, further qualified by *admission* through some means of permission and often by price. Not all attractions are equal in *importance.* Some, by virtue of both market interest and intrinsic value, are far more important as attractions for visitors than others. The factor of *fragility* needs to be considered. Some important sites cannot withstand visitor use or require special controls to protect cultural or natural resource values. Even though several attractions may be equally well located or developed, one may have much greater *popularity* than another. This popularity may or may not be earned, but it is important nonetheless.

Some researchers (Plog: 1974, 58) theorize that attractions at destinations are destined to follow a cycle of rise and fall. This is predicated on studies of airline trips. The first visitors to discover a new area are what could be labeled "allocentrics"—those who prefer nontouristy areas, novel destinations, and flying, and enjoy a different culture. Then, as the attractions become more popular, a "midcentric" audience increases. Midcentrics are defined as less adventuresome but not as conservative about attractions as "psychocentrics." it is at this point that the attractions are most successful because the widest range of markcts is attracted. Then, as the allocentrics begin to lose interest and the numbers of psychocentrics are not sufficient to fill the places once popular, the attractions decline. If this sequence is predictable, it should lend itself to correction with planning. New supply features and new marketing may reverse the trends and avoid attraction demise.

Another factor of attractiveness is *competition.* A traveler, from the perspective of home origin, now has an ever-increasing number of alternatives. Within the same time-cost range, a great multitude of attractions clustered within many destinations are available. It is likely that if two or three attractions are similar, price can become the most important variable. If, however, the attraction is unusual or indigenous to only a few places, there is less competition and less price consciousness. So, from a planning approach, thc lcss that attractions are duplicated, the better. An example was the case of historic ship attraction selection for Sault Ste. Marie and Menominee, Michigan. Local interests in these two places, some 235 miles apart, had the same concept of drydocking an ore liner for visitor inspection and education. At that time a regional plan was underway (Blank and Gunn: 1966), and the consultants, after studying Great Lakes shipping history, advised each community to build on their intended theme but select less competitive examples. Now, an old ore liner attraction is at Sault Ste. Marie and an historic schooner has been rebuilt at Menominee. Both dramatize and interpret early days of shipping without duplicating attractions.

An important part of the scope of tourist attractions is their varying dependency on natural and cultural resource characteristics. For a great many attractions, the natural resource features are important, such as topographic change, climate, wildlife, surface waters, waterfalls and other scenic features, and unusual natural phenomena such as geysers and natural bridges. Others depend more on cultural foundations, including historic buildings, sites, archeological digs and restorations, ethnic sites, and many manmade points of interest, such as manufacturing plants, convention centers, unusual scientific accomplishments, or cultural institutions such as universities. Whereas the geographic setting may have contributed to the success of Las Vegas, the cultural factor of legalized gambling is very influential.

Physical planning requires some understanding of the relationship between development and certain location factors and other land assets. If planners and developers are interested in new tourism, an assessment of these assets becomes important. Past experience shows that attractions vary regarding their dependency upon certain land assets.

Figure 5-1 shows the relationship between resource foundations and categories of attractions. These are generalized statements indicating the relative extent to which attractions in each category are dependent upon large quantity and high quality land and location assets. They are clustered into three groups— dependency upon natural resource assets, cultural resource assets, and neither strong cultural nor natural resource assets. *Natural resource* assets are defined as water and waterlife, forests and other vegetative cover, wildlife, climate, and topographic change and soils. *Cultural resource* assets are defined as historic sites, shrines, historic buildings, ethnic areas, and archeological sites. The category of *noncultural and nonnatural resource* assets refers to foundations such as access and nearness to markets and other business development.

## Influence of Ownership

In many countries, attractions are owned and managed by three sectors: government, private enterprise, and nonprofit organizations. The policies of such ownership and management can make quite a difference in what, where, and how attractions are developed. A given attraction might take on quite different form depending on whose policies were followed for its development and operation.

| KINDS OF ATTRACTIONS | DEPENDENCY UPON NATURAL RESOURCES | DEPENDENCY UPON OTHER THAN NATURAL AND CULTURAL RESOURCES | DEPENDENCY UPON CULTURAL RESOURCES |
|---|:---:|:---:|:---:|
| **TOURING CIRCUIT:** | | | |
| Roadside scenic areas | ✪ | ○ | ○ |
| Outstanding natural areas | ✪ | ○ | ○ |
| Camping areas | ○ | ✪ | ○ |
| Water touring areas | ✪ | ☆ | ☆ |
| Homes: friends/relatives | ☆ | ✪ | ☆ |
| Unusual institutions | ○ | ✪ | ○ |
| Shrines, cultural places | ☆ | ★ | ✪ |
| Food, entertainment | ☆ | ✪ | ○ |
| Historic bldgs., sites | ☆ | ○ | ✪ |
| Ethnic areas | ☆ | ★ | ★ |
| Shopping areas | ☆ | ✪ | ☆ |
| Crafts, lore places | ☆ | ★ | ✪ |
| **LONGER-STAY:** | | | |
| Resorts | ✪ | ★ | ☆ |
| Camping areas | ✪ | ★ | ○ |
| Hunting, water sports | ✪ | ☆ | ○ |
| Organization camps | ✪ | ☆ | ○ |
| Vacation home complexes | ✪ | ○ | ○ |
| Festival, event places | ○ | ✪ | ✪ |
| Convention, meeting places | ☆ | ✪ | ☆ |
| Gaming centers | ☆ | ✪ | ○ |
| Sports arenas, complexes | ○ | ✪ | ○ |
| Trade centers | ○ | ✪ | ★ |
| Service, tech. centers | ○ | ✪ | ○ |
| Theme parks | ☆ | ✪ | ○ |

✪ Highly Dependent  ★ Dependent  ☆ Somewhat Dependent  ○ Low or No Dependency

**Figure 5-1.** Resource Dependency of Attractions. The relationships among kinds of touring circuit and longer-stay developments and three categories of resources—natural, cultural, nonnatural.

Without doubt, the greatest number of tourist attractions in the U.S.A. is owned and operated by the government. Some 50 agencies of the federal government (Destination USA: 1973, 4/3) and many agencies of state governments act not necessarily as regulators but as developers and managers of land and water resources that function as tourist attractions. At present, all of the more spectacular natural resource attractions of the country are under government ownership and management.

Eighty-five percent of all outdoor recreation lands are owned by the federal government (Destination USA: 1973, 4/3). Some of the key federal agencies to hold and manage tourist attractions are the Forest Service (155 national forests with over 230 million visitor-days a year); the Corps of Engineers (3,080 recreational sites with 480 million recreation days per year); the National Park Service (298 units, over 350 million visitors per year); and 1.4 million visitor use days at sites administered by the National Wildlife Refuge System (Clawson and VanDoren: 1984). Other agencies that have recreation sites are the Bureau of Reclamation, Bureau of Land Management, Bureau of Indian Affairs, Tennessee Valley Authority, and the Department of Defense (Destination USA: 1973, 4/9-12).

Policies governing these attractions differ from those of private enterprise and vary between agencies. Most are highly resource-protection and are not primarily for recreation or tourism. Often the policies reflect more of the values of the leaders of the agencies and their mandates than recreation and tourism values. Generally, policies are applied similarly across the country.

In some instances, such as the Kentucky state parks, protection of resources is important but so is visitor use and revenue-production in lodges and other commercial-type operations. Mission Bay Park in San Diego, California, an important boating-swimming-fishing-attraction complex owned by the San Diego Parks and Recreation Department, pays for nearly all of its public operational costs from lease revenues obtained from commercial operations on 25 percent of the waterfront.

Policies of the nongovernment sector are generally oriented to either profit-making (for the commercial interests) or social welfare (for the nonprofit organizations).

But even for the commercial sector, an increasing sensitivity to the consumer has become more and more a significant part of policy. For example, self-imposed safety regulations early became a part of business for the winter sports and waterfront resort developer. The quality of construction and maintenance has improved dramatically over the years. Since negligence is generally the key to tort cases and lawsuits are bad publicity for repeat business, alert commercial attraction owners are making sure their facilities are built and maintained in good order. The private pacesetter for changing the image of the visitor experience was the construction and maintenance of Disneyland. It is reputed to have such rigid standards that no cigarette butt is allowed to remain on the ground more than 20 seconds. As discussed in greater detail in Chapter 6, profit-making enterprises must provide the quality of product and service demanded by the consumer, and consumer demand is increasing steadily.

Often forgotten but very powerful in many countries is the nonprofit segment of tourist attractions. Most of the important historic sites and buildings in the U.S.A. are owned and operated by nonprofit organizations. Their motivations are primarily specialized social good, such as preservation of history and heritage, but they frequently become involved in major investment and operational enterprises that attract travelers.

This "third sector" may hold great promise for tourism everywhere but particularly in developing countries. "To satisfy basic economic, social and cultural needs, people are experimenting more and more with local initiatives, self-help, autonomous but cooperative ventures" (Knechtel: 1985). This falls outside the traditional public or commercial sectors.

An increasingly important pattern of ownership and operation is that of mixes of organizations and agencies. For example, Texas statutes allow counties and cities to issue bonds to purchase and improve buildings or historically significant objects. These governments may, in turn, lease the site to a private organization for maintenance and operation (Historic Preservation: 1975, 14). Policies, therefore, reflect this mix.

A modern attraction complex has sprung from earlier prototypes that includes a mix of recreation activity features such as boating, swimming, golf, tennis, and personal vacation homes, either as condominium units or detached houses. Many of these complexes provide a management service so that owners can recoup some of their costs by renting to others while they are not occupying the vacation homes. Policies vary with each individual development.

## Location and Land Use

From a planning perspective, location of attractions is an important but poorly researched factor. Certainly the several generations of transportation modes have influenced the popularity of attractions at certain locations. Jet planes directly escalated the use of attractions worldwide.

But closer examination of attraction location suggests the possible influence of three important factors—ease of access from markets (distance-price), relationship to service centers (cities), and relationship to resource foundations.

*Air, land, and water* access to attractions by market sources is an essential factor of attraction location. Air routes, highways, railways, and waterways prescribe destination attraction use. This fundamental means that isolated resort or tour attraction properties, no matter how important the resource may be, are not going to enjoy heavy attraction patronage. When a geographer maps the routes across a region, a great deal of understanding of potential attraction development is obtained. The air, highway, rail, and waterway networks define those areas that are accessible and have greatest potential.

*Relationship to service centers* must be added to access as influencing attraction location. Because the businesses sought by travelers are mostly those also used by residents of cities, an urban relationship becomes important for attrac-

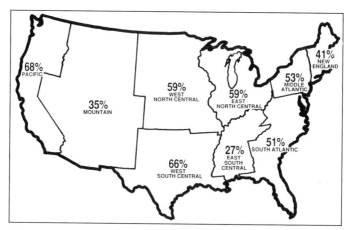

**Figure 5-2.** Origin-Destination, U.S. Travelers. Map of 1985 travel flows in the U.S.A. Percentage of travelers who originate in the region to which they are destined is shown (1985 National Travel Survey: 1986, 15).

tion use. Areas within and surrounding cities become better locations for attractions than do remote areas. Sometimes the radius from a city can be expanded with the addition of good access.

The quality and quantity of *resource foundation assets* (and lack of liabilities) is another important attraction location factor. Many tourist activities demand particular resources assets. (The relevance of this factor is described in Fig. 5-1.)

When the planner overlays these three factors, much about attraction location is defined. The best locations are easily accessible, lie within a reasonable radius of service centers, and are often supported by natural and cultural resource assets. This combination of factors can be of great assistance when evaluating the potential of a nation or region for increased tourism development. If, for example, a search identifies a reasonably accessible community and a favorable but undeveloped resource base nearby, there may be potential for attraction development.

For both touring and destination types of tourism, the origin-destination relationship of tourists is important. According to the 1985 National Travel Survey (see Fig. 5-2), domestic weekend and vacation travel distances vary. Weekend travel in 1985 averaged 660 round-trip miles (434 in 1976). Vacation travel, as one might expect, averages more miles—980 round-trip miles in 1985 (214 miles less than in 1976). This suggests that *weekend* tourism attractions are located at destinations an average distance of around 330 miles from home, whereas *vacation* attractions are spread over an average of nearly 500 miles from home.

This means generally that whereas a portion of U.S. tourists may travel either short or long distances, most seek destinations or touring travel that is not excessively far away. As illustrated in Figure 5-2, for example, in most regions over half of the visitors originated within the region. Exceptions were New England (41 percent), East-South-Central (27 percent), and the Mountain region (35 percent).

One advantage of advanced planning of attractions is to integrate the attractions that are interdependent. In today's market competition, the small and separate attraction pales before the larger complexes. Except where touring is the attraction, travelers seek to reduce the friction of many separate trips and stops for relatively unimportant attractions.

Attraction integration can create larger complexes. And larger complexes, even when the several parts are owned and managed separately, have greater interest value for visitors and are more promotable. This principle of clustering and its success have been proven by the theme park complexes.

Many examples demonstrate this principle. Cities contain many attraction complexes—clusters of entertainment, shops, museums, parks. Downtown Amsterdam, for example, contains a cluster of cultural centers and architectural grandeur—Central Station, Amstelkring Museum, Old Church, New Church, General Post Office, Royal Palace, University, Schouwburg Theater, Rijkmuseum, Stedelijk Museum, and Rembrandt's House. Canal boat trips create clusters by connecting many attractions together on a tour. The extensive U.S. and Canadian national parks are virtually clusters of attractions—wildlife, waterfalls, geysers, forests, lakes, streams, and rare plants.

Related to clustering is the planning problem of the contrived versus the authentic. Whereas planning can assist the integration of complementary attractions, there may be danger in contriving unrealistic situations with "too much" planning. Cohen (1979, 27) has identified four types of touristic situations:

1. Authentic—objectively the real thing.
2. Staged authenticity—covert (tourist unaware) tourist space.
3. Denial of authenticity—may be real but is doubted by skeptics.
4. Contrived—created settings admittedly not originals.

One may be critical of anything but "real" events and places. But landscapes are reflections of cultures, and cultures change. If one is to interpret some earlier age and all artifacts are gone, there is no recourse except to recreate the situation. Many historic sites are of this nature, and when properly and honestly presented, they provide enriching experiences for travelers. The Fortress of Louisbourg in northern Nova Scotia (Megill: 1971), consisting of more than 50 buildings rebuilt on original foundations at a cost of over $20 million, represents a "contrived" attraction. Yet it is doubtful if the 150,000 visitors a year are disappointed because these structures are replicas of the originals built in 1719. The architecture, artifacts, interpretation, period costuming, and even French-speaking attendants provide the visitor with a sense of what this community might have been in the 1740s before it was intentionally demolished (in 1760) by the British to prevent French return.

Planning of attractions at the regional scale as well as the site scale must consider the consequences of visitor volume. Whereas publicity has often exaggerated the problems of overuse and overcongestion of sites (Jordan: 1979, 31), there are concerns that must be addressed.

Generally, theories and descriptions of capacity issues fall into three groups (see Bouchard, 1974: Wagar, 1974; Stynes, 1977). For one, there are physical environmental consequences of overuse—wear and tear that publics exert when visiting sites. Natural resources can be eroded. Wildlife, plants, and soils can be eroded where too many people are given unguided and uncontrolled use of these resources. Second, there are social capacities of attractions. A wilderness hiker may believe that the experience is less than satisfactory if one other person is seen on the trail. A beach lounger's experience may be depreciated greatly if several thousand other bathers are not there also. There is a distinct difference between density (numbers per area) and crowding (feelings of too many people). Social capacity also refers to the host-guest relationship. Industrialized societies can accept much larger numbers of visitors than others. A third capacity concern is that imposed by management. Because attractions are owned and managed by someone, all sites are subject to limits of budget support for capital improvements and staff.

It is for those three factors that arbitrary limits of visitors are unworkable. Generalized standards of capacity are of no help to planners because every situation is different. Actually, most sites have great elasticity provided that good design and management have been applied.

One solution to capacity problems is to plan for a better balance among the several scales of development and developers. Based on his study of tourism in Bali, Rodenburg (1980, 194) suggests that "large industrial tourism" development may not always be appropriate. "Craft" and "small industry" developments may not benefit from mass promotion and large investment backing, but are usually more sensitive to local social and environmental capacities. Large industrial tourism developers, dominantly coming from outside (especially in developing countries), are prone to be less aware of local values and resource protection.

Another solution frequently offered for relief from oversaturation is dispersal. In theory, this appears plausible—if an area is gaining too much use, why not spread the load? Unplanned dispersal can scatter development over a larger area, thereby reducing its value as a protected resource region. The additional infrastructure of roads, water, waste, and police is much less economic than when concentrated. Hawaii's answer to congestion on Waikiki Beach was increased development and promotion of attractions on the other islands. Whereas this has increased the volume of visitor use of these other attractions,

Waikiki, merely because of its dominant and nonsubstitutable image in the marketplace, remains a very popular place.

A workable planning solution to reduce people-erosion of fragile natural and cultural resources is interpretation. Only a few visitor market segments must actually bodily occupy fragile resource sites; they are well satisfied with vicarious experiences nearby. A well-designed visitor center with exhibits, displays, audiovisual presentations, lectures, skits, and literature can often provide a richer and more entertaining visitor experience than without them. Short nature walks nearby can often provide sufficient contact with natural resources. Often, as much or more can be learned and enjoyed from an historic site by watching a film or a pageant near the site than actually walking on it.

Planners need to give careful thought to all issues relating to attraction capacity problems.

## Conclusions

Of all the factors that influence tourism, attractions are probably of greatest significance. Historical and current information about attractions can prove valuable to tourism policy and planning. Guiding the development of attractions is critical to the functioning of the entire tourism system. Attraction location and development are important throughout the regional planning process. From a study of attractions, the following major conclusions related to planning can be drawn.

*Attractions are physical place settings for experiences.* Attractions, for purposes of planning, are physical developments that in turn provide settings for meeting market needs. Although attractions cannot command user participation, they can be so located, designed, and managed to increase the probabilities of satisfying visitor travel purpose.

For most travel attractions, an abundance of high-quality resource assets and freedom from resource liabilities are essential. Physical resources can be grouped into two categories important to planning. Natural resource assets—climate, vegetation, wildlife, water, and topographic change—are strong foundations for many attractions. Cultural resource assets—historic sites, ethnic areas, manufacturing plants—are the basis for development of interest to other travel segments. Some manmade attractions such as theme parks do not depend as much as others on natural and cultural resource assets for success.

*Attractions can change over time.* The importance of physical places for attractions can change over time due to two main influences. The physical characteristics can change—cities may improve or decay; resources such as water and wildlife may improve or deteriorate in quality; developed destinations may wear out. Second, markets can raise or lower their imagery and preference for attractions over time. Popularity is as much a market as a physical factor. Influences such as international monetary exchange, fad, fashion, personal interest, public policy, and competition can change market segment interest in attractions.

*Touring attractions differ from longer-stay attractions.* Those attractions for touring circuits are more tightly aligned with the travelway and are not of kinds that can be repeated often by the same tourist. Longer-stay attractions are more tightly associated with special resource assets at destinations and must be able to withstand repetitive use by the same tourist. In many instances, land uses between touring and destination types of attractions are not compatible and therefore demand special planning attention. In other instances, they aggregate to provide a stronger market for services.

*Ownership and sponsorship varies.* Neither the public nor the private sector has a monopoly on the development of attractions for tourism. The great diversity and quantity of attractions throughout the world derive from the many development opportunities by all three sectors—government, nonprofit organizations, and commercial enterprise. Although their motivations and policies vary, they provide the singular purpose of supplying the vehicle for travel attraction participation. Public parks, forests, and beaches stand alongside commercial theme parks and nonprofit historic restorations. More and more, owners and managers of attractions and policymakers are beginning to recognize the need for integrated planning, not only for the user but in order to foster their specific organizational objectives.

*Location factors are important to planning.* Geographical distribution of attractions is not homogeneous over a region. For planning purposes, three factors are most important. First, air, land, and water access is important to link origins with destinations. Transportation routes clearly favor some locations over others. Second, all attractions are related to the nearest city as a travel service center. Transportation termini and travel services thrive best at cities. Most of the services used by travelers are also in demand by residents, who, in turn, favor urban locations for food service, entertainment, and even hotels. A third factor is that of resource foundations. The most favorable locations for attractions are those with resource assets within a reasonable radius of cities and accessible by the greatest number of market segments.

*Both urban and rural attractions have potential.* Contrary to popular belief, travel destinations are not exclusively at remote destinations; urban areas are also important. Discussed in greater detail in Chapter 11, Community Tourism Planning, are the several factors that bring both rural and urban sites into play. Certain kinds of market activity preference, such as for hunting, fishing, winter sports, and beach resorting, favor resource settings often away from cities. But even for these markets, experience has demonstrated the value of relationship to a nearby city to provide desired services such as entertainment, shopping, and especially food services. Cities are increasingly becoming travel destinations in their own right because of the many attraction amenities—convention centers, museums, art galleries, parks, performing arts centers, shopping centers, specialty foods, and sports arenas.

*Attraction capacities need planning emphasis.* Whereas good planning, design, and management can prevent many saturation problems of attractions, certain limits must be recognized. Environmental, social, and management problems have demonstrated the need for consideration of capacity. Greater dispersion, reduction in scale of development, and interpretation programs and facilities often can reduce the difficulties of capacity limits. Each destination area with its own resources and markets must be dealt with on its own conditions when planning to avoid erosion due to capacity problems.

*Attractions are clustered.* For many reasons, attraction features function best not in isolation but when clustered together. For example, most national parks represent clusters of attractions such as beautiful scenery, hiking trails, wildlife, interesting topography, swimming beaches, and historic sites. A cluster or trail bringing many historic buildings together in common context is far superior to an isolated site. Winter sports resorts are not merely ski areas; they often include other snow and ice sports as well as other recreation facilities such as swimming pools and indoor amenities. Even commercial attractions such as theme parks are clusters of many rides, shows, and other activities. Although important in the past, clustering is stronger in contemporary tourism planning because of modern transportation modes, marketing mechanisms, and higher investment of development.

# Bibliography

*America's Best, 1987.* Six Spectacular Motorcoach Tours of the U.S.A. and Canada. Hollywood, CA: Flair Tours, 1987.
*Annual Report '85.* Vienna: National Tourist Office, 1986.
Anthony, Carmen. "The New Orlando," *Corporate & Incentive Travel,* 5 (2), February 1987, 62–72.
Balsdon, J.P.V.D. *Life and Leisure in Ancient Rome.* New York: McGraw-Hill, 1969.

Blank, Uel, Clare A. Gunn and JJR. *Guidelines for Tourism-Recreation in Michigan's Upper Peninsula.* East Lansing: Cooperative Extension Service, Michigan State Univ., 1966.

Bouchard, Andre. "Carrying Capacity—Management Tools for Parks," *Recreation Canada,* 32 (1), 1974, 13–19.

Burkart, A.J. and S. Medlik. *Tourism—Past, Present and Future,* 2nd ed. London: Heinemann, 1981.

*Caribbean Sales Guide,* supplement to *Travel Agent Magazine,* Section 2, 231 (8), February 9, 1987.

Clarke, I.F. "The Grand Tour," *Tourism Management,* 7 (2), June 1986, 135–138.

Clawson, Marion and Carlton S. VanDoren, eds. *Statistics on Outdoor Recreation.* Washington, DC: Resources for the Future, 1984.

Cohen, Eric. "Rethinking the Sociology of Tourism," *Annals of Tourism Research,* 6 (1), Part I, January/March 1979, 18–35.

*Destination, USA.* Report of the National Tourism Resources Review Commission, Vols. 1–6. Washington, DC: U.S. GPO, 1973.

Dulles, Foster Rhea. *A History of Recreation: America Learns to Play,* 2nd ed. rev. New York: Appleton-Century-Crofts, 1965.

Ferrario, Franco F. "The Evaluation of Tourist Resources: An Applied Methodology," *Journal of Travel Research.* 17 (4), Spring 1979, 24–30.

Gee, Chuck Y. *Resort Development and Management.* East Lansing, MI: Educational Institute of the American Hotel and Motel Association, 1981.

Gunn, Clare A. *Vacationscape: Designing Tourist Regions.* Austin: Bureau of Business Research, Univ. of Texas, 1972.

*Historic Preservation.* Houston: Houston-Galveston Area Council, 1975.

*Japan Travel News.* February 1987, p. 1, Japan National Tourist Organization.

Jordan, Robert Paul. "Will Success Spoil Our Parks," *National Geographic,* 156 (1), July 1979, 31–59.

Knechtel, Karl. "The Role of the 'Third Sector' in Tourism Development," unpublished paper. Ottawa: Tourism, Canada, 1985.

*Liberty Centennial Travel Guide.* Albany: Division of Tourism, 1986.

Mansfield, E.D. *Exposition of Mackinaw City and Its Surroundings.* Cincinnati: self-published 1857.

Megill, Doris K. "Louisbourg: The Ill-Fated Fortress," *The Review* (Imperial Oil), October 1971.

Milligan, B.C. "Are Incentives 'Wave of the Future' for Cruise Line Bookings?" *Travel Trade,* 72 (4), January 1987, 16–17.

Murray, James H. et al., eds. *The Oxford English Dictionary,* Vol. 1. Oxford: Clarendon Press, 1933.

*1985 National Travel Survey, Full Year Report.* Washington, DC: U.S. Travel Data Center, 1985.

*1976 Travel Survey.* Washington, DC: U.S. Travel Data Center, 1977.

Plog, Stanley C. "Why Destination Areas Rise and Fall in Popularity." *Cornell HRA Quarterly,* 14 (4), February 1974, 55–58.

Rodenburg, Eric E. "The Effects of Scale in Economic Development—Tourism in Bali," *Annals of Tourism Research,* 7 (2), 1980. 177–196.

Stynes, Daniel J. "Recreational Carrying Capacity and the Management of Dynamic Systems," presentation, National Recreation and Park Association Congress, October 2–6, 1977, Las Vegas, NV.

"Tauck Tours" (advertisement), *Tour & Travel News,* 032, February 9, 1987, 7.

*Tourism in Japan 1986.* Tokyo: Japan National Tourist Organization, 1986.

Turner, Louis and Ash, John. *The Golden Hordes.* London: Constable, 1975.

Wagar, J. Alan. "Recreational Carrying Capacity Reconsidered," *Journal of Forestry,* 72 (5), May 1974.

# Chapter 6

# Services

In the minds of many the terms "tourism," the "tourist industry," and "service-facilities" are one and the same. It is through the services and facilities for tourism that local economies get their first impact from tourism. Accommodations, food services, bars, and retail sales are conspicuous business evidence of tourism development. A researcher of ancient inns of Greece and Rome observed the importance of facilities and services for travelers: "Are not a nation's inns an index to its roads and methods of transportation as well as a true reflection of the national character?" (Firebaugh: 1924, 2).

Therefore, critical to the planning of tourism for a region is identification of the planning needs of the services sector. (Because of its special function, transportation, which logically could be included within the services component, is treated as a separate component in Chapter 7).

## Historical Background

Along ancient travelways, probably the first accommodations for lodging were private homes. As travel increased, this personal hospitality became a burden and a bother. Roadside entrepreneurs then began to build special facilities for lodging and feeding travelers and their horses. However, because homes had been preferred for centuries, especially by wealthy travelers, facilities for other travelers were crude. The khans, inns, and public houses of Greece and Rome were usually "vast and miserable sheds where beasts of burden and men were herded indiscriminately into a hurly-burly" (Firebaugh: 1923, 55) and were filled with all imaginable encumbrances, such as filth, insects, adulterated wines, and bad food. In Plato's time they gave all business a bad name. A passage in Lib. XI, sec. 918, the Laws stated, "On this account (eagerness for gain) all the lines of life connected with retail trade, commerce, inn-keeping, have fallen under suspicion and become utterly disreputable" (Firebaugh: 1923, 65).

Accommodations for tourists remained primitive and fraught with risks for many centuries. During the seventeenth century, disease from infected quarters was quite common. In his tour of the continent in 1643, the Englishman John Evelyn "caught smallpox in Geneva after having slept at an inn in Beveretta in a bed that had just been used by the landlady's daughter who had not fully recovered from the malady" (Hibbert: 1969, 14). There was not much demand for hotels in this early era because horseback and horse-drawn carriage travel were limited in capacity and only the wealthy few could travel.

The U.S. carriage era fostered the tavern or inn, located about 15 miles (one day's ride) apart. The first in America, the Jamestown Inn, was built in Virginia in 1607 (American Innkeeping: 1971). These early inns often boasted food and lodging for man and beast, but provided only the minimum of comfort—often only one open room for the traveler to roll up in his own bedroll with his feet toward the fire. Service depended on the personality of the owner-host and what he offered to the tourist. The lack of amenities was tolerated because the quality of travel was even worse:

> ... the yellow flare of oil lamps and the scrape of hoedown fiddles in a crossroads tavern ... well, that made you forget the bruises of the log-ribbed trail and springless stagecoach as you hastened to the night stop (Wayne: 1942, 11).

But taverns of the stagecoach era had their problems. Supplies for feeding and lodging travelers and horses often ran short; labor was difficult, often employing children of the innkeeper; and because of low initial capital required, too many were built and not all withstood the competition. The displacement of inns with the coming of the railroads was very abrupt. Equally important to the demise of the stagecoach tavern or inn was the new ruling that separated the barroom from the inn:

> From the time of the earliest colonial regulations until the second quarter of the nineteenth century, the inn and the barroom had been legally handcuffed together. No inn could retail liquor without also making provision to care for travelers (Yoder: 1969, 178).

In the late 1800s, steamship and railway travel as well as general prosperity brought about a boom in the quantity and quality of accommodations and service. The rail lines in Great Britain fostered one of the first networks of good-class hotels—Charring Cross, Great Eastern, Grosvenor, Midland, North British, Caledonian (Jones: 1959, 370). With the addition of steamship travel to port cities, new hotels were built in London—Savoy, Berkeley, Claridges, Cecil, and later the Ritz, Strand Palace, and Regent Palace.

In the U.S.A., "The elegant hotels at Saratoga or Newport attracted people who formerly would have vacationed in the country only at exclusive house-parties. . . ." (Dulles: 1965, 160). Locations at lakeside, seaside, or mountainside became focal points for these sometimes very elaborate vacation hostelries. Sometimes resort hotels were built by the railroads, such as the El Capitan in Yosemite by the Central Pacific Railroad and the Del Monte by the Southern Pacific (Van Orman: 1966, 74).

Whereas some city hotels were opulent and favored only wealthy trade, most were modest in appointments and price. From the beginning the city hotel inherited an important function from the stagecoach tavern—"the great social center of the general public, the favorite place for balls, banquets and other affairs" (Williamson: 1975, 9). The technological advances in transportation probably had more to do with the popularity of hotels than did their quality of service. "It was not until the 1870s that steam-heat and all other facilities of modern plumbing came into widespread use" (Williamson: 1975, 65).

Around 1900, the first type of luncheon or snack service—something less than full-scale restaurant service—appeared in the drugstore. Here, the quick-service counter was born and offered sandwiches, desserts, and drinks with a carbonated water base. Important to the American travel scene was the diner—a small independently owned, short-order food service with a counter, booths, and plastic surfaces (Gutman and Kaufman: 1979). It became a hangout for locals as well as a relief stop for travelers. The style was art deco in porcelain, glass block, and stainless steel, and shaped like a railway car (the early ones). These have largely been replaced today by fast-food shops. The fast-foods concept began about 1946. This was generally defined as including high-speed preparation and service, using low-paid labor and high-speed equipment (Thorner: 1973, 7–8). The development of large-scale commercial food freezing in 1929 fostered the development of fast-food restaurants.

The modern recreation vehicle has its ancient antecedent in the horse-drawn gypsy wagons of Europe in the 1600s. The wagons served as both transportation vehicles and as homes. The Conestoga and other forms of covered wagons were the Americans' inventive answer to home mobility in the settlement of the West. As automobile travel expanded in the 1920s and 1930s, homemade trailers of plywood on old car axles provided sleeping accommodations on vacation trips.

To provide temporary housing during World War II, a new industry sprang up to produce trailers with kitchens and sometimes toilet facilities. These stimulated interest in using such vehicles for postwar vacations. Meanwhile other mutations appeared to provide rolling housing for tourists: converted buses, tents on trailers, and converted trucks. Paralleling this history has been the development of recreation vehicle parks and campsites to accommodate them. Whereas country schoolyards and churchyards (offering privies, lawns, and hand-pumped water) sufficed at the start, improved standards and technology demanded new generations of parks.

Lodging always has been closely associated with travel modes. It is not difficult to understand why motels developed in those countries where the automobile travel burgeoned into a significant segment. The motel was not spawned by the hotels but rather was an innovative response by new entrepreneurs to a new market demand. The motel in the western U.S.A. was an outgrowth of the tourist camp and cabin for new mass travelers in low-cost mass-produced Ford cars. Automobile-camping travel to the Grand Canyon first exceeded train-hotel travel in 1927. Hotels first vigorously opposed the motel concept and soon learned how competitive they were, especially where older hotels were obsolete—poor facilities and in the wrong location.

The new large corporate managers of the chains and franchises soon began to dominate the accommodations field. Their larger purchasing and promoting power together with no-cost reservations tended to put the smaller "mom-pop" accommodations out of business. Recently, however, there is a visible trend toward greater diversification of lodging services due to recognition of several market segments, from luxury suites to budget rooms. Small inns and "bed-and-breakfast" services have increased greatly in popularity. Throughout the history of travel businesses, service has remained a very important element.

In all periods of travel, the market desire to purchase curios, souvenirs, and gifts has stimulated many businesses to provide this service. Although critics demean this as a trashy frivolity of tourism, it has been and continues to be a viable service segment.

## Scope

The scope of services and facilities as used here includes all those that support tourism (except attractions, transportation, and information-direction). Services are primarily commercial and are focused in the array of first-contact businesses used by tourists.

Integrated tourism planning must take into consideration the many kinds of services used by travelers during the total trip experience. For example, for only the market segment of sports fishing in the Caribbean, Schmied (1985, 23) lists the many services used during three stages of the trip:

Pretrip planning and preparation: fishing equipment, fishing information (charts, facilities, target species, services), licenses, clothes, accessories, boat repair and maintenance, boat accessories, car repairs and maintenance, travel arrangements (passport, visa, airline, local transportation, hotel reservations), diving equipment, insurance.

Fishing excursion: local transportation, gas and oil, access (boat ramps, services, guide, dive shop), bait, tackle, ice, food, lodging, repairs to boats, supplies.

Posttrip: supplies and services for catch handling, preparation, stora‌ ping, trophy mounts, boat/car maintenance and repair, equipment replacᵤₘₑₙₜ, transportation home.

Generally, traveler expenditures are greatest on the three categories of services: lodging, food and beverage service, and retail products and other services.

## Lodging

Lodging choices for tourists are reasonably broad and include the following main categories: staying with friends and relatives; motels/hotels and motor inns, hostels, bed-and-breakfast, pensions, hiking shelters, camping facilities, and recreational vehicles; resort lodges; personal vacation homes; timesharing condominiums; and watercraft.

Today the distinctions between hotel, motel, inn, and lodge are diffuse. High among success factors for all kinds of lodging (in addition to superior management) is the market-service match—selecting a specific market segment and planning for it. The Regent in Hong Kong is reported to have paid off its $65 million cost in three and one-half years with high occupancy because it met a need for elegant facilities for upscale clothing buyers from Manhattan (Matthews: 1985).

Luxury-to-budget segmentation of lodging has been a recent trend. Haynes (1986, 42) cites the addition of deluxe suites with living room (suited to business meetings) in the newer hotels by Four Seasons, Hometels, and Hyatt. Hometels even includes a wet bar, American breakfast, and a two-hour cocktail party each evening on the patio. To better reach the business segment, Hyatt has its own video-conferencing system. At the other end of the scale, low-cost and no-frills accommodations have found their own market niche.

Pleasure travelers also now have a wide diversity of accommodations to choose from depending on interests and needs. Youth hostels, most popular throughout Europe, cater to the younger hiking, cycling, and outdoor market segments. The converted monasteries and castles of Spain fit the market need of travelers interested in history. Resort lodges include a variety of fitness and outdoor sports facilities as well as fine food and entertainment. National and state (provincial) park lodges provide fine accommodations (Banff Springs Hotel, Kentucky Lake Lodge, Ahwahnee Lodge at Yosemite National Park) adjacent to natural resources and cater to interested market segments. Club Med typifies the all-inclusive resort for those markets seeking lodging, food, and sports in a prepaid package. For gambling markets, hotel-casino combinations are popular.

With the great expansion of recreation vehicles (camper vans, motor homes, camper trailers, and pick-up campers) has come camping facilities, part of which are provided by government agencies and part by private enterprise.

## Food and Beverage Service

Travel has tended to foster both a resurgence of old-style restaurants as well as stimulate a much larger mixture of restaurant kinds than ever before. In almost any town today food service is offered in much greater diversity than in the past. Schirmbeck (1983, 8) groups restaurants of the world into several classes:

- Traditional foods in well-known setting
- Lower price level and contemporary design
- Quick service in impersonal, international setting
- Delicatessen meeting place
- Fast-food franchise hamburger and sandwich shops
- Luxury full-service in elegant or nostalgic settings

The most popular worldwide are the British pub, the Italian *locanda,* the French *bistro,* and the German *Wirtshaus.*

Food service has become a massive segment of all business in several countries. In the U.S.A., about one-third of all meals are eaten in restaurants. Food service is the largest retail employer and of great importance to tourism.

Resorts are as much attractions today as facilities and services because of the activities provided. Accommodation and food service functions, such as at resort motels and vacation home complexes, continue to be important, but resorts are now known as much for their tennis, golf, sailing, fishing, boating, swimming, and winter sports. The major business of today's cruise ships is not trasnportation; they are often called floating resorts—a mixture of entertainment, lodging, food service, and visiting attractions at port cities. A new form of resort is emerging that provides recreation vehicle space but offers the recreation activities of a resort hotel. A variety of personal watercraft today, especially the larger cabin cruisers, are used for sleeping and food service.

## Retail Products and Other Services

An essential part of tourist consumer demand is that of a variety of retail products and services. Attraction operators (both commercial theme parks and public parks) have learned that it is insufficient to develop and operate the attraction only on its own merits. The public seeks a variety of souvenir items, gifts, and mementos at the same time it seeks enrichment and participation in attraction activities. These range all the way from picture postcards and monogrammed T-shirts to costly handcrafted items, often by special national and ethnic artisans.

At fishing, camping, and winter sports sites, a great many supplies, clothing, and sports items are in demand by tourists. Visitors may have left their needed items at home or did not know of their needs until arriving at the site.

Food supplies, insect repellants, drugs, camera film, cigarettes, and a host of other personal items are often in need by travelers. They expect to find these services and products in the same or even greater abundance and availability as at home.

Many specialized activities, such as golf, mountain climbing, hunting, fishing, backpacking, charter fishing, and river running, require the services of instructors and guides. Some destination areas depend greatly upon the economic impact of such guide services.

## Influence of Ownership

The ownership and management of tourist services and facilities, as the discussion of the scope has suggested, are largely private enterprise or personally oriented. Therefore, regional planning for tourism, if it is to be effective, must include an understanding of this sector's policies and practices.

Fundamentally, commercial tourist services and facilities (as described for commercial attractions) operate with the same purpose of all other business—to make a profit. However, there seems to be increasing misunderstanding of the term "profits," some believing that in the tourism, recreation, and resource development field, profit-making is somehow evil. Many, especially those sponsoring government recreation and park areas, seek a more altruistic and a broader social responsibility from business. But first and foremost is the responsibility for private enterprise to remain economically viable. Such economic viability comes from "profits," which in reality are costs of doing business. According to Drucker (1975):

> There is no conflict between "profit" and "social responsibility." To earn enough to cover the genuine costs which only the so-called "profit" can cover, is economic and social responsibility—indeed it is the specific social and economic responsibility of business. It is not the business that earns a profit adequate to its genuine costs of capital, to the risks of tomorrow and the needs of tomorrow's worker and pensioner, that "rips off" society. It is the business that fails to do so.

In a sense, all owner-managers of tourist services and facilities (governments as well as business) have similar ultimate goals: the satisfaction of tourist needs.

Much of this satisfaction, especially in developed nations, derives from the demands of the public. It is doubtful whether a motel owner would sell rooms if travelers did not arrive at that location seeking overnight accommodation. The businessperson has to be creative enough to develop the facility and service and offer it at a price acceptable to the public.

> It is the customer who determines what a business is. It is the customer alone whose willingness to pay for a good or a service converts economic resources into wealth, things into goods. . . . What the customer thinks he is buying, what he considers value is decisive—it determines what a business is, what it produces, and whether it will prosper (Drucker: 1973, 61).

For regional planning of tourist services and facilities, it may be helpful to recognize differences among four categories of ownership management: independent, franchise, quasi-government, and nonprofit.

The *independent ownership* and managership, typical of the "mom-pop" category of business, operates on its own forms of personal enterprise policies—market segmentation, pricing, range of services, and facilities. Whereas it remains a very important category, it is much smaller today relative to the great expansion in the 1950s of small motels, resorts, and roadside restaurants. One researcher (Bevins: 1971, 3) found that the economics of operation among the small outdoor recreation owner-managers varied greatly according to their goals. He grouped them into three categories: (1) those who do not wish to maximize financial returns but are in business because of beliefs in conservation, recreational values for family members, or for retirees to keep busy; (2) those who seek supplementary incomes for unemployed or underemployed family labor, and (3) those who seek the more typical economic goals of all business—revenues that will return on the investment. The trend in recent years has been toward more of the corporate types of ownership.

The *franchise, chain,* and other multiple-establishment organizations have grown greatly in recent years. The advantages cited are: greater marketing through single image and toll-free reservations, increased buying power, and uniform standards. Arrangements vary from those in which the properties (land and buildings) are owned and managed by company employees to those that are independent ownerships but agree to certain operational standards and advertising logos for promotional advantages. A popular mode is that of the Holiday Inns, which, like other chains, use similar design of buildings, central purchasing, uniform signs and logos, and uniform operational standards. Most of the inns are owned by local people who have a franchise arrangement with Holiday Inn.

Franchising, born in the U.S.A., has promise for tourism service development elsewhere. It provides for local control but has the advantages of a larger organization. Experience has shown that there may be some difficulties in adapting franchising to other countries. Ashman (1986, 41) identifies a few: lack of understanding of its function, arbitrary governmental restrictions against it, labor requirements, property ownership laws, and quality control.

Franchising demands the right balance between centralized control and unit mangement. Kaplan (1984, 20) cautions the application of centralized manufacturing techniques to food service organizations. Too much decision making at the top with few rewards to unit managers can divorce a company from the realities of consumers and service. Production speed, product quality, freshness, speed of service, sales promotion as well as administration and training of personnel, operations, and quality control are best handled at the unit level. The more successful franchise operations recognize the importance of unit level decision making with adequate rewards at the same time that efficiencies of large scale are obtained.

*Quasi-government* commercial operations, usually called concessions, are of increasing importance. In the U.S.A., many federal and state resource and land agencies have concession agreements with private businesses to provide services to the public on government land. These include hotels, motels, trailer and other camping facilities, restaurants, stores, service stations, and marinas. Reasons cited for a concession arrangement—private profit-making operation on government land—are:

1. Private investment and arrangement reduce the need for public financing.
2. Revenues can accrue to public agencies from concession operations.
3. Innovation and economy may result from management responsibility shared with the private sector.
4. Local economies may be strengthened through profit opportunities for the private sector.
5. Greater recreation opportunity for the public may result from the provision of facilities or services that the managing agencies could not provide.

At the same time, several barriers exist to limit greater use of concession arrangements on public lands:

1. Concession interests may conflict with the management purposes for the public lands.
2. Such businesses are highly seasonal and profits are affected substantially by weather conditions.
3. Concessioners do not hold title to the land, making loans difficult to secure and tenure uncertain.
4. Federal and state civil service regulations may create difficulties in contracting for personal service.

5. Inconsistent or shifting public policies may create uncertainty for entrepreneurs.
6. A high degree of onsite supervision by the public agency is generally needed in order to ensure acceptable standards of public service (Outdoor Recreation: 1973, 82).

Because each concession usually holds a monopoly as a business, it is not subject to the same competition as other businesses outside the control of the agency. Sometimes the political and managerial constraints can perpetuate bad service. For example, a 1977 report on the concession services in Yellowstone National Park described the facilities at that time as "decrepit, unpleasant and potentially unhealthy . . . company employees are badly paid, badly trained and fed in unclean staff cafeterias." The report adds that "sanitation standards in restaurants are below par and that lodgings are in poor repair, inadequate staff . . . food is frequently left overnight at room temperature, allowing potential contamination" (Lichtenstein: 1977). In recent years, new policies and contracts with more competent concessioners have remedied this situation.

*Nonprofit organizations,* such as youth clubs and churches, often own mess halls, lodging, and campground facilities with extremely varying policies. Some are of poor quality because of the lack of financing and competent management, and they depend solely on donations for support. Others are virtually palatial resorts that are "profitable" in the sense that the revenues far exceed their immediate operating expenses. Some are oriented to conservation-resource protection; others are strongly program-oriented. Each depends upon the policies of its parent institution.

One area of contention between private enterprise and government is the problem of control. Participants of the public and private sector hearings of the National Tourism Policy Study (1977, 30) cited several problems including time-consuming bureaucratic procedures, inadequate and unimaginative strategies for implementing programs, and a lack of continuity in implementing programs. The study found that:

> Participants felt that ineffective implementation of Federal programs had exacerbated inadequacies in tourism development activities, (created) difficulties for small business survival, conflicts between environmental and developmental goals, energy constraints on development, and inadequacies in promotion of travel opportunities in the United States, both domestically and internationally.

Throughout the world private businesses dominate the tourist services. The degree to which these are "free enterprise" depends upon the extent to which the free market system is not interrupted. Whereas pure free enterprise business does not exist, even in market economy countries, the more it strives toward certain fundamentals, the more successful it is.

Allen (1979) has identified the following five fundamentals as essential to a free enterprise economy.

1. *Private property.* In a free enterprise system all property is owned by private individuals. This is based on several premises. First is the premise that individuals know best how to manage their own property. The individual is believed to have a strong interest in not littering his or her property and conserving its resources because the owner is responsible for the consequences. The property owner has certain rights:

- The owner's right to determine how his or her property is used.
- The owner's right to transfer ownership to someone else.
- The owner's right to enjoy income and other benefits that come his or her way as a result of ownership of the property.

2. *Economic freedom.* By voluntarily cooperating with each other at the same time individual interests are pursued, the freedom of individual choice is protected. No outside force, such as government, dictates this choice. The following rights are important but do not guarantee business success:

- The right to start or discontinue businesses.
- The right to purchase any resource the owners can pay for.
- The right to use any technology.
- The right to produce any product and to offer it for sale at any price.
- The right to invest in any way.

The seller and buyer make a voluntary exchange. The market, by its own selection, tells the producer what to produce and at what price. Of course, total economic freedom must be conditioned by the rights of society as a whole.

3. *Economic incentives.* When there are incentives to work efficiently and productively, business becomes more efficient and productive. Workers receive incentives through wages and other rewards for doing good work. Businesses receive their incentives through profits. The more productive and the better a business meets market needs, the more profitable it usually becomes. However, punishments, in the form of business losses or failures, can come when the questions of what to produce and how to produce are not answered properly. For this system to function properly, there must be a minimum of outside interference. Economic incentives serve to direct scarce resources to the production of goods and services that the market values the most.

4. *Competitive markets.* In a free enterprise system, the individual can choose, and people vary in their preferences. These preferences of markets are expressed to producers by means of what is purchased. This means that there must be competitive businesses rather than monopolies. Each business can then strive for its market share. If it becomes very profitable, it invites competitors who seek their market share through even better products or services. This competition stimulates greater efficiency and lower prices. Competition spreads the decision of what to produce over many producers rather than by governmental decree.

5. *Limited role of government.* The greatest role of government in a free enterprise system is to stimulate business freedom and provide only basic rules and regulations for the good of society. Governmental intervention into day-to-day economic decision making is not part of a free enterprise system. It does not interfere with what or how to produce. Its role is to keep the system free and competitive.

At the same time that tourism has provided the opportunity for many entrepreneurs to create new travel-oriented service businesses, the field has been plagued with a high percentage of business failures. Some would argue that too many amateurs are attracted to these businesses. The business appears simple and glamorous to amateurs who soon become disillusioned by the long hours, greater responsibilities, and less profits than anticipated (Lundburg: 1979).

In order to remedy the tendency for excessive failures and poor service in many tourist businesses, many educational programs at all levels have been produced worldwide. Governments and business associations provide guidelines for successful operations, such as *The Inn Business* (1982) produced by the Minister of Supply and Services, Cananda. This guide provides constructive information on pertinent topics, such as entering the business, planning and development, operation (staff, repairs, maintenance, marketing), and other sources of help. In the U.S.A., business advice and guidance are offered by agencies such as the Small Business Administration and U.S.D.A. Cooperative Extension Service.

## Location and Land Use

Business location theory as applied to tourism is sparse and inconclusive. Smith (1983, 518) states that "previous attempts in this direction have been of limited value because of too many unrealistic assumptions, insufficient empirical data, and perhaps even a naive understanding of all the forces involved in the development of a tourist landscape." Because most businesses catering to travelers also serve local markets, some general location principles of shops and restaurants apply to tourism. However, it must be remembered that these are generally based on the distribution of goods rather than of markets, as is the case with tourism. The following discussion introduces some of the issues associated with tourism business location and land use.

Land use and location principles for tourist services and facilities could be stated rather simply—carrying them out on the land is not so simple. Basically, the land needs of the tourist service businesses are twofold: (1) land large enough and of suitable quality to support the structures and their appurtenances (parking, docks, and such) and (2) location at a point of high market volume.

These simple principles become extremely complicated to follow when a land search is made. First, because tourists are less fixed in their travel patterns than residents, suitable market locations are less pinpointed. For example, shopping center feasibility is much more tightly circumscribed than that of motel, restaurant, or resort. Second, tourist service business location is contingent upon decisions of others—by airport authorities, highway agencies, park and resource managers. Third, in order to operate at volume capacity, all service businesses that cater to tourists must also cater to local nontourists, complicating their rules of location. Fourth, business staffing and operation thrive on even rather than sporadic volume throughout the year. Therefore, a good location for summer tourist trade may not be the best for year-round success. Fifth, even when a good market location is spotted, the businessperson may encounter innumerable obstacles: zoning restrictions, price, unavailability, declining neighborhood quality, poor infrastructure (sewerage, electricity, fuel, water), high taxes, insufficient size, high costs of land preparation (demolition of existing buildings, land improvement), and environmental constraints. Sixth, concession tourist services do not have complete control of their decisions. They must condition their location, selection of site, and design to meet the requirements of the owning and controlling agency.

In Spain (Barke and France: 1986), lodging diversity has developed in response to new market segmentation of domestic and foreign travelers—package and camping holidays. Growth of hotels from 1971 to 1981 was greatest in those catering to package tours—93.5 percent for four-star and 80.8 percent for three-star hotels. The second largest growth in this same period was for campsites (37.7 percent). The uneven geographical distribution of accommodations can readily be seen as linked with several important location factors: international airport and attractions of Madrid and vicinity, coastal and island areas, and adjacent to established destinations. But further refinement of lodging changes shows the importance of all four types of accommodations in Spain: hotels, pensions, guest houses, and campsites. Special governmental promotion and government-owned paradores—a system of high-class hotels in historic monasteries and castles—has stimulated a better balance between coastal and interior tourism.

Certain services and facilities are more tightly oriented to special tourist activities and therefore demand special nearby land use functions more than others. For example, snack bars, bait shops, camper grocery stores, marina supplies, and similar businesses need locations tight to the recreational activity upon which they depend. Of course, by so doing their success is entirely dependent upon the seasonality and demand for those recreational activities. This is especially true of the services and facilities on a concession basis in governmentally managed areas.

However, the bulk of commercial services and facilities for tourism is not at remote destinations but along travel routes, especially at cities because they rely on a variety of markets in addition to tourists. Long ago, Johnson (1962) observed that awesome scenery, salubrious climate, spectacular fishing, and other natural resources are economically sterile without the well-beaten path along which the necessary services can afford to cluster. Modern lodging accommodations continue to favor three basic urban locations: (1) downtown, near the cultural, business, professional, and convention activities, (2) travel termini (expressway interchanges to city, airports) offering food, beverage, and a minimum of meeting space, and (3) suburbia, providing banquet, meeting, and recreation as well as varied food service for residents and travelers. A study in Tel Aviv (Arbel and Pizam: 1977) revealed a traveler preference for suburban hotel locations, provided adequate public transportation is available. Resort hotels are located at accessible resort destinations, often in connection with or near a city.

One study of motel guest preference indicated that the following variables were most important to them, in this order:

1. Price
2. Proximity to tourist's main route
3. Convenient parking
4. Room size and decor
5. Quiet setting
6. Attractive setting
7. Proximity to tourist attractions (Mayo: 1974, 58).

Food service business locations appear to be needed at lodging locations but generally have difficulty if they depend on tourists exclusively. Generally, local trade prefers nonmotel food service (Stevens: 1976, 57). Perhaps this can be altered by site development, building design, menu planning, and quality of service that can benefit from both local and transient trade.

For urban-oriented services and facilities for food and lodging, clustering around general locations is becoming a rule. Once thought undesirably competitive, it is now considered better business to locate near competitors. For example, a major steak house owner-manager reports that "The idea behind this is to establish what might be termed a conditioned reflex so that when people think of eating out, they go in one direction" (Walker: 1968, 9). These concentrations of food services, motels and shops should exhibit efficiencies not only of markets but of costs of infrastructure.

## Conclusions

Tourism planning must take into account the many characteristics and goals of the service sector. These include the historical affinity to both attractions and transportation, responsiveness to technological change, and changing management patterns. In addition, there are problems of seasonality, cyclicality (especially vulnerable to economic shifts), labor, capricious market trends, high capital investment (high building costs), complexity, franchising risks, overconcentration, and, because many are small, all the usual problems of small business (Destination USA: 1974, 2/29). Not all of these can be solved by planning, but several could be mitigated.

Most of the negative aspects of congestion are the result of poor planning and design on how to handle large loads. (A competent trucker does not try to haul four-ton loads in an undersized truck.) Instead of selecting sites that would erode the qualities of attractions, other and more durable sites accessible nearby can be planned to support services and facilities. Much of the criticism of commercial services from park managers and environmentalists is caused by poor understanding of market demand. Visitors need and appreciate attractive and functional service centers. "Most site deterioration is caused by poor planning or administration on the part of the agency" (Craig: 1977, 156). Large complexes of motels, food services, and shops are beginning to demonstrate that large numbers of people can be handled without feelings of congestion. Furthermore, larger concentrations of services and facilities for tourists that are located near but without intruding upon important attractions can serve them well. Without planning, businesses and services can become badly jumbled with attractions resulting in reduction in functionality of both.

Much of this problem also stems from older patterns of land ownership by both public and private owners. For example, much of the congestion of Waikiki Beach today stems from the original platting in 1906, which produced a grid of relatively small lots. More flexible planning (planned unit development, for example) that would allow interspersing high-rise buildings with open space for views and access is thwarted by the grid pattern of the past. In major parks, large visitor loads were not anticipated in earlier plans (many park managers opposed the philosophy of heavy public service). Only now is there recognition of the desirability of building large communities adjacent, but not directly impinging upon, the main (and often fragile) resource assets.

The following conclusions can be drawn from a review of the historic development and current status of tourism services as they relate to overall regional planning for tourism.

*Services are essential.* In spite of moves by environmentalists to eliminate services from parks, and in spite of obvious esthetic and congestion problems often exhibited by tourist businesses, they are essential to a viable tourism system. They are essential to the basic tourism function of providing services in demand by the tourist. Therefore, all development that attracts tourists also induces the need for services. Not only commercial attractions but also public attractions, such as national parks, forest reserves, historic sites, reservoirs, and public museums, create the need for an array of supporting services.

*Competition stimulates innovation.* The planner for tourism must consciously keep plans open for diversity and competition. Because of the capriciousness of the tourism market, private enterprises must be free to innovate and react to changing trends. For example, some park concessions have difficulty responding to new market trends because of the rigid policies of the park agency. Inflexible agreements with concessioners can limit the agency's ability to invite competition or terminate services of a poor operation.

*Services lack integrated planning.* In response to burgeoning travel markets, thousands of establishments have been created but generally are not integrated at the planning stage. Even the associations representing the various segments seldom meet on common issues. In recent years a few government agencies and some private organizations have made efforts to communicate and resolve issues of interest to all segments. Opportunities are abundant for better service and better business, even in a competitive free enterprise setting, if better planning for location, design, and interrelatedness takes place.

*Services dominate tourism economics.* Of the several components of the supply side, services dominate economic acitivity. Communities and nations seeking expanded tourism economic impact must do so through more (and often better) tourist service businesses. These provide the jobs, the incomes, and in many nations provide tax revenues.

However, it must be emphasized that more and better services do not necessarily cause more travel. To the contrary, unless more attractions and better attractiveness are developed, there will be no need for more services. The kinds and numbers of services can only be determined when volumes of visitors—the market demand—become visible.

*Ownership and policies vary.* Viewed from a regional perspective, planning must go beyond the business or "tourism industry" sector. The commercial enterprise sector is only one of several decision-making groups. In a market economy, the commercial sector has autonomy over the kind, location, development, and pricing of services. Financial returns from entrepreneurship must cover not only return on investment and added wealth but also the costs of operation, payments of fringe benefits, repair, prevention of obsolescence, and compensation for risk.

Governments in most countries own land, infrastructure, and sometimes services for tourism. Generally the justification for governmental intervention is for the public welfare. Important to planning are the many bureaucratic policies that govern the development and use of lands and services for tourism. Quasi-governmental businesses, such as concessions, require special consideration because they must not only make profits but also must satisfy governmental controls.

Nonprofit organizational policies are more tightly prescribed but provide important developmental inputs to tourism. Their decisions on development and land use are very important to planning for tourism.

*Services are dominantly urban.* The dominant geographic location for tourist services is at cities. This principle does not mean that there are none in rural regions. It does mean that the farther one reaches out from cities, the higher the risk of success for tourist services.

Tourist services thrive best in cities for several reasons. This is the terminus for all travel modes. Cities offer the greatest number of attractions. Cities provide the other businesses and industries for best balance of economic support—tourism seldom is enough on its own. All the service businesses for tourism depend as much or more on local resident markets as upon travelers.

Minor services are needed at rural locations for some outdoor recreation activities, but most services thrive best in cities.

*Services are closely associated with markets.* The conclusion that service businesses for tourism must monitor changes in markets may appear self-evident, yet the practices of the several sectors of owners suggest that it is necessary to emphasize this principle. Newer research is revealing the complexity, the diversity, and the dynamism of markets and market segments. Some entrepreneurs, familiar with a singular service, may continue to provide it out of habit, not realizing that markets were gradually leaving the business behind. The increased competitiveness and greater proliferation of services worldwide demand close monitoring of market changes.

Governments, because of logically mandated preoccupation with social welfare principles of egalitarianism, have demonstrated poor sensitivity to markets. Administrators of parks and recreation attractions frequently have been led by managers with training in biology or forestry rather than marketing. Much of the friction between these agencies stems from lack of understanding of markets by the administrators.

Nonprofit organizations, with activities and programs directed primarily to their members, often have a poor understanding of the markets who use their developments.

Greater sophistication of market information is essential for all tourist services.

*Services require good site and building design.* Whereas regional planning is at a broader scale, what is planned and developed at the site scale is fundamental. The larger scale is merely an aggregation of many sites. Many of the ills attributed to services and commercialism can be traced to poor design and upkeep. With professional landscape architectural, engineering, and architectural input, tourist services can be designed to protect basic resource assets. Protection of natural and cultural resource features are fundamental to good site design. With early design input, erosion of site amenities can be avoided. Architectural design that is sensitive to markets rather than owners can provide services well adapted to tourists. When given the opportunity, designers can provide a constructive catalytic role for high-quality design and development of tourist services.

*Services demand high-quality management.* Also essential to regional planning is high-quality management of tourist services. Yet few businesses have poorer records of management quality. Whereas little research supports the observation, it appears that tourist businesses, especially the smaller lodging and food services, attract many amateur and incompetent managers.

The trend toward higher education and training for these enterprises is slowly improving this situation. The increased sophistication required for management today requires able people. Technological systems, marketing research, new labor issues, and greater competition mean that only the capably managed businesses can survive. The scope of tourism service management today includes understandings of many external factors as well as internal operational procedures. Certainly hospitality must be an essential part of all tourist service businesses. Some areas have instituted special training programs for increasing the level of hospitality.

# Bibliography

Allen, John W., David G. Armstrong, and Lawrence C. Wolken. *The Foundation of Free Enterprise.* Center for Education and Research in Free Enterprise. College Station: Texas A&M Univ., 1979.

*American Innkeeping Today.* New York: American Hotel & Motel Assoc., 1971.

Arbel, Avner and Abraham Pizam. "Some Determinants of Urban Hotel Location: The Tourist's Inclinations," *Journal of Travel Research,* 15 (3), Winter 1977, 18–22.

Ashman, Richard. "Born in the U.S.A.," *Nation's Business.* 74 (11), November, 1986, 41+.

Barke, Michael and Lesley France. "Tourist Accommodation in Spain 1971–1981," *Tourism Management,* September 1986, 181–196.

*Best Western 1975 Annual Report.* Phoenix, AZ: Best Western, 1976.

Bevins, Malcolm I. *Private Recreation Enterprise Economics.* Forest Recreation Symposium, Pinchot Institute Consortium for Environmental Forestry Research, 1971.

Craig, William S. "Reducing Impacts from River Recreation Users," *Proceedings, River Recreation Management and Research Symposium.* Minneapolis: U.S. Forest Service, 1977.

*Destination USA* Report of the National Tourism Resources Review Commission, Vols. 1–6. Washington, DC: US GPO, 1973.

Dice, Eugene F. *A Study of Expenditures and Management in the Private Campground Industry.* Bulletin E-756. East Lansing: Michigan State Univ., 1973.

Drucker, Peter F. The Delusion of Profits," *Wall Street Journal,* February 1975.

Druker, Peter F. *Management.* New York: Harper & Row, 1973.

Dulles, Foster Rhea. *A History of Recreation: America Learns to Play,* 2nd ed. New York: Appleton-Century-Crofts, 1965.

Firebaugh, W.C. *The Inns of Greece and Rome.* Chicago: Frank M. Morris, 1923, 1928, Pascal Covici.

Gutman, Richard J.S. and Elliott Kaufman. *American Diner.* New York: Harper & Row, 1979.

Haynes, Joan W. "Travelers and Their Accommodations." In: *Perspectives: Leisure Travel and Tourism,* Edward M. Kelley, ed. Wellesley, MA: Institute of Certified Travel Agents, 1986.

Hibbert, Christopher. *The Grand Tour.* New York: Putnam, 1969.

*The Inn Business.* The Minister of Supply and Services Canada. Ottawa: Canadian Government Publishing Centre, 1982.

Johnson, Hugh A. *Private Enterprise in the Development of Outdoor Recreation.* Washington, DC: USDA Economic Research Service, 1962.

Jones, A.H. "Tourism and the First-Class Hotel" In: *Travel and Tourism Encyclopedia,* H. Pearce Sales, ed. London: Travel World, 1959.

Kaplan, Atid. "Overworked and Undertrained, Unit Managers Need Attention," *Nation's Restaurant News* November 5, 1984

Lichtenstein, Grace. "Yellowstone Report Cites Poor Service," *New York Times,* April 26, 1977.

Lundberg, Donald E. *The Hotel and Restaurant Business,* 3rd ed. Boston: CBI Publishing, 1979.

Matthews, A.G. "Don't Give Them Croissants if They Want Bagels," presentation at a Tourism Marketing Planning Workshop, Pacific Area Travel Association, October 14–16, 1985, Singapore.

Mayo, J. "A Model of Motel Choice," *Cornell H.R.A. Quarterly,* 15 (3) November 1974.

*National Tourism Policy Study: Ascertainment Phase.* Senate Committee on Commerce, Science, and Transportation. Washington, DC: U.S. GPO, 1977.

*Outdoor Recreation for America: A Legacy for America.* Bureau of Outdoor Recreation. U.S. Dept. of Interior: Washington D.C.: U.S. GPO, 1973.

Schirmbeck, Egon. *Restaurants.* New York: Architectural Book, 1983.

Schmied, Ronald L. "Marine Recreational Fisheries Development in the Caribbean, Why and How," paper presented, 38th annual meeting, Gulf and Caribbean Fisheries Institute, St. Petersburg, FL, November 15, 1985.

*Small Business Enterprises in Outdoor Recreation and Tourism.* Subcommittee of House of Representatives. Washington, DC: U.S. GPO, 1974.

Smith, Stephen J. "Restaurants and Dining Out—Geography of a Tourism Business," *Annals of Tourism Research,* 10, 1983, 515–549.

Stevens, Robert E. "Community Preferences—Motels vs. Non-Motel," *Cornell H.R.A. Quarterly,* 17 (1), May 1976.

Thorner, Marvin Edward. *Convenience and Fast Food Handbook.* Westport, CT: AVI, 1973.

Van Orman, Richard A. *A Room for the Night.* New York: Bonanza, 1966.

Walker, Robert E. *Franchising and the Realtor.* Chicago: National Institute of Real Estate Brokers, 1968.

Wayne, Ralph R. "Tavern Yesterdays," *Motor News,* January 1942.

Williamson, Jefferson. *The American Hotel.* New York: Arno Press, 1975, 1930.

Yoder, Paton. *Taverns and Travelers.* Bloomington, IN: University Press, 1969.

# Chapter 7

# Transportation

Passenger transportation is a vital component of the tourism system. It provides the very critical linkage between market source and destination. Transportation between cities and attractions within urban areas and within attraction complexes requires special planning consideration. In contrast to a person's work transportation, which usually employs only one or at most two modes, it is not unusual for a modern tourist to utilize plane, car, horse, aerial gondola, train, and boat on one trip. Planning increasingly requires intermodal considerations. Except when touring is used as an attraction, transportation is not usually a goal; it is a necessary evil of tourist travel. Therefore, in the planning for tourism development, it is essential to consider all travel modes for people-movement throughout the circuit in order to reduce its friction as much as possible.

## Historical Background

Today, by observing land, air, and water shipping, it would appear that the transportation system grew out of need for commodity movement. However, in 1915 a researcher of travel in America observed that:

> The story of the upbuilding of our present methods of travel and transportation is not a record of the development of a system for the carrying of commodities. It is a history of the devices originated by the people primarily for their personal use and comfort in moving from place to place (Dunbar: 1915, 1/1).

Most ancient travel was on foot. For this reason, the traveler was vulnerable to many difficulties, such as the weather, attack from bandits, and tiring slowness. It is no accident that the word *travel* is derived from "travail" or "trouble." It was described in 1575 as ". . . nothing else but a painstaking to see and search for foreine lands, not to bee taken in hand by all sorts of persons, or inadvisedly, but such as are meete thereto. . . ." (Turler: 1951, 5). Even the use of sailcraft and horseback did not represent major improvements in comfort, convenience, or reliability over walking or canoeing.

Following Chinese and Roman road building, the first major innovation of travel was the horse-drawn stagecoach. Popular in Europe in the 1600s, the stagecoach was brought to the United States 100 years later. For localities with ice and snow, the sled, sleigh, pung, or pod became popular in winter.

The stagecoach, first merely an open wagon with backless benches, developed into a closed carriage with leather or wrought-iron springs. As with each new generation of transportation technology, the quality of the coach was ahead of its roads, which were usually no more than rough, muddy, and dusty lanes cleared of trees.

In spite of a mountain of difficulties, stagecoach travel increased greatly in the 1700s. In America, problems of Indian and bandit attack, ankle-deep mud, coach breakdown, storms, dust, uncomfortable crowding, jouncing, and sickness en route did not deter tourists from traveling, although not in the mass numbers that were to travel later. The Conestoga wagon, designed to transport household goods as well as people, was popular until about the midnineteenth century (Dunbar: 1951, 1/201).

Realizing that travel problems were as much, or more, related to roads as to the vehicles, people began to give attention to improved road construction, especially by the League of American Wheelmen. The McAdam's principle that a soil foundation that is kept dry will carry any load became significant for all future road construction (Winther: 1964, 157).

But uncomfortable and slow land travel, until the railroad, was no match for easier and faster water travel whenever it could be utilized. The original location of most major cities of the world was due to the influence of water transportation. The high demand for water transport fostered canal building in both Canada and the United States in the 1800s.

Canoes, pole-boats, flatboats, pirogues, keelboats, and sailcraft remained unchanged travel modes for many years, but were made obsolete by the invention of the steamboat. First the paddlewheel boats of the rivers and then the huge passenger oceangoing steamships opened up new travel opportunities in the late 1800s.

The river systems and canals of the U.S.A were the highways for steamboat travel for a very colorful but relatively brief period of history. The advantages of new speed and enlarged cargoes were sometimes offset by problems of hitting snags, running aground, exploding boilers, and fire. The greatest of all steamboat catastrophes occurred on April 27, 1865, when the *Sultana* boilers exploded near Memphis, killing 1,647 persons (Drago: 1967, 32). However, for several decades steamship travel along the ocean coasts and the shores of the Great Lakes dominated tourist transportation. Steamship and railway lines frequently converged at Great Lakes port cities, stimulating the development of many magnificent resort hotels.

Early railroad travel demonstrated the importance of linkage with attractions. A tourist guide of 1884 (Shearer: 1970, 300) admonishes tourists to Yosemite to choose the best coach type—Kimball wagon instead of mud-wagon or Concord coach—inasmuch as mountain roads were steep and rocky. The Coulterville Route was a favorite (Shearer: 1970, 300):

| | |
|---|---:|
| San Francisco to Merced, by rail | 151 miles |
| Merced to Dudley's, by stage | 46 miles |

A major improvement was made when the Yosemite Valley Railroad was built in 1907, connecting Merced with the border of the park, now El Portal. From 4,000 tourists in the first year, it carried over 15,000 by 1910 (Muir: 1962, 181).

Railroad tourist travel virtually closed the chapter on steamboats within the country and broadened the range of travel opportunities in three ways. Greater volumes of travelers were transported, new attraction destinations were penetrated, and railways frequently developed resort hotels served by their lines. Resorts in the lake states—Minnesota, Wisconsin, and Michigan—were often a direct result of railroad building fostered by federal government land grants.

For the first time, the pleasure of steamtrain touring—safety, comfort, beautiful scenery, appetizing menus, and access to attractions—could be promoted and delivered, especially to the upper classes. Through the 1930s, diesel-powered locomotives, Pullman cars, and special scenic cars stimulated even more tourist travel. However, railroads had become increasingly uneconomic for tourism by the 1950s because of heavy wear and tear on the equipment during the war years and increased competition from air and automobile travel. In the U.S.A., passenger miles (exclusive of commuter travel) declined from 20,207 million in 1930 to 4,332 million in 1972 (Statistical Abstract: 1969, 1976).

Amtrak, the National Railroad Passenger Corporation, a semipublic agency, was established by the U.S. Congress in 1970 to purchase new equipment, establish passenger routes, and generally stimulate a revival of interest in rail travel. But in spite of heavy governmental subsidies, only megalopolis commuter trains have been popular with markets. In Europe, rail travel remains as a viable alternative for tourist transportation.

An interesting chapter of American tourist transportation in the early 1900s was that of the streetcar or electric trolley car. For many cities during the first few decades following 1900, this opened up nearby lakes to resort, cottage, and amusement park development.

However, no other form of transportation has so personalized and revolutionized travel as has the automobile. Production and ownership continued to make
new records: 2 million in 1914, 9 million in 1921, 25 million by 1930 (Dulles:
1965, 318). As with trains, the first automobiles were open. "... 1919 was the
beginning of the modern automobile age. It marked the passing of the linen
duster visor cap and goggles" (Reck: 1950, 8). Soon automobiles were enclosed
and had self-starters, heaters, and hydraulic brakes. Automobile ownership in
the years to follow became as important, or more so, for all classes as the
possession of a home. Neither carriages, stagecoaches, steamboats, or trains ever
provided the personal flexibility of use, low cost, convenience, or comfort for
tourist travel.

Although the first concrete road had been laid in Detroit in 1908, most tourists
traveled on dusty gravel and clay roads through the depression years. Then, with
increasingly heavy input from federal and state funding sources, paved highways expanded greatly: 694,000 miles (1,120,000 kilometers) in 1930 to
3,861,934 miles (6,179,094 kilometers) in 1985 (Highway Statistics: 1973, 211
and 1985, 109).

Of major significance in the United States was the establishment of a nation-
wide divided highway network, connecting more than 90 percent of the cities in
the country with populations over 50,000. Conceived in 1956 as the National
System of Interstate and Defense Highways, it has become the network most
used by tourists throughout the country. Its increased safety, comfort, and
convenience stem from design features that prohibit all grade intersections,
provide easy alignment and grades, and separate opposing traffic flows. Further
tourist safety was demonstrated following the federally imposed 55-mile speed
limit in 1974 as an energy conservation measure. (Speed limits on some rural
and lightly traveled interstate highways were raised in 1987).

Hard rubber tires and poor roads made the early buses very uncomfortable,
but tourists rode them on special tours in parks. When railroads began to cut
passenger service in the 1950s, bus use increased by scheduled lines and charter
tours.

Airplanes, much like automobiles, were first considered only a toy for wealthy
sportsmen. It was not until the 1950s that the airlines became major passenger
movers. This was primarily due to the high cost and discomfort of trips on the
earlier propeller-driven planes. The new jet planes demanded the complete
remodeling of airport technology and management. In the 1960s hijacking
became a troublesome threat to air travel—more than 90 airlines were hijacked
in 1970. Greater security around airports, including electronic inspection of
passengers and luggage, has greatly reduced this problem in recent years. Jet
travel stimulated long-distance travel as never before. The phenomenal growth
of tourism to Hawaii—296, 517 in 1960 to 1,798,581 in 1970 (Hawaii Tourism:
1972, 9)—was due largely to improved jet transportation. Whereas in 1936 it

took the first Pan American Clipper about 17 hours to travel to Hawaii from San Francisco (Lund: 1966, A6), modern jets could make the trip in less than five hours.

This brief historical sketch serves to demonstrate the basic fundamental of human movement throughout history. This fact sets it apart from all other life on earth—the "mobile animal" (Salomon: 1985, 232). However, some are of the opinion that the new wave of telecommunications will erode this fundamental. With increased universal use of new technology, especially for business, the need for travel will be greatly reduced. Evidence to date suggests the opposite. The presently massive use of computers, telecommunications, teleconferencing, and deregulated telephone service has been accompanied by an ever-increasing amount of travel and need for more and better transportation. Some believe that teleconferencing is likely to enhance the need for face-to-face interaction. "The substitution of telecommunications for travel is of minor importance because, even if it happens, it will be offset by the human desire to exercise mobility" (Salomon: 1985, 233). This is recognized and protected in the airline deregulation acts of the United States. "There is hereby recognized and declared to exist in behalf of any citizen of the United States a public right of freedom of transit through the navigable airspace of the United States" (72 Stat. 740, 49 U.S.C., 1304, Sec. 104) (O'Connor: 1982, 189). The challenge is to continue to plan for better quality human movement by means of all modes of transportation.

## Scope

Modern transportation scope for tourism is broad and is beginning to include all modes previously made obsolete by new technology. The scope as considered here includes land, water, and air modes and those forms of tourist transportation in which attractions are an integral part.

## Water

For most of America, water transportation as a basic form of domestic passenger movement is extremely low in volume, but it is important to certain areas. Although by now many short stretches of water have been spanned by tunnels or bridges, the feasibility of ferries is increasing in some regions. Car ferries with many passenger conveniences such as snack bars and washrooms carry as many as 800 passengers and 360 cars at a time, especially in New York and the state of Washington. Hydrofoils are much faster and are in use in high-volume areas such as New York City's harbor. Air-cushion vehicles are most popularly used in Great Britain, especially on the Thames River.

Tourist transportation by ship today has all but given up to other modes. However, the "floating resort" concept that includes both on-ship recreation and entertainment and the attraction of port cities has experienced increased popularity in recent years, especially in the Caribbean and on major rivers of the world. Mississippi and other riverboats with short runs, entertainment, and food are used more as tourist attractions than for transportation. A recent innovation for touring tourism in Europe is the refitted river barge for small tourist groups. Privately owned watercraft also serve both purposes: transportation and recreation.

## Land

Today, virtually all modes of land transportation are utilized in tourist movement either to or within tourist attraction areas.

It is not enough to state that automobile travel dominates domestic tourism in the U.S.A. For planning purposes one must understand characteristics of automobile ownership and use. In terms of raw transportation data—strictly people-movement—private automobile tourist transport is a bargain. A tourist's cost calculation is usually only on the basis of cash-trip expenses—gasoline and oil—because fixed costs, such as insurance, interest on investment, and depreciation, would continue even if the car remained parked.

An automobile is not bound to fixed routes; it can be made to travel partly by freeway, by secondary road, and by car ferry. It can carry more luggage, sports items, and extras at close personal range than can planes and trains. Automobile travel is not bound to schedules set by government and business; deterred only occasionally by weather, one can travel at any time of day or night and in any season. This versatility also means that a driver can quickly and easily shift from one trip plan to another and make as many stops on the way as desired.

Furthermore, no other mode of transportation offers the pride of ownership and personal possession offered by the private automobile. Size, color, style, and accessories are highly personalized, placing car ownership equal to and above that of clothing, jewelry, and even housing. It is for these reasons that both capital investment and operation costs have great elasticity, as has been documented (McCool et al.: 1974).

Recreation vehicles—pickup covers, truck campers, travel trailers, motor homes, camping trailers—have become of increasing importance. As against the disadvantages of parking problems, low gas mileage, and relatively high cost, motor homes provide housing (and often kitchen and bath facilities) as well as transportation. They are especially popular with retirees. As parking and service facilities have increased, there are more opportunities for using recreation vehicles to sightsee and visit a variety of attractions.

Bus travel is of two basic types. Scheduled bus transportation between cities and to attractions is almost universally available. Although slower than air travel, buses reach many more destinations and are less expensive. They are largely patronized by those under 24 years of age or over 65 and by a large percentage of low-income customers. Travel purpose is seldom for business but rather to visit friends and relatives and to reach attractions.

Motor coach tours, popular in Europe for many years, have experienced rapid growth in the U.S.A. in recent years. These provide attraction stops for narrations, entertainment, and other features as well as transportation. Modern coaches with heat, air conditioning, and washroom facilities also provide a group sociability not available by automobile. One study (Cunningham and Thompson: 1986, 11) showed that users preferred motor coach tours because someone else (travel tour broker, bus company) made all the arrangements. Other advantages cited were economy, risk aversion, and gregariousness. These qualities are preferred especially by older patrons. There was no correlation with automobile ownership, a clue that the lack of owning a car had nothing to do with selecting the motor coach tour mode. Buses also offer important linkage between air terminals and destinations.

Cable cars, providing vertical transportation, are generally of two types. Some utilize rails, such as those of San Francisco and Mt. Washington, New Hampshire. Others are carried on cables suspended between towers, popular in winter sports areas. Capacities depend upon the size and number of cars or gondolas as well as speed. They are particularly well adapted to mountain and gorge attractions and make little environmental impact.

Because of the vast network of highways and high per capita car ownership and general popularity, rail passenger travel in the U.S.A. has dwindled to a relatively small proportion of the total. In 1981 passenger car miles represented 89.6 percent of the total, whereas bus, rail, and water combined were 2.4 percent (Homburger: 1984, 3). In Europe and in several other locations around the world, rail passenger travel remains the dominant mode. In the U.S.S.R., over 30 million passengers a year travel on a journey of more than 2,000 kilometers (1,240 miles) (Ambler et al.: 1985, 63). But even in these areas, the demand for more automobiles and better roads is increasing rapidly in spite of government policy against developing a car-owning democracy. In the Soviet Union, between 1966 and 1980, motor vehicle production increased 224 percent while hard surface roads increased only 64 percent (Crouch: 1985, 168).

Within major parks and attraction complexes, a variety of people-movers are being utilized. Minibuses or minitrains with closed or open passenger trailers pulled by a tractor often provide linkage between parking lots and features of attractions. Modern theme parks and other attraction complexes favor walking over mechanical devices for reasons of cost and liability.

For certain hunting, fishing, and wilderness uses, horseback or mule travel is utilized. Although slow, uncomfortable, and adapted only to those who can ride, this mode provides access in locations where other modes would not be feasible or appropriate.

Wilderness transportation is usually limited to hiking or canoeing. These modes are increasing in importance.

## Air

Although the zealots of the 1930s, who predicted a small plane for every home owner, have been proven wrong, air travel is an increasingly important mode for tourism. For international travel, it has virtually killed overseas ship travel.

Generally, all three forms of commercial airlines—scheduled, supplemental, local service—and private planes are involved in air transportation for tourism. Whereas intercity tourist air travel is dominated by the scheduled airlines, the charter and nonscheduled forms of supplemental air travel have increased in popularity primarily because of lower fares. Because of the major investment in modern sophisticated airports and larger capacity planes, only the larger volume destinations—usually larger cities—can be served by scheduled airlines. Therefore, for many resort and vacation destinations, local feeder air, land, or water commuter service becomes very important. For some tourists, either for sightseeing or trips to attractions (often resorts or private vacation homes on islands or remote locations), private plane transportation is their favorite mode.

Tourism involves nearly all kinds of airports, both land and sea. General-use airports provide service for scheduled, nonscheduled, and seasonal use. For personal use, often at resort destinations, limited use or even restricted use airports may be important. Services vary considerably between the extremes of those able to handle large volumes of people and the most recent generations of aircraft to simple grass strips suited only to small craft and limited use.

For tourism planning, it is becoming increasingly important to recognize all modes and their interdependence. In the U.S.A., the common carrier share (air carriers, buses, rail) of total intercity passenger miles continues to rise. It made up about 17 percent of the total in 1975 and grew to 24 percent ten years later (Frechtling: 1986, 19). The private vehicle (auto, truck, RV) continues to dominate American domestic travel. For many other countries, where people depend more on international visitors than domestic, air transportation is dominant.

## Intermodal

One of the most difficult aspects of travel for the traveler is to connect travel modes—major air, commuter air, rental car, bus, train. For many reasons, and under several political idealogies, there is much uncertainty and confusion for the traveler to integrate several modes with one trip.

Ownership of each mode tends to separate. Whereas there are advantages to the several forms of public and private management, the separate cells tend to fragment travel flows. Schedules, pricing, convenience, quality of equipment, service, and location of terminals seem to be decided so independently that the traveler must make a special effort to put the several pieces together.

Of increased popularity is the "fly-drive" modal mix, using air travel for the major trip to the destination and rental car at the destination. A study of the fly-drive travel to Tasmania (Henshall: 1985, 25) demonstrated that convenience— the flexibility of travel routing, scheduling, and visiting attractions in the destination area—was the main reason for choice. Some 87 percent drove between prearranged hotels.

Gradually, there is increased interest in intermodal continuity. Owners, managers, and politicians are recognizing the traveler's dilemma in making the fragments into a system. Travelers to the London area appreciate the railway connection in Gatwick airport. In California, the Santa Ana Regional Transportation Center integrates several travel modes, such as Greyhound, Trailways, Camino Real Express, Amtrak, Orange County Transit District bus service, airport limousine and taxi service, and personal car. Other cities have added intermodal transportation integration—Kalamazoo and Pontiac, Michigan; Harrisburg, Pennsylvania; Charleston, South Carolina; Sacramento, California.

Although planning after the fact is difficult and costly, these changes suggest that tourism planning, if it is to meet market needs, must cross jurisdictional boundaries of transportation agencies.

## Package Tours

Of increasing importance from a planning perspective is the significant increase in the development of "package tours." A package tour "is often understood to mean two or more elements of a trip, such as transportation, accommodations, meals, or sightseeing, purchased together and prepaid" (Frechtling: 1984, 1). In 1984, package tours accounted for nearly one-third of all U.S. trips to foreign destinations. Domestic package tours in the U.S.A. dominantly use buses for travel, whereas overseas travel is usually by air. That portion of the tour at destinations is frequently by train, bus, rental car, and cruise ship. Domestic U.S. tours are generally shorter, more business-oriented, and use more ground transportation than foreign ones. Typical of package tours advertised in the U.S.A. is a "Costa Del Sol" trip to Spain (Spain's Costa Del Sol: 1986, sec. 4). Included in one price is air travel from one of seven American cities, accommodations for seven nights, round-trip transfers, and a variety of attractions, including beaches, shopping, dining, and side tours to historic sites.

Greater traveler demand for package tours is one more clue to the need for integrated planning of travel development, especially intermodal transportation.

## Influence of Ownership

Both governments and private enterprise are heavily involved in transportation planning and development throughout the world. The two bodies are so intimately intertwined that it is sometimes difficult to clarify roles. In most countries the public role is dominant.

For example, in the U.S.A., there are some 64 agencies in addition to the Department of Transportation, more than 30 congressional committees and subcommittees together with many nongovernmental and professional constituencies that have a hand in shaping transportation policy—and this is only at the federal level (Smith: 1986, 28). Whereas issues were simpler years ago, decision making today is compounded by environmental, state, city, and private enterprise issues.

U.S. transportation policy has been strongly influenced by political changes in the decades since the 1960s (Hazard: 1986). Many separate agencies and policies were consolidated in 1966 with the formation of the Department of Transportation (DOT). Although greater coordination has been accomplished, intermodal and multimodal policies have not met all expectations. It has been most successful in deregulation and transferring many responsibilities to the states and localities. Issues of social regulation have increased—highway beautification, safety, environmental impact, emissions, social welfare, and handicapped. Funding at the federal level has been caught in the issue of increased budget demands from defense and social welfare. Research and development expenditures at the federal level have decreased. Future policy may need to emphasize more careful selection of political managers, longer range planning, and better communication with the many constituency and agency groups (see Fig. 7-1) and reorganizing along more functional lines.

Within the U.S.A., there is no indication that highways will not continue to dominate personal and business travel. As a consequence, governmental support of highway improvement is essential. Bad roads are not only inconvenient—they are costly. Bad roads increase fuel consumption by two to five billion gallons annually, double the likelihood of accidents, and increase construction costs by 160 percent when capital improvements are deferred (Smith: 1986, 132). Key issues are traffic congestion, maintenance and rehabilitation,

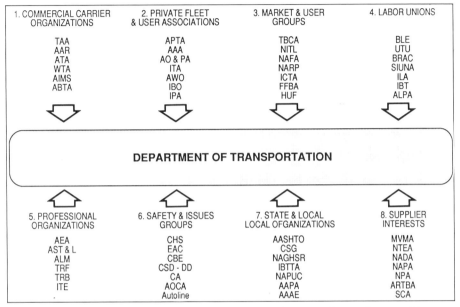

**Figure 7-1.** Transportation Associations and Constituency Groups, U.S.A. A diagram of the many nongovernmental private associations and organizations that exert considerable influence on the federal Department of Transportation (Hazard: 1986).

personnel, and management. Needed are new thrusts in research, planning, and legislation.

Similar issues of governmental responsibility for transportation appear throughout the world even though the emphasis on mode may vary. In the Soviet Union, for example, a national policy of avoiding overdependence on an outside world forces emphasis on interregional transportation. But the division of responsibilities into branch ministries tends to set up barriers to communication (Shaw: 1985, 3). Whereas state control avoids cut-throat competition, it faces problems similar to others worldwide—the need for better roads, improved air controller systems, improved safety, more funding, and integrated bureaucratic pluralism (Ambler: 1985, 175).

Air transportation equipment in the West—planes, computerized communications, and maintenance —is usually privately owned. However, the airports and weather-reporting services are usually publicly owned. Automobiles, buses, and taxis used by tourists are most often owned privately, whereas the highways, control systems, signs, bridges, ferries, and information centers are usually publicly owned. Ships are most often provided by private firms, whereas ferry services, seen as highway connectors, are usually governmentally developed. But even these divisions are diffused because fee systems and lease arrangements from private enterprise have much to do with the financing of the publicly owned operations.

As a consequence of U.S. policy toward strengthening private enterprise, governmental deregulation became a policy trend and was particularly applied to airlines and buses. However, because of governmental sovereignty within European countries and many legal agreements within and between countries, deregulation is viewed differently (Quigley: 1986, 401). Existing main carriers enjoy a high degree of protection from competition, and past policies are not easily changed. But consumer group pressure toward greater diversity and less regulation is being felt by governments worldwide. The success of the independent charters suggests stronger private sector development of air travel everywhere. The deregulation issue is further evidence of the need for recognizing the differences in tradition and political ideology when planning for better integration of tourist travel.

Many controls and regulations of transportation for tourists are carried out at the local and state levels. For example, county (and sometimes township) highway departments have several categories of road design and maintenance. Whereas roads developed to these standards provide for widespread movement of tourists by personal auto, rental auto, or bus, they are designed primarily for volume rather than tourist experience. As a consequence, excessive clearance of right-of-way has sometimes destroyed much of the appeal of scenic roads unless special legislation designates different standards for scenic routes.

Generally in the West, private enterprise policies have centered on: (1) equipment design and manufacture (automobiles, airplanes, trains, ships, buses, recreation vehicles), (2) operation of travel system (scheduled and charter airlines, shipping firms, bus lines), (3) mining and manufacture of transportation fuels, and (4) related transportation services (travel agents, service stations). Private enterprise competition has provided increasing comfort, speed, and efficiencies of equipment over the years. Passenger compartments of buses, airplanes, and automobiles are far more comfortable than ever before. Private enterprise competition has developed attractive and efficient service stations as well as travel agencies and other supporting systems of help to the traveler. Airline, bus, and car rental reservations are done quickly and efficiently on computerized systems.

It seems that transportation goals are the same no matter if governmentally or privately owned, causing much vacillation between owners. In many countries airlines are state owned, whereas in the U.S.A they are privately owned. Intracity buses are frequently publicly owned. But competitive private bus companies have taken over with reasonable success in many places—Hong Kong, Kuala Lumpur, Buenos Aires, Calcutta, Manila, Istanbul, Cairo, Singapore, Nairobi, Belfast (Glaister: 1985, 65). However, in Korea (Won: 1986, 280) chaos resulted from deregulation of buses in several cities. Cooperatives, a modified form of private enterprise, has worked reasonably well since. In Daejon, accidents have been reduced, route restructuring has been successful, operational reliability has been increased, there is better tailoring to demand, and the several companies are showing profits. Cooperatives have been implemented also in Gwangju City, Daegu, and Incheon.

Critics of private enterprise roles in the U.S.A., however, point to several problems. Manufacturers are criticized for not improving passenger safety and operational efficiencies of travel equipment. Those owning and managing travel systems are criticized for not providing adequate schedules, adequate routings, and better intermodal systems—air to bus and the like. For example, some blame poor food service and lodging linkages for the low popularity and social status of tourist bus travel. Oil producers are criticized for not keeping pace with demand for fuels and for not investing more in research and development for new oil and oil-substitute sources.

It can be concluded that ownership and control are directly related to the span of that ownership and control. Therefore, one cannot speak of a transportation system as a whole without full understanding of the ownership and control of individual parts of that system.

## Location and Land Use

Patterns of transportation land use are critical to all tourism planning and development. Locations of highways and air terminals provide the skeleton upon which most tourism thrives and grows. However, the spatial network of present transportation modes was based upon a complicated array of factors that generally were nontourism-oriented.

For example, present highways, very important to tourism, have been located primarily on industrial point-to-point linkage criteria. Therefore, routings of shortest mileage have precedence over other factors, such as scenery or access to attractions. Many state highway systems reflect rural-city linkages, such as the "farm-to-market" roads of Texas. In recent years, some highway engineers have begun to consider both tourism needs and commodity movement in locating highways.

Displacement, caused by highway relocation, has been a major problem of tourist service business location and land use in the past. Hotels once located for railway trade were rendered obsolete by new highways unless the highway happened to skirt the old location. Motels built along two-lane highways were rendered obsolete by the new freeways. Perhaps no change in the U.S.A. in recent years has had greater impact on the location of tourist business activity than the establishment of the interstate highway system, especially near attraction resources and near cities. It has been difficult to shift habits of tourists from a "visual" selection to an "informational" selection pattern of attractions and services. Expressway speeds and limited access tend to dilute the extent to which tourist establishments can depend upon visual roadside marketing.

As with most other technological advances in America, the new freeways were first accepted, carte blanche, as being the salvation to growing highway problems. And similar to other single-minded engineering solutions, freeways encountered many land use obstacles in the 1960s. Because of relatively lower costs of land acquisition and easier demolition, urban freeways were routed through urban ghettos, parks, and historic areas, causing considerable social and economic displacement problems. Plans that called for ruining the great tourist attraction of New Orleans, the French Quarter, with an eight-lane elevated expressway directly through its heart, were finally halted.

By 1966, land use concerns over highways in the U.S.A. stimulated the passage of an act that states that the secretary of the Department of Transportation shall not approve any program or project that requires use of any land from a park, recreation area, wildlife and waterfowl refuge, or historic site unless there is no feasible alternative (Whalen: 1968, 58).

Because of the resurgence of interest in urban downtown areas by both residents and visitors, planning attention is increasingly focused on better pedestrian use. Until traffic is solved, improvements in pedestrian movement are difficult. However all of the amenities downtown—theaters, entertainment, museums, restaurants, plazas, ethnic specialties—must be approached on foot to be enjoyed. Before the automobile, the social and cultural assets of streets exceeded traffic movement in importance. In the U.S.A., as early as 1971, according to a survey (Appleyard: 1981, 3), pedestrian values had been sacrificed to traffic in most cities. The highest priority problems cited by residents were street noise and heavy traffic. Any visitor to major cities around the world, such as Mexico City, can attest to the health hazard to pedestrians from automobile fumes and congestion.

Airport land use ranges all the way from relatively small areas with mowed grass strips to runways as long as 14,000 feet (4,170 meters). Besides, runways must have clear zones of unobstructed air space at either end, sometimes requiring a total length of several miles. The largest U.S. airport, Dallas-Fort Worth, covers 17,000 acres (7,082 hectares). This demand for special qualities and quantities of land has often dictated locations that are not necessarily best for linkage with other modes of transportation, such as personal automobile, bus, and trains.

Because air transport termini are located at cities, they must fit the overall planning of cities. However, rapid suburban expansion has frequently led to conflict of land use—residential, business, schools. In Virginia, a 630-foot skyscraper (60 stories) tower was proposed within a waterfront development project along the Potomac River near Washington, DC. Because the Potomac is a flyway for approaching Washington National Airport, the proposed tower was rejected by the Federal Aviation Administration. It was reported to have had a significant adverse impact on airplanes and helicopters approaching the area (Post America: 1986, 3).

What are the most critical tourism criteria concerning transportation? One way of grouping the criteria would be according to the following five categories: tourist transportation needs and desires (users), tourism and the transportation industry, tourism, government, and transportation, tourism environmental considerations of transportation, and tourism transportation as related to overall transportation systems.

## Users

Tourists as transportation users have three major functional demands upon transportation. First, they need movement from home to destination and within destinations, basically a commodity shipping procedure. From this standpoint, the shorter and less costly the better. The tourist would prefer, in this instance, one ticket, one capsule throughout the entire trip, and integrated accommodations and food along the way.

Second, tourists often seek pleasurable travel. This is more than commodity movement and includes an attraction function. The extent of pleasure en route is dependent upon the tourist's objectives. If the tourist seeks only to get to a weekend cottage, the characteristics of the trip are of little consequence. If, however, the travel and its amenities *are* the attraction, the opportunity to participate in travel attractions along the way is very important.

Third, no matter the mode, tourists seek several personal travel factors and will opt for the best combination. Included are comfort (freedom from fatigue, discomfort, poor ridability), convenience (absence of delays, cumbersome systems, roundabout routings), safety (freedom from risk, either from the equipment or other people), dependability (reliable schedules and conditions of travel), price (reasonable, competitive), and speed.

From a land planning and development point of view, these functions require consideration of linkages between all modes of transportation for the traveler and linkages with attractions, both along the way and at the destination.

A major highway travel issue in the U.S.A., important to tourism, is the conflict between large trucks and passenger cars. Both car drivers and truckers are faulted. In Michigan, the trucking faults were described in three main categories (Kulsea: 1986/I, 22): (1) unsafe driving practices including excessive speed, tailgating, driving in all lanes, excessive hours, and lack of proper training, (2) unsafe vehicles such as worn tires, poor brakes, and leaky fuel tanks, and (3) property damage from uncovered truck cargoes, wear and tear on highways, and scattered trash.

Whereas the majority of truck drivers are competent professionals, the increased safety problem is borne out by statistics. Trucks caused 81,647 accidents in 1984 in Michigan, killing 452 persons and injuring 32,630. Inspections show 25 percent of trucks to be unsafe (Kulsea: 1986/II, 28). In 1985, police ordered 10,932 trucks off the road as unsafe. One insurance company stated that in the U.S.A. annually, over 2,000 accidents and $13 million worth of windshield and headlight damage are caused by trucks (Kulsea: 1986/IV, 18).

Car drivers are not without their faults: (1) speeding to pass and then slowing down too quickly to get off a ramp, (2) lack of proper signaling, (3) unawareness of trucker "blind spots," (4) expecting trucks to stop quickly, (5) risking accidents with unsafe driving, and (6) lacking understanding of driving procedures with trucks (Kulsea: 1986/IX, 19). This issue demonstrates how the several forms of transportation, in this case trucks and cars, need to be planned and managed in an integrated manner.

## Transportation Industry

As now constituted, those private enterprise components of transportation seek profits from commercial establishments in the several modes used by tourists and others. Profits demand volumes of users (sometimes freight and nontourists as well as tourists) at prices that can meet operational costs and provide adequate return on investments.

From the industry's point of view, technology and management suggest that the most efficient operations are those of individual business establishments for each mode. In other words, airlines are best able to run and maintain air travel; trains, train travel, and so on. Although some integration has taken place (within antitrust constraints), most modes in the U.S.A. are separate and many competitive units are operating within each mode.

From a land use point of view, this means that routes and facilities and equipment are developed and managed by a great many separate and competitive businesses for each mode. This provides for competitive innovation and initiative on the part of the entire industry. However, if the functional needs of tourists are to be considered, it also means the need for integrated place and time considerations—location of joint terminals for airlines, trains, buses, and information centers, and joint scheduling and joint preparation of information.

## Governments

First, it must be repeated that for tourism, governments perform two important roles: regulator and developer-manager. From a regulatory function, tourism planning requires consideration of joint action on transportation regulations rather than only by mode. In other words, regulatory functions for air travel need to be coordinated with automobile and bus travel. This requires new functional relationships between existing regulatory agencies at all levels.

As developer-managers, government agencies are seldom responsible for more than one mode of transportation for tourists. Airport, highway, ferry, and harbor agencies are usually separate, although the consequences of their development and management are uniformly critical to tourists. This suggests that from a tourism planning point of view, existing systems and certainly proposed projects of each agency need to be reviewed for continuity regarding these services to tourists. This does not mean that tourism always has priority, only that tourist functions be considered alongside others.

Because of the close relationship between tourism and governmental transportation agencies, projects and policies must be articulated and coordinated in all tourism planning.

## Environment

Basic to all private and governmental decisions on tourist transportation is the impact on both social and physical environments. Decisions that in the past were based primarily on technological and marketing factors must now be modified to reflect environmental factors.

Social environments are often very sensitive to transportation decisions and therefore need to be considered in decision making. The power of these decisions to remodel cities and countrysides is great and, as has been pointed out, can be destructive as well as productive. It may be necessary to place severe constraints on the location of new airports, highways, canals, and harbors because of potential disruption of the existing social environment or introduced conflicts with new social groups—tourists.

Increasingly, researchers and planners (McHarg, Eckbo, Lewis, Lynch) have called attention to the need for physical environmental (ecological) considerations in transportation location and development. As yet, however, these studies and plans have not been universally adopted by decision-makers. Thus far, transportation schemes and policy have been influenced primarily by work, home, and point-to-point engineering. To these must be added, for tourism planning, objects and locations for personal travel and physical characteristics of all transportation routings between home and destinations. Movement is an integral part of the modern person's ecosystem and therefore must be considered in a physical environmental context. The commercial market and economic factors and the many governmental development influences of tourist travel need to be modified to adapt to climates, mountains, plains, deserts, forests, rivers, lakes, and other environmental factors.

# Conclusions

Because an integral part of any tourism system and plan is travel, transportation planning has high priority. However, most passenger transportation systems are the result of separate histories, developments, and management policies for the several modes. Therefore, tourism planning must integrate these modes to best serve tourism functions. The following conclusions are drawn from the study of transportation background.

*Technology determined modes in the past.* Until today, the rule of transportation practice has been the adoption of the latest technology. Technology of a new travel generation (more than the travelway) over history has been the prime factor to make a mode obsolete. Better carriages did it to walking, horseback, and wagon travel; steamboats did it to carriages. Railroads made obsolete steamboats on the continent. Auto and jet plane travel made obsolete steamboat, railroad, and propeller plane travel. Along with these shifts in mode have come major changes in travelways. Perhaps the major planning lesson to be derived from this history is that potential technological change must be considered when planning locations for tourism developments.

*Modern tourism utilizes all modes.* Today, however, there appears to be a turnaround. All present and past modes are utilized to some degree within tourism. Jet planes and automobiles may take passengers to tourism destinations, but steam trains, cruise ships, helicopters, buses, and rental cars may be used at destinations. Within attractions, small trains, cable cars, horse-carriages, bicycle trails, horseback trails, foot trails, and minibuses may be utilized. So each mode, even older historically, may have merit for special tourism situations.

*Ownership and control are complicated.* No matter the governmental ideology of a nation, it appears that transportation throughout the world is complicated by jurisdictional management and control. Freight and passenger movement have different criteria, needing different policies and regulations. The several modes—air, bus, train, ship, rental car, taxi, and even personal car—are administered by different public and private entities. Even in market economy countries, where much of transportation is performed by private enterprise, governments have strong involvement with function and regulation. This diversity of ownership and control has often stimulated greater opportunities and better quality transport for tourists. But it presents a challenge for integrated tourism planning.

*Intermodal travel requires new planning cooperation.* Tourist demand is seldom directed toward a single transportation mode. Increased availability (price, scheduling, airline options) of air travel has introduced many more destination choices to the prospective traveler. But access to the specific attractions and circulation within a destination frequently put several other modes into play. Increased popularity of package tours forces greater integration of travel modes. If any one travel link fails to provide the quality of service desired, the entire trip may be spoiled. The planning of intermodal transportation centers is needed for domestic local as well as outside visitor markets.

*Transportation is more than movement—it is an experience.* Whereas providing access to destinations distant from residential market areas is a prime function of transportation, travelers cannot be treated solely as freight. For many travelers, the several environments, social and physical, inside and outside the vehicle, are of interest and concern. Although levels of service quality have increased generally in recent years, the fragmented ownership and control of travel modes can often produce a ragged sequence of travel experiences. Planning cannot be directed solely to physical access. Quality of total traveler experience from home to destination and return requires collaborative planning.

*Transportation is essential but not a cause of tourism.* There is no question about the essential role of transportation in tourism. However, it is an error to believe that new transportation alone will create a new tourism destination. Only if there are resource foundations for attracting a variety of services and sufficient information and promotion will transportation be effective in fostering tourism. Even after favorable factors are known to exist, much care must be exercised in locating roads, port facilities, and airports. In single-minded zeal to increase tourism, highways and airports have been so poorly located that the cultural and natural resources—much of the attractiveness for tourism—were destroyed. Transportation system planning must be integrated with transportation site planning for successful tourism.

*Origins-destinations are important.* Closely related to quality of travel provided by each mode is its origin-destination planning implications. For example, jet planes brought Hawaii into mass travel range, allowing its expansion as a major tourist destination. Air-conditioned autos and buses have changed destinations from dominantly northern travel to include the South. In spite of energy threats to travel in 1973, it was rumored that some new owners of gas-economy autos were able to reach more distant resort destinations than ever before. Regular monitoring of origin-destination patterns for the several market segments is needed for effective tourism planning.

# Bibliography

Ambler, John, Holland Hunter, and John Westwood. "Soviet Railways—Lethargy or Crisis?," Chap. 2, In: *Soviet and East European Transport Problems,* John Ambler et al., eds. New York: St. Martin's Press, 1985.

Ambler, John, Denis J.B. Shaw, and Leslie Symons, eds. *Soviet and East European Transport Problems.* New York: St. Martin's Press, 1985.

Appleyard, Donald. *Livable Streets.* Berkeley: Univ. of California Press, 1981.

Crouch, Martin. "Road Transport and the Soviet Economy," Chap. 6. John Ambler et al., eds. *Soviet European Transport Problems.* New York: St. Martin's Press, 1985.

Cunningham, Lawrence F. and Kenneth N. Thompson. "The Intercity Bus Tour Market: A Comparison Between Inquirers and Purchasers," *Journal of Travel Research,* 25 (2), Fall 1986, 2–12.

*Destination U.S.A.* Report of the National Tourism Resources Review Commission, Vol. 1. Washington, DC: U.S. GPO, 1973.

Drago, Harry Sinclair. *The Steamboaters.* New York: Bramhall House, 1967.

Dulles, Foster Rhea. *A History of Recreation.* New York: Appleton-Century-Crofts, 1965.

Dunbar, Seymour. *A History of Travel in America,* Vol. 1. Indianapolis: Bobbs-Merrill, 1915.

Frechtling, Douglas C. "Travel Trends in 1985," *1985–86 Outlook for Travel & Tourism.* Proceedings of the Eleventh Annual Travel Outlook Forum. Washington, DC: U.S. Travel Data Center, 1986, 5–19.

Frechtling, Douglas C. *U.S. Market for Package Tours.* Washington, DC: U.S. Travel Data Center, 1984.

Glaister, Stephen. "Competition on an Urban Bus Route," *Journal of Transportation Economics and Policy,* 19 (1), January 1985, 65–81.

*Hawaii Tourism Impact Plan,* Vol. 1, "Statewide, Tourism in Hawaii." Honolulu Department of Planning and Economic Development, 1972.

Hazard, John L. "The Institutionalization of Transportation Policy: Two Decades of DOT," *Transportation Journal,* Fall 1986, 17–32.

Henshall, Brian D., Rae Roberts, and Andy Leighton. "Fly-Drive Tourists: Motivation and Destination Choice Factors," *Journal of Travel Research,* 23 (3), Winter 1985, 23–27.

*Highway Statistics.* U.S. Dept. of Transportation. Washington, DC: U.S. GPO, 1973 and 1985.

Homburger, Wolfgang and James H. Kell. *Fundamentals of Traffic Engineering.* 11th ed. Institute of Transportation Studies. Berkeley: Univ. of California, 1984.

Kulsea, William C. "Cars and Trucks—They Must Mix Safely," a series in *Michigan Living,* Part I, 68, (9), 22+; Part II, 68 (10), 28+; Part IV, 69 (1), 18+; Part IX, 69 (5), 18+, 1986.

Lund, Kay. "5 on First PAA Pacific Passenger Run to Return," *Honolulu Star-Bulletin,* October 19, 1966.

McCool, Stephen et al. "The Energy Shortage and Vacation Travel," *Utah Tourism Recreation Review,* 3 (2), May 1974.

Muir, John. *The Yosemite.* New York: Doubleday, 1962, 1912.

O'Connor, William E. *An Introduction to Airline Economics,* 2nd ed. New York: Praeger, 1982.

"Post America Skyscraper Rejected," *Transportation Policies,* 40 (2), December 1986, 3.

Quigley, Hugh. "Deregulation or Liberalization: The Quandry Facing Europe." In: *Perspectives: Leisure Travel and Tourism,* Edward M. Kelley, ed. Wellesley, MA: Institute of Certified Travel Agents, 1986.

Reck, Franklin M. *A Car Traveling People.* Detroit: Automobile Manufacturers Association, 1950.

Salomon, Ilan. "Telecommunications and Travel." *Journal of Transport Economics and Policy,* 19 (3), September 1985, 219–235.

Shaw, Denis J.B. "Branch and Regional Problems in Soviet Transportation," Chap. 1. In: *Soviet and Eastern Transport Problems,* John Ambler et al., eds. New York: St. Martin's Press, 1985.

Shearer, Frederick, ed. *The Pacific Tourist.* New York: Crown, 1970.

Smith, Wilbur S. "Current Highway Transportation Interests," *Transportation Quarterly,* XL. (2), April 1986, 131–141.

Spain's Costa Del Sol, advertisement in *Travel Holiday,* 166 (5), November 1986, SDC4.

*Statistical Abstract of the United States.* Bureau of Census. Washington, DC: U.S. GPO, 1979, 1976.

Turler, Jerome. *The Traveiler.* Gainesville, FL: Scholars-Facsimiles and Reprints, 1951, 1575.

Whalen, Richard J. "The American Highway," *Saturday Evening Post,* December 14, 1968.

Winther, Oscar O. *The Transportation Frontier: Trans-Mississippi West, 1865-1890.* New York: Holt, Rinehart and Winston, 1964.

Won, Jaimu. "Buscooperative Systems in Korean Cities," *Transportation Quarterly,* 40 (2), April 1986, 227–287.

# Chapter 8

# Promotion/Information

Increasingly, all those programs and physical developments that attract and inform tourists comprise a very important component of the functional tourism system. Although promotion and information are not the same functions, they are often administered by a single governmental agency. Communications of all types are becoming more and more important to link the consumer to the product. Simply, if tourists do not know about travelways, attractions, services, and facilities, and do not know how to get to them, tourism can be less than satisfactory for both consumers and suppliers. Certainly, the planning for tourism must include understanding of the essential component of promotion/information.

## Historical Background

The history of tourist promotion and information materials and methods is virtually a history of communication and travel. The desire to travel and see the habitat of animals on the other side of the hill may have come from the first animal pictographs on the cave wall or from tales of a returning hunter in prehistoric times. One of the earliest forms of communication was that of asking questions of travelers. Throughout history, visual graphics and audio presentations have had much to do with stimulating, guiding, and describing travel.

Even the ancients had guidebooks. As early as the second century BC., there were titles available such as *The Athenian Acropolis, Spartan Cities, Guidebook to Troy,* and others (Casson: 1971, 56). However, because these were handwritten on leather or papyrus, they were too bulky and too valuable for the tourist to take on travels.

So, then as now, the tourist depended greatly on guides who personally showed and talked about points of interest. And, then as now, guides often preferred to break the boredom of their routines by telling jokes to the travelers, who would much rather hear about the things they were seeing. Often, guides seemed to run on and on with their narrations. In a sketch Plutarch wrote: "The guides went through their standard speech, paying no attention whatsoever to our entreaties to cut the talk short" (Casson: 1971, 56).

There must have been a degree of travel enticement and enrichment as well as religious thrust from the herbals and illustrated manuscripts and calendars of medieval Europe. The landscape paintings of the seventeenth century brought the qualities of exotic places to the attention of many viewers. Travelers of the Grand Tour brought home sketchbooks and notebooks filled with information about distant places to show their friends and relatives.

Throughout history, signs have been an important part of travel. A sign in Pompeii, buried in lava from Mt. Vesuvius in A.D. 79, carried this invitation (Presbrey: 1929, 9):

> traveler going from here
> to the twelfth tower
> there Sarinus keeps a tavern
> this is to request you to enter
> farewell.

In eighteenth-century England and the United States, colorful signs with carved and painted maidens, lions, dragons, and many other subjects adorned the street shops, inns, and restaurants. The first massive electric sign in the United States was oriented to tourism. It appeared on Broadway in New York City 1891, advertising vacation homes on Long Island, "swept by ocean breezes," and promoting a dance band and fireworks (Presbrey: 1929, 618).

With the inventions of printing came atlases and maps, which served as information pieces for travelers, even though they were inaccurate and incomplete at times. Newspapers in England in the seventeenth century (and later in the U.S.A.) began to print stagecoach timetables and itineraries, probably the first step in tourist information.

With trains in the U.S.A. came elaborate tourist guidebooks. A guidebook issued by the Union and Central Pacific Railroads in 1884 boasts that it contains:

> full description of railroads routes across the continent, all pleasure resorts and places of most noted scenery in the far west, also of all cities, towns, villages, U.S. forts, springs, lakes, mountains, routes of summer travel, best localities for hunting, fishing, sporting, and enjoyment, with all needful information for the pleasure traveler, miner, settler, or business man (Shearer: 1970, 1).

For generations, the word "Baedecker" has been synonymous with superior guidebooks. In 1824 Karl Baedecker established his business on the assumption that the traveler was sophisticated, literate, and curious, and therefore needed more than merely accurate information about trains, food, and hotels. In his introduction to a recent reprint of a U.S. *Baedecker,* Commager states that: "The Baedecker guidebooks therefore provided authoritative essays on the geography, history, government, art, and culture" of European countries (Baedecker: 1971, v). Not until seventy years later did Baedecker believe that there were sufficient attractions and tourists (not immigrants) coming to America to develope a guidebook for the United States. In 1893 he published (Baedecker: 1971, v.) a massive guidebook

> for those who came to see and buy and enjoy—those who needed to know what trains to take (imagine having a choice!), what cities to visit, what hotels to patronize, what luxuries to carry home, what historic monuments to contemplate (and probably sneer at), what scenic wonders to compare with the Alps and with the Lake Country.

The quality of research, description, and maps used in Baedecker guides has not been equaled since.

Without doubt, the informative power of literature has had much to do with describing and stimulating travel. Both fiction and nonfiction about exotic places produced vivid images among those who never dreamed they might some day visit them. Surely for the American, the adventure travel books of Richard Halliburton—*The Glorious Adventure* (1927), *New Worlds to Conquer* (1929), *The Flying Carpet* (1932), *Seven Leaque Boots* (1935)—and those of Lowell Thomas—*With Lawrence of Arabia* (1924), *Kabluk of the Eskimo* (1932), and *Back to Mandalay* (1951)—may have been more effective than paid advertising for influencing travel.

Tour guides and park interpreters have, since their inception, represented a special segment of tourism communication. With roots in the activities and nature writings of John James Audubon, Alexander von Humboldt, Charles Darwin, and many others, interpretation of parks and zoos expanded greatly in the early 1900s. Esther Burnell, one of the first professional interpreters in the U.S.A., was licensed by the federal government in 1917 as a nature guide in what is now Rocky Mountain National Park. She was responding to the need for telling visitors about the special natural history characteristics of the region (Sharpe: 1976, 10). In 1932 guided auto caravans in Yosemite became so popular that they had to be stopped because of the traffic jams. Emergency federal funds gave support to expanded interpretation programs in the national parks in the 1930s.

The history of tourism is intimately linked with that of promotion important to transportation, products, and services ever since the early Greek criers. George Eastman's 1891 slogan for his camera, "You press the button; we do the rest" and the 1896 steamship advertisement, the most costly of its day, "To Far Away Vacation Lands," offering appeal to Germany, Switzerland, France, and

Italy, were the beginning of the modern tourist advertising era. Soon automobiles, bicycles, hunting guns, vacation homes, steamship trips, and hotels became popular topics of advertising copy in newspapers and magazines. Throughout its long history, one of the most travel-influential magazines in the U.S.A. has been the *National Geographic* magazine because of its advertising and journalism, especially its color photography.

One of the most amusing experiments in roadside advertising dominated American automobile travel for three decades following 1930. Many a boring landscape was enlivened by road sign sequence of humorous jingles, often admonishing highway safety while advertising a shave cream. Rowsome (1965, 105, 115) cites some examples:

| | |
|---|---|
| The safest rule | Little Bo-Peep |
| no ifs or buts | has lost her jeep |
| just drive like | it struck a truck |
| everyone else is nuts! | when she went to sleep. |
| Burma Shave | Burma Shave |

But these ads came long before the days of consumerism and "truth-in-advertising," and one viewer demanded fulfillment of an offer that was made as a spoof. In answer to the following jingle on the roadside (Rowsome: 1965, 56), he collected his 900 jars and was ready to go.

Free—free
a trip to Mars
for 900
empty jars
Burma Shave

The company gave the collector and his wife a free round trip to Moers (pronounced Mars) in Germany!

Certainly not all tourism promotion and information has come from the private sector. Many public agencies, motivated and justified as educational information for visitors, were entering the field by the late 1930s. The publication of maps, guides, descriptive literature, and periodicals about natural and cultural resources became common practice.


## Scope

Today, promoting destinations and informing tourists encompass a wide array of communications. Some believe that the recent revolution in communications has had greater impact on society than any other factor. Platt (1975, 268) predicted that "One of the most powerful methods for bringing pressure to bear on national elites and business elites may be the power of TV, with its instant-outrage and consciousness-raising power, which has already been shown to cross national boundaries."

The scope of promotion includes not only all forms of travel advertising, publicity and public relations, and special incentives (discounts, give-aways), but also the many other ways in which we are influenced about travel and travel objectives. The topic of information includes the full range of descriptions about places and modes of travel. And it includes the many ways in which guidance and direction are given to tourists in their travel.

## Information

Today's more sophisticated travelers are no longer satisfied with generalized descriptions, exaggerated phrases, and incomplete or inaccurate guidance. They seek helpful insight concerning their travel destinations. A frequent travel annoyance is to arrive in a hotel and not be able to find out when a certain attraction is open, how much it costs, and how to get there. Travelers want more and better information before they travel and also at destination sites.

For example, motorcoach tour managers well understand how important information is to touring circuit travelers. All exploring is done efficiently and rapidly (Cooper: 1981, 369). Travelers continually wish to minimize the risk of disappointment. Information through interpretation techniques can encourage use of prescribed routes and avoid misuse of resources at the same time experiences are enriched. These planning techniques can also reduce visitor-resident conflicts.

The scope of information is extremely broad and almost defies classification. However, for planning purposes it is helpful to distinguish between directed and nondirected information for tourists.

*Directed* information means that the sponsor has intentionally prepared messages for tourists. This is overt tourist-oriented material that includes interpretation, personal guiding, guide books, magazine articles, travel atlases, travel clubs, trip counseling, and many other forms.

A growing form of information dissemination is interpretation—"a service for visitors to parks, forests, refuges, and similar recreation areas" (Sharpe: 1976, 3). Whereas the primary purpose is to provide information about natural and historic resources, interpreters are sometimes expected to promote the agency and to modify outdoor recreational behavior to protect fragile resources (Sharpe: 1976, 4). Generally, two types of interpretive processes are used: personal (information duty, conducted tours, interpretive talks, living history, and demonstrations) and nonpersonal (audio presentations, audiovisuals, exhibits, centers, trails, signs, publications, and self-guided trails and tours) (Sharpe: 1976).

Foreign visitors especially need adequate sources of information and interpretation. For many years, the U.S. National Park Service (NPS) has operated information and interpretation programs in the 673,575-acre Grand Canyon National Park in Arizona. However, because of the great increase in foreign visitors, a special study regarding new policy and practices was conducted in 1981 (Machlis and Wendroth, 1982). It focused on organized tours to the South Rim and included 1,440 observations and 908 questionnaires covering 40 tours. Among the recommendations to NPS were better informational literature at entrance stations, foreign language information at trailheads, bilingual employees, and international road signs. Also included were foreign language presentations at several visitor centers, and foreign language exhibits and publications. Special training for interpreters was stressed so they can better understand foreign customs and behavior. Needed were better ways of informing foreign visitors regarding the regulations for safety and protection of the environment. Better planning of routes, parking, and scheduling of tours was recommended.

The study advised the concessionaires in the park to provide better information in foreign languages at shops and accommodations, and on tours. Fewer but better (descriptions) stops were suggested. Better service could be provided if the concessionnaires performed market studies to determine the needs of specific segments.

Research was advised at other parks to compare results, which might reveal the potential for interconnected park circuits whereby cooperative efforts could improve information exchange. Study of the park's relationship to the Grand Canyon community was recommended.

Recommended cooperative efforts were: sharing the costs of publications, regular cooperative monitoring of tours, joint language training program for interpreters and drivers, better coordination of scheduling to meet other park needs, and joint funding of research. This study illustrates the need for research in communication.

Many guidebooks and maps are issued by oil companies, travel clubs, state agencies, credit card companies, and guide publishers. An important criterion of evaluation for all such publications is how well they relate to the actual physical development. Do the descriptions fit the places?

Important in the scope of tourist information is the literature produced by governmental tourism and travel agencies. Whereas these agencies are primarily promotion and advertising oriented, they publish great quantities of informational literature on attractions. Sometimes, as in the case of information literature for Kentucky, tourism data are primarily about the state parks. One study in Texas showed that state-produced information services were rated highest in credibility compared to other media. Furthermore, the majority of those responding to advertising coupons ". . . are probably seeking this information after they have decided to return to Texas. . . . The advertising is functioning as a service to the respondents and not as a stimulus to the decision to travel in Texas" (Nolan: 1974, 44).

Research is showing that just as the markets have become segmented, so must the design and development of information. In order to reach more discriminating market segments, better understanding of the differences in planning horizons and traveler reaction to information must be obtained (Gitelson and Crompton: 1983, 7).

As new computer communications are devised, travel information may become more readily available. American Airlines Sabre Vision is a "user friendly" system that allows the traveler to make plans, check availability, and book his or her own vacation reservations (Drenick: 1987, 1).

Although research has not yet proven the point, *nondirected* information may have even greater impact on the tourist. Throughout our lives we are bombarded by a variety of messages about places. These nondirected messages are very powerful even though they were not prepared and disseminated for tourism purposes.

One of the most important nondirected sources of information comes from friends and relatives. Katz and Lazarfeld (1955) showed that person-to-person communication remains the most influential means of transferring information. When friends return from a vacation trip, we may be invited to home movies, slide shows, and displays of albums and souvenirs from the vacation. No matter how poor the quality of photographs or how biased the information may be (certainly not a scientifically representative sample of all tourists), it is absorbed and often believed without question because the friends or relatives obtained the information firsthand. All the comments—about the attractions, services and facilities, transportation, and the characteristics of the hosts—are probably given greater credibility and weight than other information.

Another very effective form of nondirected information comes from general literature. Both prose and poetry provide descriptive data about places that make impressions upon potential travelers. Certainly as people read the folktales of Paul Bunyan, they develop a curiosity about the North; and the historical poetry of Longfellow, such as *Evangeline,* offers place interest in Acadian locations—Nova Scotia, New Brunswick, Maine, Louisiana. Michener's geographic novels, such as *Sayonara, Tales of the South Pacific, Hawaii, The Source,* and *Centennial,* did much to publicize Japan, the Pacific, Israel, and the U.S. Great Plains. Nonfiction books and magazine articles, such as those in *Science, Smithsonian, Sports Illustrated, Heritage,* and *National Geographic,* often contain vivid place descriptions important in tourist information.

Youngsters in school are exposed to geography, government, history, and economics books that identify and describe characteristics of places around the world. Who knows how much these impressions provide basic understanding about places everywhere?

Today even magazines not devoted to travel carry many columns of copy about people and places throughout the world. Undoubtedly, these make impressions and help to form attitudes and opinions as well as provide factual information about places, accommodations, transportation, and attractions everywhere.

All mass media of communications, such as newspapers, radio, TV, and movies, also carry many powerful messages about travel and places incidental to news reporting. Parts of the world, such as Vietnam, Sinai, Rhodesia, Moscow, Peking, Uganda, Korea, or Iran, would have little meaning to most people except for what we learn of them through news reports. Movies such as *South Pacific, Lawrence of Arabia, Ryan's Daughter* (Ireland) and *Witness* (the Amish country of Pennsylvania) have stimulated travel to the locales of these plots (Butler: 1986, 113).

Although little research has been performed on the success of transmitting essential directional messages for tourists, general observation suggests that some is effective and much is not. Highway departments have greatly improved coordination of directional signs between the governmental unit sponsors, but some confusion remains in linking secondary routes, particularly those of special agencies with recreational resource development responsibilities. Perhaps the greatest confusion results from the breakdown of direction between the travelway and the many travel objectives: attractions, services, facilities. Sometimes literature and maps are incorrect or misleading. Proper directions, perhaps in several languages, are needed for foreign visitors. Attempting to obtain proper directions from local residents who may be unfamiliar with tourist objectives and routes may not be successful. In an effort to be courteous, misdirections are given more freely than an admission of no knowledge.

As great growth in all tourism development has taken place, it would appear that very little attention has been given to physical and program development that assures proper direction for travelers. This may be one of the greatest of challenges for the regional tourism planner.

Because recent research of host-guest relationships has exposed some social problems and issues, there is great need for better information and guidance on travel etiquette. Both hosts and guests need better understanding of how behavior and attitudes can be improved. Because travel has become such a dominant force in society, it would seem mandatory to include travel etiquette in all educational systems for children. Travelers need to understand that even though they are *allowed* to enter another region, perhaps another country, they are visitors there and must behave accordingly. Even though the vacation traveler may have as an objective the shedding of inhibitions, the impact on a local population should be anticipated. Dress and manners can be modified if the traveler knows local customs and standards of behavior. It is equally important for hosts who have promoted tourism in their community to be more tolerant of the sometimes strange and offensive ways of visitors. Again, planning policy should anticipate these social issues and build educational and informative programs to lessen the conflict.

## Promotion

Without doubt, the greatest commitment to tourism by the governmental and private sectors is promotion of tourism. Every National Tourism Organization (NTO) spends great amounts of money on tourism promotion, which generally takes on four forms: advertising, publicity, public relations, and incentives.

*Advertising* is defined as persuasive marketing efforts whereby the promoter wishes to communicate a given sales message to a specific customer audience. It is paid for and includes print, direct mail, broadcast, and billboard. Advertising is mainly intended to gain attention, interest, and awareness, and to stimulate action to purchase (Davidson: 1987, 475). Davidson explains further that advertising must be stated in terms of the mind-set change desired, such as for a vacation market segment:

- To increase awareness of the destination/hotel/carrier as an option for the next decision.
- To increase the number in the target market who hold favorable attitudes toward the destination/hotel/carrier.
- To broaden the knowledge of the specifics of the destination/hotel/carrier and what it offers to the vacationer.
- To increase the number in the target market who say they will seriously consider the destination/hotel/carrier the next time the vacation decision arises.

Another form of promotion, *publicity,* is not paid advertising space but other programs to stimulate interest and awareness. Familiarization tours are sponsored by several airlines and tourism organizations. These are tours paid by the sponsor and offered to travel agents, tourist information counselors, and writers. When these practitioners are back on the job, they understand the merits of destinations they have visited and hopefully will influence travelers to visit these same destinations.

Whenever a tourism official appears or makes a presentation at a conference, there is promotional spinoff in the form of *public relations.* When such an official calls on travel businesses or gives talks to organizations, there is promotion implied.

In recent years, *incentives* have become popular for travel promotion. Hotel room discounts, prizes and gifts, and specially priced packages are used to promote trips, especially by the business sector.

Tourism promoters more and more are performing research and increasing their levels of sophistication in order to meet the new market needs. Travel behavior and segmentation (described in Chapter 4) are being utilized to refine promotional techniques and approaches. For example, application of the VALS market segmentation was applied to travel promotion for Hong Kong (Skidmore: 1987, 111). It was found that past promotion had been satisfactory for "achievers," but the greater effort should be made to attract "belongers" and the "socially conscious."

Further research is needed to measure relationships between promotion (and information) and quality of experience. From a planning perspective, it may be well to ask, was the promoted product truly available? With the increased ease and proficiency of creating images of destinations through the several media, the visitor may be disappointed in the experience because the promotion was overstated. For example, an advertiser may hire an agency to produce literature on natural or historic sites. The photographer waits until clear weather and assumes a special vantage point that required permission from an owner to create an ideal and artistic image. The average tourist who is more likely to see the site on a cloudy and rainy day and from a far less dramatic overlook may feel cheated.

## Influence of Ownership

Perhaps the most important fact regarding the entire component of promotion/information in tourism is that it is not owned and directed by any single agency or enterprise. In most countries it is performed by diverse and unconnected individuals and agencies of government and private enterprise. Probably the majority—those who are nontourism identified such as those who produce radio and TV programs, publish news, write geography books and novels, publish magazines, and those who tell their friends and relatives about their experiences—do not realize they are a powerful element in informing and directing tourists because that is not their objective. All the same, what they say and write are extremely influential.

Because of the power and widespread use of nondirected activities in promotion and information, it is even more important that those agencies and organizations that directly spend moneys and operate programs for tourism recognize their relationship to the nondirected. Furthermore, it is incumbent upon them to exercise their directed functions clearly, fairly, and effectively and in the best interests of the tourist and the environment.

Governmental tourism ministries and agencies at all levels throughout the world are heavily engaged in many forms of promotion and information for tourism. Advertising is influenced by both general agency policies to entice more visitors and the commercial advertising policies of the contractor hired to carry out the advertising.

Other agencies, such as health departments and park agencies, issue literature and news releases on hazards, rules of safety, and regulations affecting lodging, food, and beverage service, and many recreational activities. Agencies on history, culture, and the arts produce guides and descriptive literature on historic sites, museums, pageants, and crafts. Highway departments, in addition to providing directional routing signs, are beginning to experiment with informational signs and display boards for travelers along the highway and in special turnouts. Whereas their policies generally prohibit listing of commercial services, facilities, and attractions, they often display "tourist services" or "gas, food, lodging" signs. Unfortunately, the many agencies that relate to tourism promotion and information often fail to integrate their efforts.

In the U.S.A., many federal agencies, including the National Park Service, Fish and Wildlife Service, U.S. Forest Service, and the Army Corps of Engineers, operate under policies that allow the publication of literature, production of TV and radio shows, and often the placing of signs in areas under their jurisdiction. Policies on the design, use of graphics, and marketing stratification vary with each agency. There is no federal attempt to integrate the policies of these many agencies.

Likewise, private interests, both commercial and nonprofit, are very active in many forms of information and direction functions for tourism. Their policies are generally directed toward profit-making or promoting their own programs (youth, church groups), promoting products (oil company guides), promoting services to members (AAA, Chamber of Commerce), promoting tourist services and products (tourist businesses and associations), and identifying places or giving directions through the use of markers and signs. These are as varied in policy as the individual enterprises and there is little evidence of integrating policies on promotion and information by private organizations and enterprises. Whereas innovation by the several actors should be protected, planning policy should require awareness of their interdependence.

## Location and Land Use

Most promotional and informational functions for tourism are program-oriented and do not directly involve land use (folders, TV, magazine advertising), but certain elements are critical to all tourism land use. In most cases, travel folders or spot radio announcements can do a better job of communicating than can billboards. Therefore, it is important for the planner to have knowledge of the land use implications of the component of promotion/information. How is highway and street signage implemented and what are the consequences for the traveler and the environment? What is the impact of the visibility of the landscape, building design of motels, or car service stations at highway exits on informing the public?

The use of billboards has been controversial ever since they were introduced. Advertisers and businesses point to billboard promotional effectiveness, whereas environmentalists and designers point to their landscape degradation and ineffectiveness. States and local governments have reacted in various ways to arguments from both sides.

Perhaps Hawaii has the longest tenure and Oregon the shortest regarding legislative control of billboards in the U.S.A. Hawaii's efforts toward conservation of the natural beauty of the island of Oahu and ridding the city of Honolulu of billboards began with the organization of the Outdoor Circle of Honolulu in 1913. By persuasion, and particularly by an "antibillboard" rubber stamp campaign, all but one firm took down its signs. The stamp was used on checks, receipts, bills, letterheads, and envelopes to attract attention. The final holdout was purchased by the Outdoor Circle in 1927. In the same year territorial legislation was passed prohibiting billboard use (Campbell: 1962) and it remains in effect today.

In response to the U.S. Federal Highway Beautification Act of 1965, Oregon passed its own motorist information act in June 1975, which brought down nearly all commercial roadside signs. State-supported directional "logos"— signs identifying specific gasoline, food, and lodging brands—are replacing the old billboards at important exits (Motorist Information Act: 1975).

But not everyone in Oregon is satisfied. In the opinion of one state official, "The ultimate in highway beautification and traveler information is no signs at all" (Calcaterra: 1976). However, a research team concerned with highway signage states that "any arguments unfavorable to billboards based on aesthetic reasons are counterbalanced by the public need for information" (Claus and Claus: 1976, 92). Thus, the ambiguity between advertising and information continues.

## Conclusions

Several conclusions about tourist promotion and information are important to tourism planning. Throughout history travel behavior has been influenced by these activities. Today, with the availability of more rapid and more vivid communication techniques, characteristics of places become a part of everyday knowledge, even without concerted tourism efforts. In addition, places are being promoted as never before. Modern modes of transportation make promotion, guidance, information, and direction more necessary and more complicated. Several conclusions can be drawn regarding the relationship of the component of promotion/information to tourism planning.

*Information is a growing need.* One of the weakest links in all tourism is that of imparting information to the traveler. Because governments and the private sector have been preoccupied with promotion, information gets relatively little attention and budget support.

More traveler information is needed at both the origin and destination. Today's great proliferation of destinations and travel packages requires more information on choice-making. More information on transportation modes, services available, and especially the attractions to be visited is needed by today's market. Before leaving home, the traveler needs adequate information on social customs of the destination so that behavior can be tempered to avoid embarrassment or conflict. Whereas the spontaneity of travel experience must be maintained, on-site interpretive programs can enhance the experience. Natural and cultural resource visits can be much more memorable when the visitor has a better understanding of the place and its characteristics. Capacity problems are reduced greatly by informative visitor centers at the edge of resource areas. Tourism policy must increasingly speak to the issue of greater information for travelers.

*Better quality information is needed.* Better, not just more, information is needed. Few agencies and organizations have been established for this purpose. Although the number of travel journalists has burgeoned in recent years, most have been engaged in promotion, not information. Journalists need greater understandings of travelers and travel. Opportunities abound for more accurate, more relevent, better market-targeted, and more interesting travel writing. The travel critic is yet to appear alongside those serving the public in art, music, and drama. Here may be a great opportunity for a new form of journalism. Better quality signs, exhibits, displays, and presentations at visitor centers require a new level of professionalism. Tourism planning policy must include stronger emphasis on improving the quality of information.

*Direction requires integrated planning.* How the tourist finds the destination is also a mix of printed literature and physical development. Probably no other aspect of tourism planning and development has received so little attention. State tourism information centers are increasingly helpful, but they represent only one small aspect of the problem of directing a tourist from home to the desired attractions, to the supporting services and facilities, to the proper travel routings, and back home.

Much of the direction problem is being solved by better highway marking. The U.S. interstate system now has matured to the level that signage is increasingly informative. However, the questions of detail and completeness are yet to be resolved. How much directional information can and should be placed on a highway sign? Perhaps more dependence will have to be placed on other media—maps, atlases, tape narratives dispensed at regular travel stops such as car service stations, restaurants, and accommodations. The multitude of separate administrative units of transportation planning and management (county, city, state, federal), lodging (hotels, motels, campgrounds), food services (restaurants, snack bars, drive-ins), and attractions are subject to the influences of all directional guidance. The need for collaboration on all directional programs and physical development is great.

*Markets require segmented promotion.* Perhaps the greatest research accomplishment in travel in the last decade is the recognition that tourists are not all alike—the market is proliferated into many segments. But technological transfer sometimes is slow. Laments a veteran hotelier:

> The conventional wisdom was to offer what was convenient for the operator to provide, or what was better than the competition, rather than to define what the customer wanted and then be sure he received it. Accordingly, many hotels have competed on the basis of trying to out-towel, out-robe, or out-chocolate-mint one another without regard to the customer's purpose of travel or how the hotel can help ensure the success of the trip (Goeglein: 1986, 3).

As more statistical data are obtained for tourism, there is danger in translating averages into average travelers, who do not exist. Only when planning policy recognizes the several travel market segments can appropriate information and promotion be effective.

*Evaluation of promotion and information is needed.* In the past, millions of dollars have been spent on information and promotion in blind faith that they were effective. Legislators and businesses are increasingly asking whether these expenditures actually "work."

For advertising, "evaluation research is that body of systematic, scientific procedures employed to isolate, define, measure, and understand the relationship between advertising efforts and the influence of such advertising on the marketplace" (Davidson: 1987, 473). When performed by competent researchers, guidance for improved promotional methods and applications can be obtained.

Similar measures can be applied to effectiveness of informational efforts. But even though techniques are available, change will not take place until they are applied more liberally. Planning policy must incorporate means of monitoring the effectiveness of human and financial input into travel information and promotion.

*Management must accept greater responsibility for promotion/information.* There is ample evidence to demonstrate the need for management at all levels to accept greater responsibility in the promotion/information component of the tourism system. Policy must reflect the great proliferation of public and private managers. Those responsible for transportation could greatly improve their developments and effectiveness for travelers if more and better information was prepared. Those in charge of the many public and private attractions could enhance the experiential satisfaction of visitors with more and better promotion and information. Certainly the service sector—lodging, food, entertainment, shops—could improve business with better promotion and information. And those charged with the development of promotion and information need to reach toward higher levels of professionalism for the good of all.

*Enticement must be linked with actual developments.* Advertising that promises more than can be delivered is misleading. Knowledge of a destination, particularly how the tourist is able to see and do things at that destination, must not be overstated or the consumer will be disappointed. This fundamental of advertising and consumer behavior is well known but often violated. For two reasons, the quality of the experience at a destination place (related to its planning, design, and management) is critical to future promotion and enticement of new travelers. First, there is a high degree of repetition—going back to the same place—if experience equaled of exceeded those promised. Second, because of the great impact of comments on friends and relatives, the tourist's satisfaction with meeting expectations is important for future referrals.

> Advertisement and selling messages, then, should be designed to create expectancies that will be fulfilled by the product insofar as is possible. The extent to which this seemingly common-sense precaution is violated through use of "cute" exaggerations and other forms of creative "gimmickry" in advertisement is, at times, appalling (Engel, and others: 1973, 534).

Planning for tourism must include solutions to promotion and information problems related to delivery of tourism products.

# Bibliography

Baedecker, Karl. *The United States—With an Excursion Into Mexico. A Guidebook for Travelers.* Introduction to 1971 reprint by Henry Steel Commager. New York: De Capo Press, 1971, 1893.

Butler, R.W. "Literature as an Influence in Shaping the Image of Tourist Destinations: A Review and Case Study." In *Canadian Studies of Parks, Recreation and Tourism in Foreign Lands.* John S. Marsh, ed. Peterborough, Ontario: Trent Univ., 1986, pp. 111–132.

Calcaterra, Patty, personal correspondence, Travel Information Council, Salem, OR, 1976.

Campbell, A. N. "How Hawaii Erased a Blot." From the archives of the Outdoor Circle. Honolulu: The Outdoor Circle, 1962.

Casson, Lionel. "After 2000 Years Tours Have Changed But Not the Tourists," *Smithsonian,* 2, (6), September 1971.

Claus, R.J. and Karen E. Claus. "The 1965 Highway Beautification Act—Part II," *Cornell H.R.A. Quarterly.* 17 (1), May 1976.

Cooper, C.P. "Spatial and Temporal Patterns of Tourist Behavior," *Regional Studies,* 15 (5), 1981, 359–371.

Davidson, Thomas L. "Assessing the Effectiveness of Persuasive Communications in Tourism," Chap. 40. In: *Travel, Tourism and Hospitality Research,* Ritchie and Goeldner, eds. New York: John Wiley & Sons, 1987. pp. 473–479.

Drenick, E.D. "AA Tests Computer for Clients to Select Their Own Vacations," *Tour & Travel News,* 032, February 9, 1987, 1.

Engel, James F. et al. *Consumer Behavior.* Hinsdale, IL: Dryden Press, 1973.

Gitelson, Richard J. and John L. Crompton. "The Planning Horizons and Sources of Information Used by Pleasure Vacationers," *Journal of Travel Research,* 21 (3), Winter 1983, 2–7.

Goeglein, Richard J. "Technoloy and Tourism: A Growing Partnership," *Tourism and Technology: A Growing Partnership,* proceedings of 17th Annual Conference of Travel and Tourism Research Association, Memphis, June 15–18 1986, pp. 14.

Katz, E. and Lazerfield, P.F. *Personal Influence: The Part Played by People in the Flow of Mass Communications.* New York: Free Press, 1955.

Machlis, Gary E. and Ellen Wenderoth. *Foreign Visitors at Grand Canyon National Park: A Preliminary Study.* CPSU/UI S82-2, College of Forestry, Wildlife and Range Sciences. Moscow: Univ. of Idaho, 1982.

*Motorist Information Act.* Oregon Travel Information Council, Salem, 1975.

Nolan, Sidney D. *Response Evaluation of Selected Texas Tourism Advertising.* Unpublished dissertation. College Station: Texas A&M Univ., 1974.

Platt, John. "The Future of Social Crises," *The Futurist,* October 1975.

Presbrey, Frank. *This History and Development of Advertising.* New York: Doubleday, 1929.

Rowsome, Frank, Jr. *The Verse by the Side of the Road.* Brattleboro, VT: Stephen Greene Press, 1965.

Sharpe, Grant W. *Interpreting the Environment.* New York: John Wiley & Sons, 1976.

Shearer, Frederick E., ed. *The Pacific Tourist.* New York: Crown, 1970, 1884.

Skidmore, Sarita. "Lifestyle Segmentation of Vacation Visitors to Hong Kong." *Tourism and Technology A Growing Partnership,* proceedings of 17th Annual Conference of Travel and Tourism Research Association, Memphis, June 15–18, 1986, pp. 111–117.

# Chapter 9

# Regional Potential Planning

This chapter takes the preceding chapters one step further—into the process of planning. Chapters 1, 2, and 3 offer some historical background of tourism planning and describe the functional tourism system of highly interdependent components. Chapters 4 through 8 provide some depth regarding the function and planning issues for each of these components. How these building blocks for tourism development can be given better integration and how the developmental problems can be alleviated through planning processes are now addressed (a condensation and updating of Chapters 8 through 12 in the first edition).

## Models

Although planning of tourism in several forms and at several levels has taken place for many years, only a few system models have been created. As more is learned about how the many elements and components of tourism function, the greater becomes the need for knowing how the system can be manipulated, particularly to improve its functioning, to better meet objectives, and to minimize problems and obstacles.

One early systems model grew out of research and observation of tourism development in Michigan in the 1950s (Gunn: 1965). After application to Michigan's Upper Peninsula (Blank and Gunn: 1966), the model was refined and elaborated (Gunn: 1972, 1979). Getz (1986, 24) lists this model and others developed during the same period—Bargur and Arbel (1975), Arnott (1978), Lawson and Baud-Bovy (1977), and Mill and Morrison (1985).

### Recreation Areas

Romsa (1981, 343) has described a planning model prepared by Kiemstedt (1967) that measures and maps three sets of factors to determine areas best suited for recreation development. Factors investigated are: inherent physical attributes, available leisure facilities, and the cultural milieu of the region. Each set is summarized to arrive at an index of attractivity. The measurement problem was solved by use of a delphi technique evaluation.

The process includes several steps, beginning with measuring the relevant subcomponent variables for a given quadrant of land. From an attractivity function, the attractivity value for each variable is calculated. This value is then multiplied by a weighting factor and the resultant values form an index. Kiemstedt describes this function in the following formula:

$$\text{Ar I} = \sum_{i=1}^{m} i \ \sum_{j=1}^{n} j \cdot aj \cdot wj$$

Ar I = attractivity indext of region I
    i = quadrant
   aj = the attractivity value of variable j
  wj = the weight assigned to variable j

Overlay maps are then prepared for each component. Those land areas that fall into the upper categories for all components are deemed the most attractive for recreation development.

Further modifications were made by others in an attempt to gain greater insight into analysis for planning.

## Pasolp

Baud-Bovy (1980, 1982) has elaborated on his earlier experiences of applying some of his concepts and principles of tourism planning in several countries. He stresses integrated planning—planning that breaks from the traditional technical planner's approach. By integration, he asks that tourism planning be integrated with the nation's policies, with the physical environment, with the related sectors of the economy, into the public budget, into the international tourism market, and with the structure of the tourist industry.

His experimentation with an updated PASOLP approach in Niger resulted in the concept shown in Figure 9-1 (Baud-Bovy: 1982, 312). The central part of this approach is the flow-by-flow elaboration and analysis of tourism products. These analyses take into consideration the:

Resources of the country (established and potential).
Specific requirements of each tourism market, of each flow, each category of
    present or potential tourists.
Country's structures, policies, and socioeconomic constraints.
Existence of competing destinations.

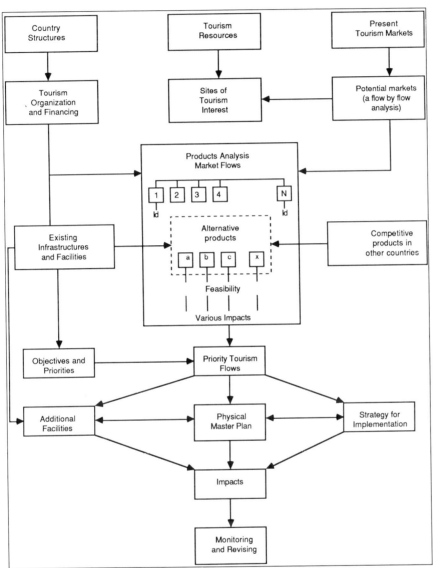

**Figure 9-1.** Pasolp Model. Product Analysis Sequence for Outdoor Leisure Planning (PAS-OLP), a planning model that encompasses socioeconomic, political, sociocultural, and environmental factors (Baud-Bovy: 1982, 312).

Baud-Bovy emphasizes that planning should be a continuous process because of the vagaries of tourism over time—economy, politics, fashion. Required is a regular monitoring system.

## Canada

In the last few years, Canada has continued to build on its earlier precedent of tourism planning at both federal and provincial levels. Because of the political ideology of concern over the welfare of its people and because of the long and linear distributions of population between two oceans, a study was commissioned to establish the foundation and guidelines for fostering interprovincial travel for cultural exchange. *Intercan: Understanding Through Travel* (1977) offered recommendations for subsidizing travel to other provinces and special programs of family-to-family host-guest relations.

The federal-provincial agreement program providing financial assistance to development within provinces stimulated increased study of destinations. A special study in 1979, *The Tourism Destination Concept* (Ruest), elaborated on the nature of tourism destinations.

In 1981, the Canadian Government Office of Tourism (CGOT) commissioned two special investigations. One was to focus on the market-product mix; the other on federal criteria for destination zones. Taylor of CGOT in 1980 published his recommendations concerning the need for a better market-plant relationship (Taylor: 1980, 56). He based this need on his observations that "the characteristics of tourism demand are changing rapidly and these changes outstrip the present ability of the plant to adjust and that a measurement system can be devised that will permit the plant to adapt to changing demands in a rational manner." The physical plant—all the goods and services used by the traveler—were related to the experience sought and satisfied by a trip.

As an experiment, Taylor applied market analysis in several countries to the Canadian tourism plant. It was found that of the six segments of Swedish travelers, the Canadian plant could satisfy only one segment; West German markets were matched by only two. All American market segments, however, matched the available Canadian plants.

A special contract to develop a plant/market model for Canada was offered by CGOT. The objectives were to:

Assemble and store key data on tourism market and plant.
Find where markets should be directed to meet expectations.
Show gaps in tourism plant.
Show Canada's position in world plant.
Relate CGOT'S program to market/plant (Plant/Market: 1981, 1).

The diagram in Figure 9-2 illustrates the basic concept of this market/plant match model.

Running parallel with this investigation was a second study directed toward elaborating a federal perspective on tourism destinations throughout Canada.

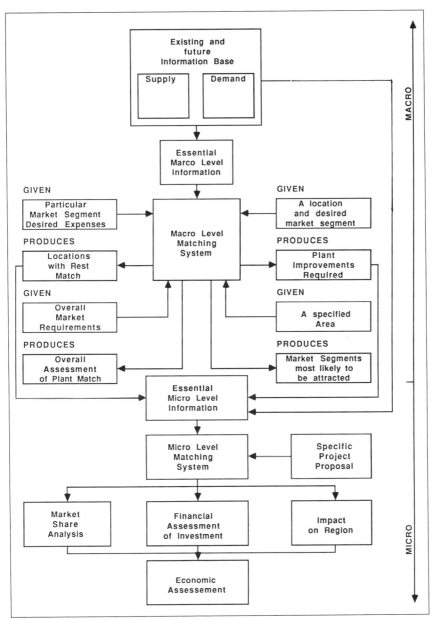

**Figure 9-2.** Plant/Market Match Model. A systems model that guides planning toward matching market segments with the most appropriate supply development (Taylor: 1980, 58).

Particular reference was to be made to four markets: the United States, overseas, interregional Canada, and intraprovincial. The investigation was limited to review of existing documents on planning and zones. The study (Gunn: 1982) resulted in two main conclusions. First, several functioning destinations have developed even though they are not necessarily the result of plans. Market data

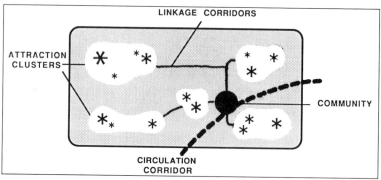

**Figure 9-3.** Destination Zone Concept. Key elements that make up a destination zone: attraction clusters, service community, circulation corridor, and linkage corridor.

for the four market thrusts of the project were applied to known Canadian resources, access, and developed destinations. The "ideal" criteria used for new delineation of zones were: access, service centers, natural resource base, cultural resource base, other resource factors, and program management factors.

The second conclusion was that in the effort to delineate zones based on these criteria, several observations suggested that a new approach to planning was needed. The existing basis for destination delineation seemed to have several weaknesses:

The provinces used different criteria.

The definitions of destination zones were not uniform.

The dates of basic data of provinces and the federal government varied considerably.

## Destination Zones

As a consequence, further study produced *A Proposed Methodology for Identifying Areas of Tourism Development Potential in Canada* (Gunn: 1982). The intent was to provide a uniform approach to replace the great diversity of provincial methods that defined destination zones in many different ways. Conceptually a destination zone was defined as illustrated in Figure 9-3. Four essential elements of a destination zone are identified: (1) *attraction clusters,* groups of things to see and do, (2) *community,* providing services, facilities, attractions, products, (3) *circulation corridor,* main access by land, air, water, and (4) *linkage corridors* between the supporting service center and attraction complexes. All elements require integrated planning for a viable destination zone.

It is no coincidence that this destination zone concept agrees with contemporary views of urban development. At one time, and particularly in response to apparent problems of urban congestion, the principle of dispersion was put forward. Conservationist proponents of national parks still support this principle. In some parts of the world, this was seen as an egalitarian socialist solution. Instead, today the concept of combining a major central city and its periphery into a single unit seems to have more merit. Gradus (1980, 418) calls this a "regiopolis." The regiopolis combines the advantages of social and human as well as economic development. This system of a cluster of smaller cities and town linked by adequate transportation and communication with a central city provides for interdependence at the same time that it protects geographic and human differences.

The destination zone concept with a central service community is compatible with better schemes for social and income improvement in developing countries. The objective of alleviating poverty, dominantly in rural areas, required not only increased productivity but also greater access to services and facilities for health and education. Sponsored by the U.S. Agency for International Development, a concept of urban function in rural development (UFRD) was created and applied in several regions, including the Philippines, Bolivia, the Cameroons, and Ecuador (Rondinelli: 1985, 435). By analyzing the resource distribution and extent of services, the need for strengthening urban centers and the services to be added can be indentified. More equitable economic and social development can be guided in this way.

Further refinement of the destination zone concept suggests three destination subzone types, as illustrated in Figure 9-4. Subzone A recognizes the urban area as a very important target for many travel objectives. Subzone B identifies linkage between a surrounding area and a community service center. Subzone C responds to a significant percentage of travel patterns that include first arrival at a major travel terminal (usually a large city) and transfer to some other area as a final destination.

The Canadian Government Office of Tourism elaborated this approach into policy (Gunn: 1982, 21). Also recommended was that CGOT accept the responsibility for three important efforts, each resulting in a popularly written publication issued annually. The purpose would be to provide *all* actors and publics with base information to stimulate better local and regional planning. The first publication would be a "Research of Physical Factors" summary. Descriptions and computer-generated maps would be aggregated to a composite map showing areas where basic resources were of best quality and quantity. The resources to be analyzed and mapped were water-waterlife, vegetative cover, wildlife, climate, topography, soils, geology, history, ethnicity, archeology, esthetics, institutions and existing attractions, service centers, and transportation access.

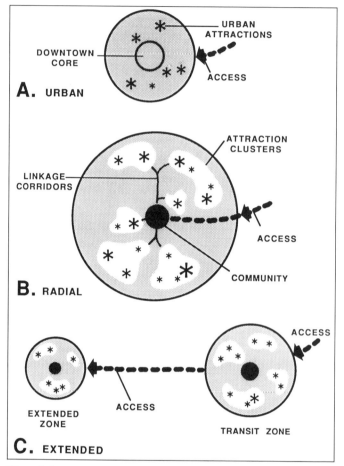

**Figure 9-4.** Three Kinds of Destination Zones. Three popular patterns of tourism destination zones: (A) urban, (B) radial, and (C) extended. All contain the key elements shown in Figure 9-3.

A second publication could summarize "program factors." This would include a component of information on international market trends—key market sources, political stability, economic stability, international currency exchange, preference trends, and traveler characteristics. A second component would be national market trends, including federal/provincial/territorial policies, economic conditions, destination trend preferences, changing characteristics of Canadian travelers, and federal marketing strategies. A third element would identify national development trends and changing policies of government, the private sector, and the nonprofit sector.

The third publication would be a guide to "destination development" based on a market resource match. The physical factors would be evaluated in terms of several key market segments. Each market segment has preferences of activities and locations that could be tested against the resource factor distribution.

Destination zones could be delineated wherever there was a good market resource factor match.

The technical foundation for the mapping of physical factors utilizes a computer aggregation of nine resource maps whose influences have been weighted subjectively.

An adaptation of the concepts outlined by Gunn for Canada was prepared by Matheusik (1985). This proposal would strive for an even closer product market match. He outlines a nine-step planning process:

1. Select market segments to be used.
2. Identify the key tourism components or product characteristics to be mapped.
3. Collect tourism product data on each of the tourism factor components.
4. Convert tourism product characteristics to base map format.
5. Compare market segments with tourism product potentials.
6. Weight each product segment for each market segment.
7. Develop composite map for each market product match.
8. Delineate appropriate destination zones for each market segment.
9. Delineate key aggregate destination zones for combined markets.

Table 9-1 illustrates suggested weighting of the several factors for each of four market segments of U.S. vacationers to Canada.

**TABLE 9-1**
WEIGHTED FACTORS FOR FOUR MARKET SEGMENTS

Weighted index shows the relative significance of each factor as a basic foundation for tourism supply development that would be necessary for four segments of U.S. markets to Canada.

| | WEIGHTED INDEX | | | |
|---|---|---|---|---|
| TOURISM FACTORS | OUTDOOR | TOURING | RESORT | URBAN |
| Transportation | 25 | 30 | 25 | 20 |
| Natural resource base | 33 | 19 | 21 | 12 |
| Service centers | 16 | 20 | 18 | 28 |
| Attractions/events | 10 | 18 | 20 | 25 |
| Other Resources, programs, management | 16 | 12 | 17 | 15 |
| | 100 | 100 | 100 | 100 |

(Source: Matheusik: 1985?, 8)

It should be noted that these efforts were within the current Tourism Development Branch of CGOT objectives, which called for:

Elaborating a federal tourism development policy and strategy.

Improving the governmental framework policy environment for development of the Canadian tourism industry.

Stimulating investment in facilities and capital flows to the Canadian tourism industry.

Assisting the private sector in productivity improvement and facilities/services upgrading (Tourism Program: 1981).

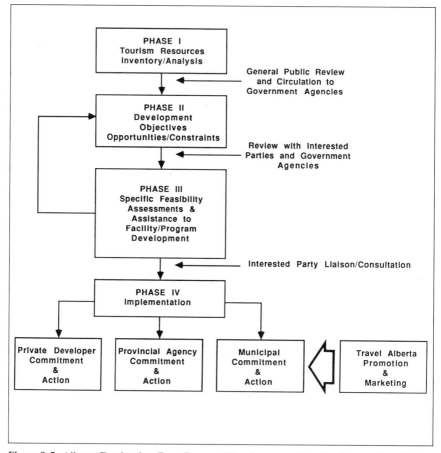

**Figure 9-5.** Alberta Destination Zone Process. The phases used in planning tourism destination zones in Alberta. Promotion is added only after development action has taken place by private developers, provincial agency, and municipal agency (Southeast Alberta: 1982, 4).

A planning process for destination zone development was recommended for the province of Alberta (Southeast Alberta: 1982). This approach is diagrammed in Figure 9-5. Phase I is an inventory of all existing attractions and potential opportunities. Included in the process is issuing a publication of the results and holding public hearings for discussion and information purposes.

Phase II identifies opportunities and constraints as derived from phase I and then sketches development concepts for future tourism. This step also results in a document that is given widespread circulation in the destination zone. The document includes policy statements as well as physical development ideas. This is reviewed by publics and the government agencies involved.

Individual site projects are then reviewed and feasibilities are determined in phase III. Consultations between interested parties form an integral part of this phase. Also at this point potential implementation measures are delineated. Priority is given to new attractions that support and reinforce one another.

An implementation phase follows. It involves three principal actor groups: private developers, provincial agency, and municipal agency. This is the critical phase of commitment and action, where plans become realities. Only after these are in place will Travel Alberta, the provincial agency, initiate promotion and marketing.

Although this proposed process is valid theoretically, the province has reduced it to a less structured and detailed approach to planning tourism.

During the same period, Statistics Canada, an arm of the federal government independent from CGOT, attempted to elaborate destination zones based on population (Chadwick: 1979). It was found that adaptation of the Standard Geographical Classification (SGC) used by the agency was not uniformly accepted by tourism interests in the provinces, who had formed their own subdivisions. Through compromise, the SGC system was adapted to the provinces to establish census subdivisions of provinces of value in maintaining tourism statistics.

## Capacity Assessment Planning

Whereas much polemic discussion has blamed tourists for environmental degradation, most researchers agree that design and management can still allow for mass tourists without major environmental disruption. In addition to the discussion of carrying capacity in Chapter 5, other sources of capacity considerations can be found in: Young (1973), Turner and Ash (1975), Bosselmann (1978), Stankey and Lime (1973), and Hendee et al. (1978).

A suggested modification of a traditional planning process for tourism development was advanced by Getz (1983, 252). Figure 9-6 illustrates his model of an eight-step planning process that allows for capacity assessment:

1. Describing and modeling the system and its environment
2. Forecasting and choosing alternatives
3. Evaluating planned development

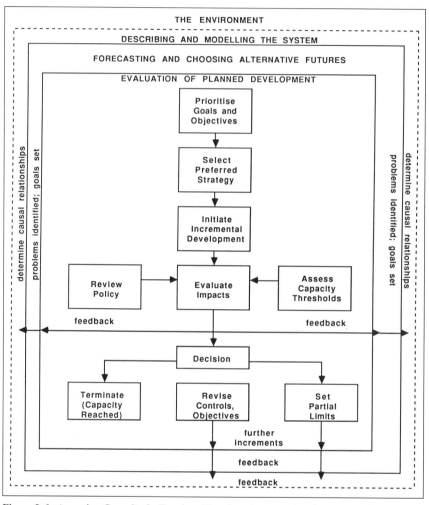

**Figure 9-6.** Assessing Capacity in Tourism Planning. A comprehensive approach to tourism planning with special emphasis on identifying capacity thresholds. The sequence begins with goals and research and ends with controls that allow but limit development (Getz: 1983, 252).

4. Prioritizing goals and objectives
5. Selecting preferred strategy
6. Initiating incremental development
7. Reviewing and evaluating
8. Decisions

# Spatial Interest

At the same time that planning consultants and governments awakened to the need for tourism planning, several geographers began to take interest. Pearce

(1981) has built on earlier work of Gilbert in England, Miege in France, and Poser in Germany to identify structures and processes of tourist development, evaluation of resources, and analysis of impact of tourist development.

Pearce (1981, 6) identifies five elements of tourism supply: attractions, transport, accommodations, supporting facilities, and infrastructure. He divides attractions into three types: natural features, man-made objects, and cultural, such as music, folklore, cuisine, and such. Transportation for travelers has shifted, causing spatial changes in tourism development. Accommodation patterns have diversified and a wider range of other services are in demand. Whereas infrastructure is not revenue-producing, it is essential to tourism. "Successful tourist development depends in large part on maintaining an adequate mix, both within and between these sectors." Developmental factors of interest to geographers include location, land tenure and use, carrying capacity, and analysis of tourism impacts. With these foundations, several spatial planning issues are identified: areas with greatest developmental potential, the need to foster growth in dispersed areas, and the need to ameliorate local cultural disruption.

Pearce (1981, 83) concludes that there are many constraints on planning for tourism. One problem, particularly in developing countries, is the temptation to use models from elsewhere due to lack of data and expertise. This may lead to highly inappropriate development. Implementation is also an issue, especially when roles of the several sectors are not clearly understood. Coordination between agencies is often difficult. Planning, as a process, is preferred over a plan due to the dynamics of tourism. Feedback, monitoring, and flexibility to meet changing conditions are needed.

Because tourism involves elements of great interest to geographers—spatial differentiation and regularities of occurrence (Pearce: 1979, 247)—few disciplines have contributed as much to the literature of tourism development. Pearce identifies interest topics such as spatial patterns of supply (Thompson: 1971; Wolfe: 1951; Piperoglou: 1966; Pearce: 1979), spatial patterns of demand (Wolfe: 1951; Deasy and Griess: 1966; Boyer: 1962, 1972), the geography of resorts (Pearce: 1978; Pigram: 1977; Relph: 1976), tourist movement and flows (Williams and Zelinksy: 1970; Guthrie: 1961; Archer and Shea: 1973; Wolfe: 1970; Campbell: 1966; Mariot: 1976), impacts of tourism (Christaller: 1954, 1964; Coppock and Duffield: 1975; Archer: 1977; Pearce: 1978; Odouard: 1973; White: 1974; Smith: 1977), and models of tourism space (Miossec: 1976, 1977; Yokeno: 1977). Van Doren and Gustke (1982, 543) analyzed shifts over time (1963–1977) of the growth of hotel development with particular reference to the "sunbelt" of the U.S.A. This sampling of scholars and topics emphasizes the fact that economics and promotion, although dominating political interest in tourism, are not the only topics of study important to planning.

Other geographers have studied special aspects of tourism. Demars (1979, 285) has traced the development and distribution patterns of resorts in North America and Britain. Murphy (1979, 294) investigated the special imbalance in travel patterns and the place of camping in market development strategies in Canada. He concluded that marketing plans must be much more aware of the specialized behavioral preferences for destinations. Britton (1979, 326), in his study of Third World tourism, concludes that "Host governments must convince an arrogant and powerful industry that local citizens *and* tourists would benefit from the honest representation of places." Geographical studies in developing countries have been made by many geographers, including Helleiner (1979, 330), Hyma and Wall (1979, 338), and Collins (1979, 351).

## Integrated Development Planning

The theme of this book is planning so that there is better integration between the diversity of demand and the multiplicity of supply. Integration is an obvious need but not so easily accomplished. Each element has its own identity and must respond to its own objectives. But individual self-interest objectives can be met in even a better manner if integrated with other elements. The issues—social, environmental, economic—that have become more visible in recent years also endorse an integrated approach.

Lang (1986) has provided some fundamentals of integrative planning that incorporate broad goals beyond physical development. He warns that development is not mere growth but is a "process of learning, adaptation, and purposeful change capable of releasing new potentials" (Lang: 1986, 5). He proposes planning that has two key dimensions: interactive and strategic. Integrated planning should involve not only the stakeholders responsible for action but also those who are affected by the outcome. Interaction must include information feedback, consultation, collaboration, and negotiation. (A comparison between interactive and conventional planning is shown in Figure 1-1 in Chapter 1.)

An admonition to planners is Lang's (1986, 24) identification of three perspectives important to all planning:

> *The organizational perspective* which sees the world from the viewpoint of effecting and affected organizations, whether formal or informal, permanent or ad hoc. Primary concerns are the survival and stability of the organization and achievement of its goals. Organizations are often characterized by narrow objectives, short time horizons and a healthy fear of failure.
>
> *The political perspective* which is the way elected representatives and those who work directly with them perceive events, processes and decisions. This view

emphasizes the exercise of power and the allocation of resources through bargaining and negotiation. Time horizons typically are short, concerns are highly focused and specialized, and balance is sought in the short run rather than in some distant future.

*The personal perspective* filters the world through the individual's perceptions. Such factors as intuition, self-interest and ethics, personality traits and leadership qualities play an important part in determining how each of us constructs our unique reality and makes our decisions.

## Tourism Planning Goals

A review of the social and environmental aspects of tourism shows that sole support of economic advantage is incomplete and at times even damaging as an exclusive planning goal. In today's context there are four goals of tourism planning: satisfactions to users, rewards to owners, protected utilization of environmental resources, and local adaptation.

Tourism begins with the desires of travelers to travel and ends with the satisfactions they derive from such travel. But as has been pointed out, the complicated characteristics of modern tourism tend to reduce these satisfactions from the possible level desired. Mere volumes of mass participation, a popular method of evaluating "success," do not necessarily translate into satisfaction. Visitors often waste much time, money, and energy in attempting to weave the myriad of development into a meaningful whole. They are too often frustrated in finding their way about, seeking the appropriate services, and understanding the attractions when they do find them. Furthermore, developers' lack of understanding of the tourist market results in less than satisfactory services, such as poor accommodations, food services, and interpretive tours. Fragmentation of transportation services creates irritation and sometimes disappointments, especially when packaged tours are aborted midstream.

> The amateur tourist needs help. He needs confidence that the strange world to which affluence admits him deals candidly with him and that there are standards of price, service and facility he can rely upon. He needs assurance that if somehow things do not turn out as expected, he has a recourse (Destination USA: 1973, 1/95).

Planning should not only eliminate the problems outlined above but also provide the positive mechanism whereby land acquisition, design, development, and management have the greatest chance of providing user satisfactions. In this sense, planning is not only user problem solving, it is user problem avoiding. Planning should provide a check on interrelationships of development to make sure the participant's desires, habits, wishes, and needs are satisfied insofar as physical development and management can do so. The worth of the planned development is not to be judged solely by the planner's own value scaling but by the user's. This standard demands a flexible rather than a fixed planning policy.

The importance of tourism to the individual and society was identified in the "Manila Declaration on World Tourism," (Records: 1981, 118) resulting from the World Tourism Conference of 1980. The "spiritual" elements that must be given high priority were cited, including:

Total fulfillment of the human being.
Constantly increasing contribution to education.
Equality of destiny of nations.
Liberation of man in a spirit of respect for their identity and dignity.
Affirmation of the originality of cultures and respect for the moral heritage of
    peoples.

This expression by 107 national delegations and 91 observer delegations from international, governmental, and nongovernmental organizations was evidence of the need for considering the human dimension in all tourism planning.

Therefore, one major goal of collaborative tourism planning is the *provision of user satisfactions.*

A second conclusion that can be drawn from the review of tourism development ills today is that of owner-investor rewards. Lack of communication and even antagonism between governmental agencies and private enterprise often does not allow private investors adequate freedom to get an important commercial job done. When private development becomes "so involved in red tape and so expensive that it has no appeal to a potential investor," something is wrong (Brown: 1975, 24). Successful private enterprise must make profits. Likewise, successful bureaucratic development, as in local, state, and federal parks and reserves, must meet those objectives of common weal important to public goals. This kind of "success" is as important to the not-for-profit sector as profits are important to business. Only adequate rewards of this nature can provide the incentive to develop. The desired social and economic impact then follows. Fragmentation and isolation of policies, regulations, and managerial practices tend to reduce greatly the potential of rewards.

Planning should not only address itself to the elimination of the above problems but also to the provision—the insistent provision—of positive rewards to those who identify, design, develop, and manage areas for tourism. In other words, cooperation, collaboration, and coordination must foster, not destroy, individual creativity and innovation in development to meet new needs. It must be self-serving at the same time that it is socially responsible. Private enterprise should be guided into locations and programs in which it can be *more,* not less, successful. Public agencies should be guided into locations and programs that meet their special governmental mandates.

Therefore, another goal of collaborative tourism planning is the provision of *increased rewards to ownership and development.*

Third, the many environmental problems of tourism development are increasingly critical. A major criticism of tourism is its esthetic damage to natural beauty. Massive resort and vacation home complexes create major waste disposal and runoff problems. It is seldom that these developments are located within a regional context that respects physical and esthetic resources. Piecemeal development by governments and the nonprofit sector provides little integration between their establishments and the resources of an area. Equally critical, because of increased demand for heritage development, are the historical and archeological resources. These are nonrenewable resources because once a site is destroyed or used for other purposes, it is difficult to restore it. Tourism demands much from the social and economic environment wherever it takes place. Whereas it has many positive qualities, it can do irreparable damage to established communities and resources. According to Dasman et al. (1973, 115): "The more local people benefit from tourism, the more they will benefit from a commitment to preserve the environmental features which attract tourism."

Therefore, a third goal of collaborative tourism planning is the *protection of environmental resource assets.*

The above three goals are appropriate from a tourism developmental perspective. However, tourism is not an isolated phenomenon; it is an integral part of all other aspects of an area's noneconomic as well as economic life. The comprehensive nature of tourism forces incorporation into the social, traditional, political, physical, and economic goals and objectives of localities. Tourism must interact with decisions on infrastructure and amenities sought by residents. Whereas development remains a goal for most areas, the impact of that development demands equal attention.

Therefore, the goal of collaborative tourism is *integration into total community and area social and economic life.*

These goals of tourism planning provide the framework for setting more precise objectives. The distinction between goals and objectives is not merely semantic. It is made in order to provide good planning flow from overall policy directions to achievable action. Goals are continuous; they are abstractions that never are reached, or, rather, are always being reached for. For a nation or any region, the four goals above could be made policy that could continue, year after year. For tourism, these could be adopted by both the public and the private sectors.

## The Planner and Publics

Perhaps in no other field of endeavor is there more skepticism about planning than in tourism. This situation was demonstrated by the results of hearings held in 1976 for the National Tourism Policy Study (1977, App. A.). Although some mention was made of the need for planning, most of the testimony was devoted to image-building, increased promotion, and deregulating tourism controls. Private enterprise generally views planning within the very narrow context of someone else imposing unwanted controls upon them. The image is usually that of an autocratic and insensitive planner, coming from outside their own ranks. This view was stated very succinctly by Trecker (1950, 232):

> The notice of planning is frequently obscured by our feelings about it. It is strange but true that many people have a fear of planning because they envision someone else making the plan for them to obey or execute. The fear is not of planning per se but rather a fear of *how* the planning is done. In the last analysis it is fear of control rather than of planning.

Although planning frequently must exercise some changes in the usual routine of development (otherwise it will not accomplish more than the present helter-skelter of development), planning has much broader meaning. For one thing, there is no reason why tourist business operators might not begin their own broad-scale planning for their own self-interest. By means of collaboration among the many segments of commercial tourism (within legal limits, such as antitrust laws) businesses could provide a much stronger systems approach to tourism development. Furthermore, planning authority need not be vested in only a high-level government planner. More likely, many individuals within the many segments, both public and private, could and should perform planning roles.

Whereas much experimentation is taking place in order to meet public pressure and legal mandates for public involvement in planning, little is fully known about how effective these experiments are. Some agencies believe they have developed new rapport with the public, whereas others are not sure.

Although advantages of public involvement may seem obvious, the pitfalls and obstacles are numerous but less visible. Bury (1977) has grouped these problems into three categories, depending on the viewpoint of the agency, individual, or group. From the agency point of view, public involvement may be seen as not worth the effort: it may encourage distraction from mandates; it may encourage waste of time and may even stimulate alienation. Furthermore, the effectiveness has not been proven. Professional and administrative power may be weakened and studied reason may give away to heated emotionalism.

From the perspective of the citizen, involvement is costly, sometimes futile (because no one listens or power groups take over), appears to threaten authorities, is often rigged or only tokenism, and frequently is loaded with complicated jargon. Organized groups object that the effectiveness of their public involvement is unknown: no one reports subsequent change in policy or practice; presentations are personal rather than the consensus of the group; financially wealthy groups overpower poor groups; minority opinions are ignored; and group proposals are demeaned.

In a way, all these efforts toward greater public involvement represent an indictment of past planning. One may well ask why management hasn't always been involved with or at least been sensitive to publics in tourism. For example, why haven't the recreation land management agencies in the U.S.A., such as the Army Corps of Engineers, U.S. Forest Service, and the National Park Service, always involved publics in their planning?

Perhaps the answer lies in the type of planning that has been utilized by these agencies. A popular mode was that of a tight organizational hierarchy operating on a site basis after land boundaries were set by Congress or by the agency. It was a "batch" and "inhouse" process rather than a continuous and open process. In other words, the planning was primarily on a site basis using a landscape architectural approach, which assumes a "program"—a definition by management regarding the needs (type of use, capacity, quality, other special considerations such as budget). Then the designer creates a site "solution" based upon his concepts of visitor use and interpretations of the program furnished by management. In this process, the assumption is that the manager has full knowledge, by means of training and experience, of all the publics involved. But even later programs of "supply" and "demand," as contained in state comprehensive outdoor recreation plans, did not contain adequate information on the behavior, attitudes, and opinions of the many publics involved in park development and use. Therefore, the need arose for reassessment of planning that is more sensitive to these publics.

With regard to tourism planning, it is likely in the future that new regimes of planners and influential publics will occur. Although sporadic, universities are beginning to recognize the need for providing educational options that involve tourism planning curricula. State governments are considering the need for tourism planners, and planning may become an important part of the federal role in tourism. Edgell (1977, 35) has pointed out that several factors—American lag in tourism policy, growth of both domestic and foreign tourism, and increased recognition of the worth of tourism—will

> bring about—indeed, will force—a greater degree of tourism planning within the United States. Similarly, it is probable that those states with tourism-based economies will, like Hawaii, develop plans for assuring the quality of life for both tourists

and their host communities. Tourism is simply too important an industry to permit to develop without planning and policy direction.

## Intersector Planning

A review of the tourism planning experience suggests that the greatest opportunities for achieving desired goals may lie in voluntary integration among and between the many sectors as their own plans are made. This approach is the extreme opposite of centralized planning. Rather than bureaucratic planners at the federal level demanding that each part, such as hotels, integrate their decisions on location, size, and market segments with other key actors, such as in transportation, this interaction would be voluntary, based on self-interest.

The concept of self-interest, the very foundation of a market economy, could be utilized more effectively throughout all parts of tourism. This concept is based on the motivation to achieve and to reap the rewards from such achievement. In the past, this has been directed only internally—within the firm, agency, or organization. But, as has been dramatized by the model of the functioning tourism system, an exclusively internal perspective is no longer valid. Many externalities influence success. Therefore, when an external perspective is added, it will be seen that integration with other parts is not only desirable but virtually mandatory. This, together with guidance from expert planners, may produce the best tourism planning. An approach of intersector planning could also be identified as *organic* tourism planning because it is founded in the organic optimal functioning of each individual part, recognizing that this functioning is dependent upon successful functioning of all the other parts.

For example, decisions on sites, structures, and management of hotels are likely to be more productive if done in concert with decisions by others on resource protection. Agency policies on management and protection of forests, wildlife, and water resources have considerable influence on the tourists who seek these resources as well as lodging services nearby. Lodging decisions should also be made in concert with those on transportation, traffic control, and visitor information. Public decisions on infrastructure must reflect the needs of tourism as well as local citizen development if a community has adopted tourism growth as policy. Developers of attractions—parks, entertainment, historic sites, theme parks—will gain when their plans are integrated with those of the service businesses and transportation planners. Both governments and private enterprise will gain when relations are more cooperative than adversarial. A proactive rather than reactive stance by private enterprise can foster greater self-interest, which, in turn, can be better for visitors and for business.

However, in many countries, several barriers prevent or constrain the exercise of adequate networking between the many parts and sectors.

Enterprise often has a misconception of the tourism product. Hotels often believe their product is the hotel and its services, which is only partly true. Certainly, the bulk of effort, time, and money must be spent on internal operations—housekeeping, engineering, sales, food service, maintenance, personnel. But an important part of its product is the travel purpose and the attractions that brought the traveler to the hotel. It is important that hotel planning be integrated with attraction planning to foster a more satisfying travel product.

All enterprises and government agencies have great difficulty in justifying outreach. Capital and operating expenses generally are not budgeted for this function. So, it is not assigned to anyone. When such outreach is seen as a part of normal operation, greater planning integration will take place.

For all businesses and government operations related to tourism, a major barrier to integration is turf protection. Other agencies and businesses are seen as competitive. Employees often are made to feel that any contact with outsiders is taboo. But if the tourism system is to function more smoothly and if each part is to gain, greater integrated planning, on a voluntary basis, will be required.

When tourism economic impact is described, it is popularly assumed that the business sector is exclusively oriented to tourism. This is not true. Restaurants, entertainment, shops, service stations, and even hotels cater to local markets as well. This fact of dual markets divides the attention of businesses, often diluting their support of tourism programs. When business practice and policies are directed toward integration of both local and visitor markets, they will gain.

Unfortunately, educational institutions and training programs tend to foster divisiveness rather than integration. Hotel schools teach hotel management; park schools teach park management; restaurant schools teach restaurant management. Seldom are students taught how their segment is interrelated to and dependent upon many others. Broader education and training programs are needed to alert new trainees regarding planning integration.

Intersectoral as well as intrasectoral planning will be fostered when each individual part recognizes its highly interdependent role with many other parts—even from a self-interest perspective.

## Conclusions

From investigation of several planning approaches to tourism development, some conclusions can be drawn. Perhaps the main conclusion is that tourism planning models are few in number and relatively recent. As more planning experience is accumulated, better models most certainly will appear. Viewing tourism development beyond the site and into the regional scale does offer some conclusions that should be useful in future applications.

*Planning models tend to include similar components.* Independently, several authors and planners have developed tourism planning models. Some are more comprehensive than others. However, all identify similar key components that are essential to the overall tourism system. All recognize the basic components of a demand (market) side and a supply (destination) side linked by transportation and communications. All endorse the need for planning in order to accomplish a better demand-supply match. The components of market interest and ability to travel, attractions, transportation, services, information, and promotion, although expressed diffrently by the several experts, appear to be essential for tourism functions.

*Special relationships are essential to planning.* A tourism destination zone utilizes many of the same elements of traditional area planning but places them in slightly different context. A simple urban-rural dichotomy does not apply. Rather, four elements—attraction clusters, community, circulation corridor, and travel linkage between the services of a community and the attraction clusters—require integrated planning. This four-part pattern is an important overlay of value in the search for potential. This pattern also suggests important relationships between big city and small city. Such a regiopolis (central large city surrounded by smaller cities) is compatible with desirable social and economic development generally.

*Integrated planning is organizational, political, and personal.* Past "muddling through" has shown to be inadequate in today's tourism development world. Instead, planning must integrate the functional parts in ways that protect their individual integrity and also complement one another. Integrated planning has been identified as a process of learning, adaptation, and purposeful change capable of releasing new potentials. Whereas increasing order integrative planning releases new energies for better tourism development.

*Economic development cannot be an exclusive goal.* Traditionally, strong emphasis by communities and nations has been placed on the economic advantages of tourism. Ample support for this overall goal has come from many examples throughout the world where greater income, more jobs, and greater tax returns have resulted from expanded tourism development. However, experience has also demonstrated that the pursuit of this goal is hollow without consideration of three other parallel goals. The goal of visitor satisfactions must also guide all public and private planning and development of tourism. Although this goal may have been assumed in the past, it must become more explicitly declared or the economic goals will never be accomplished, An equally important third goal is that of resource protection. Short-sighted exploitation of natural and cultural resources now demonstrate how sensitive all sectors must be to the principle of resource protection. Finally, planning must integrate tourism into ongoing area social and economic life. Otherwise severe conflicts may arise. Without a balance of all four goals—economic gain, visitor satisfactions, resource protection, and local integration—planning and development is destined to be less successful.

*Tourism planning must involve publics.* The statement that tourism planning must involve publics suggests that centralized planning is wrong, but this is not the intent. Rather, this principle means that tourism is so complicated and so comprehensive a social and economic phenomenon that many publics are involved. An all-publics approach is contrary to the view that tourism is simply one overlay upon a community affecting only the primary businesses of hotel and food service. Local citizens, educators, workers, merchants, professionals, politicians, and many separate constituencies will be impacted whenever tourism development takes place. Planning demands their early involvement so that negative consequences can be minimized from the very start.

*Self-interest fosters integrated planning.* Contrary to popular belief, self-interest can foster integrated planning, provided that the several barriers can be eliminated. The barriers of narrow product definition, lack of commitment, turf protection, single-market perception, and narrow educational programs often obstruct the realization of the many opportunities through networking. Every business and public development for tourism is so related to many others that internal management practices limit opportunities for success. As each enterprise and development reaches out and networks with many others, greater self-interest is fostered, not diminished.

# Bibliography

Arnott, A. "The Aims and Methodologies Used in a Study of Tourism," *Planning Exchange,* (11), 1978.

Bargur, J. and A. Arbel. "A Comprehensive Approach to the Planning of the Tourist Industry," *Journal of Travel Research.* 14 (2), 1975, 10–15.

Baud-Bovy, Manuel. "Integrated Planning for Tourism Development," presentation at CAP/CSA seminar, Colombo, Sri Lanka, May 8–18, 1980. Madrid: World Tourism Organization.

Baud-Bovy, Manuel. "New Concepts in Planning for Tourism and Recreation," *Tourism Management.* 3 (4), December 1982, 308–313.

Blank, Uel and Clare A. Gunn. *Guidelines for Tourism-Recreation in Michigan's Upper Peninsula.* East Lansing, MI: Cooperative Extension Service, 1966.

Bosselman, F. *In the Wake of the Tourist: Managing Special Places in Eight Countries.* Washington, DC: Conservation Foundation, 1978.

Britton, Robert. "The Image of the Third World in Tourism Marketing," *Annals of Tourism Research,* 6 (3), 1979, 318–329.

Brown, D.R.C. "The Developer's View of Ski Area Development," *Man, Leisure and Wildlands,* Proceedings, Eisenhower Consortium, September 14–19, 1976, Vail, CO. Springfield, VA: National Technical Information Service, 1975.

Bury, Richard L. "Obstacles and Pitfalls of Public Involvement," Proceedings, Ninth Recreation Management Institute, Texas A&M Univ., March 31, 1977.

Chadwick, R. "Tourism Regions for Domestic Travel Statistics," paper presented at annual meeting of the Canadian Association of Geographers, Victoria, British Columbia, May 28, 1979.

Collins, Charles. "Site and Situation Strategy in Tourism Planning: A Mexican Case Study," *Annals of Tourism Research,* 6 (3), 1979, 352–366.

Dasman, Raymond F., John P. Milton and Peter H. Freeman. *Ecological Principles for Economic Development,* London: John Wiley & Sons, 1973.

Demars, Stanford. "British Contributions to American Seaside Resorts," *Annals of Tourism Research,* 6 (3), 1979, 285–293.

*Destination U.S.A.,* Vol. 1. Report of National Tourism Resources Review Commission. Washington, DC: U.S. GPO, 1973.

Edgell, David L. "International Business Projects for the Rest of the Century: International Tourism and Travel," presentation, Key Issue Lecture Series, Univ. of Texas, Dallas, 1977.

Getz, Donald. "Capacity to Absorb Tourism: Concepts and Implications for Strategic Planning," *Annals of Tourism Research,* 10 (2), 1983, 239–263.

Getz, Donald. "Models in Tourism Planning," *Tourism Management,* 7 (1), March 1986, 21–27

Gradus, Yehuda and Eliahu Stern. "Changing Strategies of Development: Toward a Regiopolis in the Negev Desert," *APA Journal,* 46 (4), October 1980, 410–423.

Gunn, Clare A. *A Concept for the Design of a Tourism-Recreation Region.* East Lansing, MI: BJ Press, 1965.

Gunn, Clare A. *Vacationscape: Designing Tourist Regions.* Austin: Bureau of Business Research, Univ. of Texas, 1972.

Gunn, Clare A. *Tourism Development Potential in Canada* and *A Proposed Methodology for Identifying Areas of Tourism Development Potential in Canada,* for Canadian Government Office of Tourism, Ottawa, 1982.

Gunn, Clare A. *Tourism Planning,* 1st ed. New York: Crane Russak, 1979.

Helleiner, Frederick. "Applied Geography in a Third World Setting: A Research Challenge," *Annals of Tourism Research,* 6 (3), 1979, 330–337.

Hendee J., G. Stankey and R. Lucas. *Wilderness Management.* Washington, DC: U.S. Forest Service, 1978.

Hyma B. and G. Wall. "Tourism in a Developing Area: The Case of Tamil Nadu, India," *Annals of Tourism Research,* 6 (3), 1979, 338–350.

*Intercan: Understanding through Travel.* Report of Interdepartmental Task Force for Canadian Government Office of Tourism, Ottawa, 1977.

Kiemstedt, H. "Zur Bewertung der Landschaft fur die Erholung," *Bertrage zur Landespflege.* Sonderheft 1, Stuttgart, 1967.

Lang, Reg. "Planning for Integrated Development," paper presented at conference on Integrated Development Beyond the City, Mount Allison Univ., Sackville, New Brunswick, Canada, June 11–14, 1986.

Lawson, F. and M. Baud-Bovy. *Tourism and Recreation Development,* Boston: CBI, 1977.

Matheusik, Mick. "Toward a Federal Tourism Strategy for British Columbia," unpublished paper, 1985, Vancouver.

Mill, Robert and Allistair Morrison. *The Tourism System.* Englewood Cliffs, NJ: Prentice Hall, 1985.

Murphy, Peter. "Tourism in British Columbia: Metropolitan and Camping Visitors," *Annals of Tourism Research,* 6 (3), 1979, 294–306.

*National Tourism Policy Study: Ascertainment Phase.* Senate Committee on Commerce, Science, and Transportation and National Tourism Policy Study. Washington, DC: U.S. GPO, 1977.

Pearce, Douglas G. *Tourism Development.* London: Longman, 1981.

Pearce, Douglas G. "Towards a Geography of Tourism," *Annals of Tourism Research,* 6 (3), 1979, 245–272.

"Plant/Market Match Model," unpublished progress report, November 24, 1981. Ottawa: DPA Consulting Ltd.

*Records of the World Tourism Conference.* Madrid: World Tourism Organization, 1981.

Romsa, Gerald. "An Overview of Tourism and Planning in the Federal Republic of Germany," *Annals of Tourism Research,* 8 (3), 1981, 333–356.

Rondinelli, Dennis A. "Equity, Growth and Development," *APA Journal,* Autumn 1985, 434–448.

Ruest, Gilles. *The Tourism Destination Concept,* for Canadian Government Office of Tourism, Ottawa, 1979.

*Southeast Alberta Tourism Development Concept,* discussion paper, Tourism and Small Business. Edmonton: Travel Alberta, 1982.

Stankey, G. and D. Lime. *Recreational Carrying Capacity: An Annotated Bibliography.* Washington, DC: U.S. Forest Service, 1973.

Taylor, Gordon D. "How to Match Plant with Demand: A Matrix for Marketing," *Tourism Management,"* 1 (1), March 1980, 56–60.

*Tourism Program and the Canadian Government Office of Tourism.* Ottawa: CGOT, 1981.

Trecker, H.B. *Group Process in Administration,* rev. ed. New York: Woman's Press, 1950.

Turner, L. and J. Ash. *The Golden Hordes: International Tourism and the Pleasure Periphery.* London: Constable, 1975.

Van Doren, C.S. and Larry D. Gustke. "Spatial Analysis of the U.S. Lodging Industry: 1963–1977," *Annals of Tourism Research.* 9 (2), 1982, 543–563.

Young, G. *Tourism: Blessing or Blight?* Harmondsworth: Penguin Books, 1973.

# Chapter 10

# Three Levels of Planning

The general concept of tourism planning seems too abstract unless it is brought into implementable levels. Tourism covers all scales of development from national to local. Even though the goals of planning may be the same, the approaches and processes may have to be different for several geographical and political levels.

A review of planning applications suggests that three levels of planning processes are worthy of consideration (Figure 10-1). More and more, planning is seen as a *continuous* process whereby changes in many factors can be given planning and implementation response. Thus, planning takes place at all times, not just when an agency requests that a plan be prepared. Because of the many factors in tourism decision making, this approach is very complicated. Yet it is essential if a market-product match is to be kept current.

As an adjunct to continuous planning, there is merit in devising a specific *regional strategic* plan for tourism development. Such a plan, developed and updated every few years, provides for specific objectives. It sets forth specific projects to be accomplished by target dates. A regional scope (an entire country, state, or province) allows more than community concerns to enter into the process. The regional policies and overall relationships are included. A physical planning component can identify destinations with the greatest potential for tourism development.

If based upon continuous planning processes and regional strategic plans, tourism planning at the *local* level can be accomplished most effectively. Sporadic and independent community tourism planning lacks integration with the region. Thus, it may lack valuable inputs at the regional level and even may be counterproductive. Because of the great importance of community tourism planning, Chapter 11 is devoted to its discussion.

## Continuous Planning

Increasingly in recent years, planners and scholars of planning theory have been giving attention to planning as a continuous process. Much of this activity is a

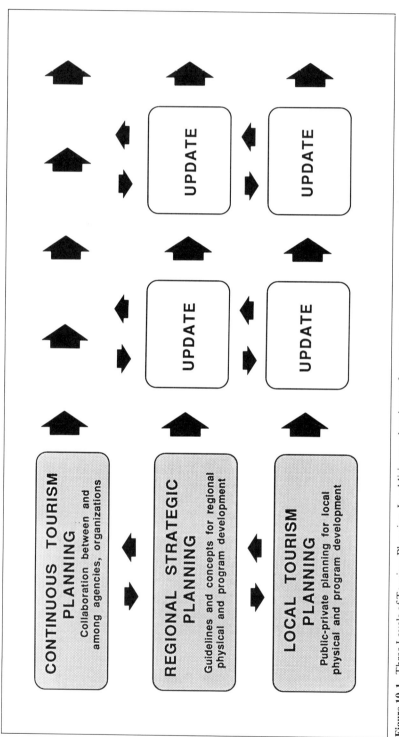

**Figure 10-1.** Three Levels of Tourism Planning. In addition to planning performed by the firm and agency, overall tourism planning requires three levels—continuous, regional strategic, and local community planning. Integration of these levels is the key to successful tourism.

reaction to the inadequacies of the project or plan approach (sometimes called the master plan), "which gave a detailed picture of some desired future and state to be achieved in a certain number of years" (Hall: 1975, 269). The project approach grew out of landscape architecture, which dealt with specific buildable site development, such as for the development and opening of parks within a specific time period. Today, it is increasingly recognized that in addition to this approach, planning as an ongoing process has great merit.

The process that checks back upon itself has grown out of the science of cybernetics, coined by the mathematician, Norbert Wiener, in 1948. This process was presented as a means of controlling complicated mechanisms by interrelating important information. It was applied not only to internal control exercised by the nervous system in an animal but also to the engineering control of equipment such as guided missiles. An important aspect of cybernetics is feedback, in which corrections are made as necessary in the functioning of a system such as the path of a missile. From this, the concept of systems planning developed, meaning the integrated and operational planning of the entire system as a whole composed of interrelated parts. Today, systems planning is yet to be initiated at both private and public levels for tourism development and management. The concept of continuous planning is an application of systems planning to existing agencies, organizations, and the private sector.

Within each component of the tourism functioning system is massive involvement by public agencies. In addition, many organizations outside government exercise great influence on functions of each component. Seldom, however, are these ever integrated. In fact, they are frequently counterproductive. Certainly, for the sake of diversity and countercheck, it is desirable to have such an array of agencies and organizations. But there are instances, particularly at the planning stages, when even a small amount of collaboration and cooperation would be constructive.

Not only must the tourism system receive better continuous planning, but also tourism must be integrated with all other planning for social and economic development. Many governmental agencies at the federal-to-local level are engaged in programs fostering new jobs, housing, and general social welfare. However, seldom do these programs include tourism. A review of structure plans (official planning documents) in England (White: 1981, 40) revealed that tourism was seldom mentioned. Tourism agencies and enterprises apparently have not yet raised tourism to the needed level among the body politic. Undoubtedly, this general lack of integration of tourism into overall social, economic, and environmental planning is typical the world over, not just in England.

A continuous tourism planning function could be modeled as an interactive system whereby each sector is not subjected to a superior level of planning. Instead, each sector, on its own initiative, interacts with all others in its own decision making. Since this is not a legislated planning model, it does not depend on a planning bureaucracy or heirarchy. It capitalizes on its own self-interest to benefit from communication and interchange with other sectors.

For example, the accommodations sector (entrepreneurs, managers, organizations), in order to stimulate greater occupancy, would on its own initiative interface with the historical restorers, festival backers, park-owned recreation interests, and entertainment sectors because they are the ones to stimulate more visitors. Because accommodations can be affected by the policies and decisions of the other sectors, the leaders would open up communications with them. Although some overall committee, council, or other structure may be needed to integrate action, it would seem that the greater each sector increases its own sophistication regarding tourism integration, the more it will contribute to overall integrated planning for tourism.

Such an approach may be getting closer to Lang's call for better integrated planning—"there is need to define shared interests and values and to work out ways that these may be pursued for mutual benefit" (Lang: 1986, 30). He utilizes the concepts of "domain" (Trist: 1985), used to describe a set of interdependencies among stakeholders in a transactional (shared) environment. "Collaboration begins by stakeholders recognizing that they have mutual interests and that their problems are too complex and too extensive for organizations to go it alone" (Lang: 1986, 32). Each one may be willing to engage in a give-and-take exchange on the basis of mutual gain. Only when each sector sees the advantages of interactive functioning will it reach out beyond its traditional and mandated turf.

In order to activate such interagency and intersector cooperation, a detailed review of existing practices, policies, and legal mandates may be necessary. If, for example, a federal agency's enabling act makes no mention of tourism in spite of its actions impinging greatly on tourism development, it may be necessary to amend the legislation. To illustrate further, a highway department may have great competence in engineering construction but lack planning data on tourism trends that influence future highway planning. Until the department's official mandate is amended to include greater responsibility to traveler needs, it may be expecting too much of officials to reach out and cooperate with tourism developers on a voluntary basis.

Even though a continuous planning process ideally would encompass great integration of all factors, it may be necessary to empower a central tourism agency at the highest level of government to be the catalyst for continuous planning. Most tourism agencies today at the federal level are promotion-oriented and do not have powers of coordination and integration of overall tourism. When the key tourism agency is given responsibility for more than promotion, more effective planning and decisions can be made by both public and private sectors.

## Regional Strategic Planning

Nothing as complicated and comprehensive as tourism can be planned by one or two strategies, models, or processes. Because few mechanisms presently exist for interagency and interorganization collaboration, it would seem that continuous planning might have merit. However, it must be supplemented by other planning approaches for potential solution to tourism problems, from local to national.

Described here is an approach for experimentation and study purposes that is adapted to the regional level (state, province, nation). It contains elements of planning that have been well tested and therefore represents a traditional approach to planning. However, since tourism problems of development require special adaptation, the special needs of tourism planning have been incorporated. Experiments in portions of this approach have been made in a few places, such as reported in *Comprehensive Outdoor Recreation Plan: State of Hawaii* (1970); "Tourism Planning for East Texas" (Gunn: 1973); *Tourism Development—Assessment of Potential in Texas* (Gunn: 1979); "Tourism Potential in Central Texas," (Gunn: 1982); and *Guidelines for Tourism-Recreation in Michigan's Upper Peninsula* (Blank, et al.: 1965). With the new trend toward interest in tourism planning, it may find increased use.

For want of a more descriptive title, this approach is called *regional strategic planning*. Whereas the purposes of regional strategic planning are the same as those for continuous planning—visitor satisfactions, rewards to owners, environmental protection, and integration into community life—the scope is slightly different. Considering a region, such as a state, it provides generalized regional information and guidelines that can foster tourism growth to meet these goals. As such it does not solve site problems except as they relate to other sites. It is designed to provide recommendations on both physical and program tourism growth and development, especially the identification of potential destination zones.

Because regional strategic planning is focused on a specific region and because it is a definite project with a completion date for recommendations, it demands special technical planning inputs by tourism planners and others who can provide special expertise. This type of planning is called "intervention for change" and is defined by Friedmann (1973, 347) as ". . . the introduction of ways and means for using technical intelligence to bring about changes that otherwise would not occur."

However, it must be emphasized that technical information and processes alone will not solve the problems of tourism. The effectiveness of technical planning functions in guiding social, economic, and environmental change will depend upon:

(1) the clarity of system objectives, (2) the extent of consensus about them, (3) the relative importance that politicians attach to them, (4) the degree of variance relative to objectives expected in the performance of the system, and (5) the extent to which a technical (as contrasted to purely political) approach is believed capable of making system performance conform to these objectives (Friedmann: 1973, 353).

## Regional Development Factors

Since planning involves predicting the consequences of the manipulation of many development factors, it is necessary to identify these factors. Based on past tourism experience, a dependency hierarchy, such as diagrammed in Figure 10-2, can provide a useful foundation for planning.

Regional development of tourism (A in Fig. 10-2) generally must have an increase in the volume of participation (B in Fig. 10-2). More people must go to a region and spend moneys on tourism activities in order to generate new jobs, new incomes, and new tax revenues. However, increased participation depends on two very important factors.

First, there needs to be a heightened demand to visit the given region. In this context it means that more people, at their home origins, must be able to exhibit both the *desire* and the *ability* to travel to the region and participate in its offerings. If prospective visitors do not have a desire to visit the given region, it is doubtful that they will. In addition to desire, they must have the time, money, transportation, and equipment necessary to make the visit.

Second, if more people are to do this, changes in present levels of offerings—the supply—must take place (C2 in Figure 10-2). Either the capacity of the present physical plant or the total number of offerings must be increased. In other words, either more people or shifts of markets must be accommodated at more attractions, lodging, food service, transportation, and retail sales and services. Furthermore, if the region has a reputation of low attractiveness or poor service, this image must be reversed. And, from an economic point of view, the local system servicing tourism should have the highest "export" ability; that is, it should import the fewest services and goods. Finally, whatever changes are made must be appropriate to both national and local political and social goals. For these to happen, there need to be changes in both markets D1 and resource development D2 (Fig. 10-2). Some of these changes can be manipulated from the standing of the region; others cannot.

## Expanded Markets

Changes in D1 (Figure 10-2) are often influenced by overall cultural and economic trends of the nation (for domestic tourism) and the world (for foreign

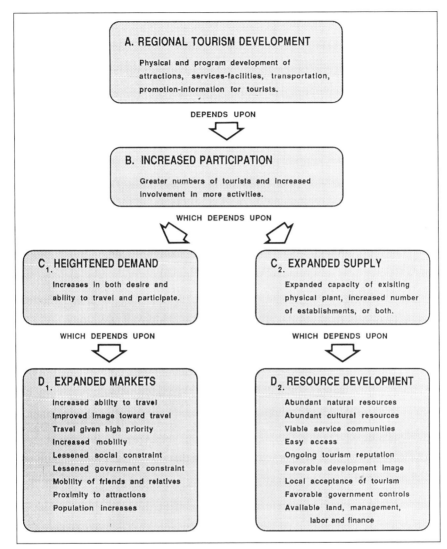

**Figure 10-2.** Tourism Development Dependency Hierarchy. If tourism is to be increased, greater market stimulation and greater development of tourism supply must take place. The sequence of factors upon which demand and supply depend is shown in this diagram.

tourism) and are not subject to easy manipulation by a tourism region. Some, however, can be influenced.

A review of some of the important factors within markets D1 (Figure 10-2) may provide some insight into the opportunities and limits of market manipulation.

*Ability.* Shifts in the ability of people to travel and to spend moneys on travel objectives can have a very important impact on a region. Factors such as increased incomes, greater job security, and discretionary incomes can also have great impact. These, however, are not easily manipulated by the tourism region as they are general conditions of a total society and economy at the source of markets. Sometimes the lowered pricing of transportation and destination services can increase a market's ability to travel to the destination.

*Image.* Markets have either negative, neutral, or positive image values toward regions. These image values can have a good or bad influence upon the popularity of a region. Their creation is extremely complicated and not easily manipulated, as has been discovered by tourist agencies who have been disappointed with massive image-changing programs. In the long run, better supply development at destinations can enhance its image.

*Priority.* People must place high priority on travel in their expenditure of time and money if tourism is to thrive. If society shifts its priorities into other consumer targets—homes, automobiles, education—tourism will lose. Although advertising and promotion may have some impact, it is difficult to change the priority ratings of families and individuals. Whereas some modification of priorities on personal expenditures can be determined from the standing of a region, it is very limited.

*Mobility.* Obviously, mobility of the market—opportunity to travel from home to the destination—is very important. The numbers of people able to travel to tourist destinations may be drastically affected by continued changes in costs of automobile, bus, and air travel. To some degree, mobility both within and relating to a region can be changed by influence from a destination area. However, increased ownership of automobiles and controls of gasoline and plane ticket costs are difficult to influence from the standing of a destination region.

*Social constraint.* Over history, social controls have been increasingly liberalized. Changes in dress, conduct, and an increase in the variety of recreational activities have influenced travel land use in recent decades. However, such cultural shifts come from market sources and are not easily manipulated by the destination region. A local society may even have values in conflict with those of visitors. Better information and educational programs can be used to lessen social conflict between hosts and guests.

*Governmental constraint.* Generally, there are few direct governmental constraints to travel. But government policy and regulations do have an impact on travel and tourism development. Taxes on gasoline, air travel, and properties exert constraints on certain recreational activities. Governmental distribution of subsidies to highways, airports, reservoirs, and recreation areas may favor some activities over others. Blue laws place constraints on purchases of recreational goods on Sunday. Land use regulations and environmental controls are of increasing importance. Political agreements (or disagreements) between nations, such as for control of disease or terrorism, can foster (or restrict) travel. To some extent tourism regions can exert some control over governmental constraint.

*Friends and relatives.* A major category of the tourism market activity is the visiting of friends and relatives. The habitat shifts of these people make a difference in the destinations of travelers. Knowing the locations and changes in locations of friends and relatives is important, but it is difficult for a tourism region to have any influence over this factor.

*Proximity.* Generally, the shorter the distance between home and tourism region the better. Markets nearby or within a region are usually the ones most productive. One way a region can affect this proximity is by favoring destination development at locations most accessible from markets. Another way is to tap new market segments within the market area. Improving the time-distance relationship between home and destination can sometimes improve the effect of proximity.

*Volume.* Given today's burgeoning environmental issues, it is fanciful to suggest increasing the population in market areas in order to increase tourist business. However, for certain destinations, it may be within a region's power to tap increasing volumes of new market segments in cities of origin within their market range, especially those that exhibit elements of growth.

## Resource Development

Although market manipulation is not the subject of this book, it has been introduced to more clearly identify the relationship to the role of resource development (D2, Figure 10-2) in changing the economic and social impact of tourism. In other words, if anything major is to take place within a tourism region, the changes in the supply—providing more for people to see and do— become very important. If numbers of attractions, transportation facilities, accommodations, food services, and retail sales and services—those economic impact generators—are to increase, upon what factors do they depend?

*Natural resources.* The given natural resource assets that lend themselves to development are important if growth of development is to take place. If a region has an abundance of usable surface water, esthetic and game-laden forests, interesting topography, buildable soils, and favorable climate, it has greater potential for tourism development than one without these assets. Of course, reservoirs can be built, forests can be planted, wildlife can be increased with controlled management, and hostile climates can be ameliorated (enclosures with heat in winter and with air conditioning in summer). However, these come at a cost, a cost that often can be lessened by avoiding areas needing environmental modification and by selecting areas with most suitable natural resource assets. Certain rare natural resource assets cannot be replicated and therefore demand special consideration. Many developed attractions, such as natural resource parks, scenic overlooks, resorts, scenic tours, hunting and fishing areas, depend greatly upon natural resource assets.

*Cultural resources.* Many participants seek their travel objectives less from developed natural resource assets than from those that are the result of human cultural imprint. Religious, scientific, and educational institutions, trade centers, national shrines, engineering feats, and manufacturing processes, as well as historic and archeological sites, buildings, and artifacts, are examples of the wide range of cultural resources important to tourism development. These usually already exist within a region.

*Viable service communities.* Generally, the cities that lie within a region being considered for tourism development serve two functions. First, they frequently contain attractions or resources with potential for attraction development. Second, they provide many needed services, facilities, and products. The primary ones, of course, are lodging and food services, but equally important for many tourists are other services, such as police, communications, medical aid, and banking. In addition, cities offer the infrastructure—water, waste, fuel, electrical power—necessary for tourism development. For example, a water shortage on California's Catalina Island in 1977 forced hotels to use disposable dishes for meals and salt-water in toilets, and guests to bring their own linen. (Saltus: 1977, 1G). Tourism expansion, therefore, depends upon the distribution and viability of cities. Generally, the larger the city, the more complete its ability to provide these functions.

*Easy access.* Tourism expansion depends heavily on access, and not all regions are equally served by transportation and access. Within the continental United States, the highway system is very important because automobile travel dominates. However, for certain localities and activities, air (and even ship) travel are important access factors for tourism. Even when expansion is considered, existing routes generally are favored over new ones.

*Existing markets.* An area that already possesses an ongoing tourism development has a stronger factor in its favor for future expansion than does raw and undeveloped land. The existing development may have established a reputation that is well known in the marketplace. Existing development, such as public parks, theme parks, historic sites, and beaches, can provide many clues to the relative importance of the resource potential of an area.

*Favorable development image.* Image is a product of both the supply and user attitude and therefore cannot be dealt with only on a resource development basis. However, if an area, no matter the reason, now has a reputation of poor (or excellent) quality of tourism participation experience, it can deter (or favor) further expansion greatly. Changes of development can alter this image but not without massive change.

*Local acceptance of tourism.* Expansion of tourism depends greatly on the local attitude toward expansion. If the local electorate and leadership fully understand the implications of tourism and favor its development, further expansion has support. However, if attitudes are antagonistic or hostile, it will be difficult to develop tourism.

*Favorable government controls.* Tourism development can be accomplished best with the least governmental constraints. If too many legislative controls are enacted against it, certainly development is restricted. Legislation, however, is a human act and can be changed. Jurisdictional problems can sometimes limit full opportunity of developing legislation that gives tourism greater chances of growth. Care needs to be exercised in evaluating controls. Many of the recent environmental controls may appear to work a hardship on some development, but, upon deeper examination, may be protecting and perpetuating tourism attraction assets.

*Available land for development.* Tourism development certainly depends upon space for development. Some segments use more land than others. Beach use may be very intensive, whereas hunting and wildland recreation are probably the most extensive. Vacation home development uses relatively large areas of land. Probably a greater constraint is not being able to purchase land that is properly located. Since modern concepts of ecology do not allow any land to be classified as "waste," new tourism development must be a tradeoff from existing or other potential land use. Land price and ability to purchase are important. Therefore, tourism development depends on the availability of suitable land for expansion—suitable in both quantity and quality.

*Availability of entrepreneurs and managers.* The tourism development of a region will depend on the availability of entrepreneurs and managers. If a region does not have these resources, they will have to be imported. However, this can cause conflict with local aspirants who expected to qualify for the positions. The greater the supply of the several types of developers and managers—for attractions, transportation, lodging—the more favorable a region is disposed to development.

*Availability of labor pool.* Tourism employs a wide range of job categories from highly skilled to unskilled. The nature of development will determine the labor needed and whether the region will have to import labor or supply its own. If new labor is imported, local underemployed or unemployed people may resent their coming and resist tourism expansion. Therefore, source of labor is an important factor in tourism development.

*Availability of finance.* Tourism development demands great amounts of capital investment. Perhaps it would appear that the means of finance would not matter as long as development takes place. This may be true in the economic sense that some additional development is better than none. However, investment sources often carry with them contingencies that may or may not be compatible with local interests. For example, local residents may realize the value of certain land use controls to protect resource assets. An outside investor may refuse to invest under such conditions. Availability of finances is important from many standpoints.

This hierarchy of dependency identifies a number of variables, any one of which can make considerable difference in opportunities for tourism development. Some are geographic; others are not. Some are slow to change; others change very rapidly. Some are subject to legislative control; others are determined—sometimes permanently—by given environmental conditions. Some are subject to the caprice of society—either that population who lives in and controls land use of a region or that coming in as visitors. Some can be maneuvered from the standing of the region; others cannot.

Study of this hierarchy can reveal a number of factors that are within the realm of regional strategic planning possibility.

## Regional Strategic Planning Process

From this description of the various factors upon which tourism development depends and from the identification of tourism development problems, it can be observed that some are maneuverable at the regional level and some are not. It would appear that the developers—commercial enterprise, nonprofit organizations, and governments—could gain by having regional (national, provincial, state) guidelines that provide facts, policy guidelines, concepts, and recommendations for tourism development. Whereas it would be difficult for any one state to alter the national economy, to change its tourist image quickly, to change national priorities toward the purchase of travel rather than goods, it is possible to plan tourism in a systems manner that utilizes resources prudently and integrates the many fragments into a viable whole (Figure 10-3).

Together with the continuous planning process, a regional strategic planning process could provide tourism-oriented recommendations and guidelines as a foundation for regional policy and local implementation. Conceptually, the following steps would be necessary: (1) setting objectives, (2) researching the regional tourism development factors, (3) synthesizing these facts and drawing conclusions about the region's potential, (4) conceptualizing ways of developing tourism, and (5) making recommendations for regional development. The project would bring recommendations to the point of identifying concepts and general zones of development that would then form a foundation for local planning and individual feasibility studies for specific establishments. Five steps are involved, detailed below.

Step 1. *Setting objectives.* Whereas regional strategic planning for tourism is directed toward the same four goals of all tourism planning, it consists of a specific project that terminates at reaching specific *objectives.* Even though updating from time to time will be required, this form of planning can provide recommendations on a number of tourism development issues and projects important to the overall goals of ideal tourism system functioning. The objectives should be broad enough to be comprehensive but precise enough so that they can be accomplished within a given time frame. Certainly, they should provide regional organizations and local interests with systems guidelines for tourism development.

Step 2. *Research.* This second step serves two purposes. It provides basic data important to tourism planning. It also is a familiarization step by getting prime actors of the planning process—the sponsors, the specialists, and the publics— started in their mutual assessment of tourism development potential. The emphasis is primarily fact-finding, but this research should also spark new ideas for development. Implementation and public involvement are integral parts of the process right from the start. Included in this step are two sets of information search: on physical resources and on program resources, especially markets.

From a tourism development point of view, not all areas have the same *physical assets.* If service businesses (the producers of local economic impact) are to thrive, they need locations where tourist volumes can be obtained. Volumes will occur at destinations where the physical factors are the best. Therefore, a basic principle is: *the greatest tourism development potential can be*

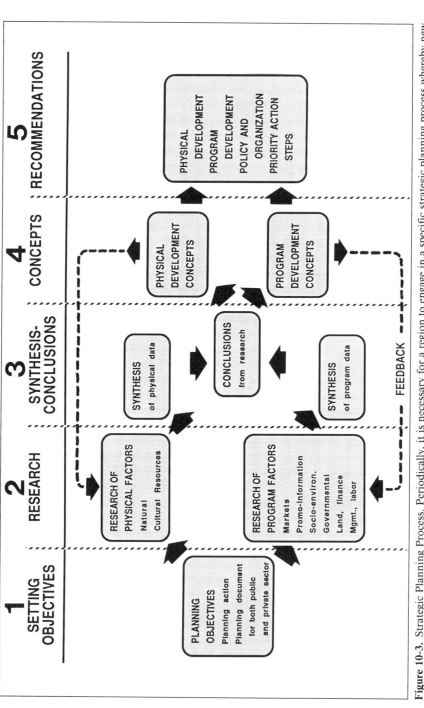

**Figure 10-3.** Strategic Planning Process. Periodically, it is necessary for a region to engage in a specific strategic planning process whereby new market-plant match can be accomplished. However, this five-step process should be closely integrated with continuous and community planning for tourism.

*found in those areas where the several factors of greatest value are the most abundant and of highest quality.*

Experimental work (Gunn: 1979) has shown that map overlays of the several physical factors can display graphically those locations that have the greatest tourism potential. For ease of application, the universe of influential physical factors could be grouped into nine classifications:

1. Water and waterlife
2. Vegetative cover and wildlife
3. Climate and atmosphere
4. Topography, soils, geology
5. History, archeology, legend, lore
6. Esthetics
7. Existing attractions, industries, institutions
8. Service centers
9. Transportation

The technique of "multiple map overlay" by computer is useful in performing this task. "The multiple map overlay is an analytical process involving the systematic weighting and rating of desired attributes and displaying their locations" (Mutunayagam and Bahrami: 1987, 100). The formula for this process is:

$$O_{ij} = (1/W) \times \sum_{k=1}^{n} A_{ijk}$$

$$W = \sum_{k=1}^{n} w_k$$

where:

$O_{ij}$ = color code of the cell in row $i$, column $j$ of the overlay product map
$n$ = number of maps to be overlaid
$w_k$ = weight factor for the $k$th map
$W$ = summation of all weight factors
$A_{ijk}$ = color code of the cell in row $i$, column $j$ of map $k$

Mutunayagam and Bahrami describe the process as consisting of two procedures. First, the *off-line* procedure is for the researcher cartographer to create a hand-drawn map to scale based on study of the particular attribute and its distribution over the region. For example, a map could display coded zones in which water resources are found. For tourism purposes, the zones would be larger than the precise boundaries of the attribute to indicate the importance of lands adjacent to water. A second part of the off-line procedure is weighting each map to reflect the relative importance of each attribute factor. Because this step does not lend itself to engineering or scientific precision, the best method to date is the use of a panel of experts. A delphi approach, using several iterations, can provide consensus in the relative importance of the several attribute factors for future tourism development, considering the full range of market preference and activities needed.

The second procedure, *on-line,* involves translating the maps into computer overlays. Several mapping software packages are available today, such as GRI-DAPLE, PC ATLAS, STATMAP, PCMAP, SYMAP, and others. The hand maps are digitized for computer input for each overlay. For visual check, each factor overlay can be printed out on hard copy with key shadings. For overall evaluation of the locations where all factors are of greatest significance, all overlays are aggregated by computer. The resulting hard copy display provides a range of shadings that then can be interpreted by the cartographer and analyst for synthesizing the information and deriving conclusions.

Research of *program factors* is equally important, especially *market* analysis. If adequate secondary data on market characteristics are available, they may suffice. However, it is likely that primary market surveys will be required to provide more current, more definitive, and more complete data. In recent years, market researchers have increased the sophistication of analysis greatly, providing much greater insight into locations, interests, and propensity to travel.

Study of the extent and quality of existing travel *information* will be needed in order to assess deficiencies. Throughout the world as travelers face greater destination alternatives and as activity horizons broaden, the need for information is greatly enlarged. Guidebooks, booklets, maps, atlases, taped travel guides, and interpretive information will be part of regional planning research.

Certainly, part of a regional plan will be understanding the effectiveness of existing *promotion.* Research of promotion should cover all programs of advertising, publicity, public relations, and incentives. This research can provide facts about existing destinations and their attractions and guidance on new programs that will be needed after further development of supply takes place.

Examination of the *socio-environmental* climate of the region is very important. Study of local attitudes toward tourists could be revealing. Several constituencies other than the business sector should be polled. Environmental impacts can be complete studies unto themselves, revealing the fragility or capacity limits of certain resources.

Review of the *governmental* roles is an essential part of the research. Both the regulatory and the developmental policies and practices of all levels of government require review and evaluation. Excessive or insufficient regulation for tourism should be identified. Overlapping, competitive or insufficient roles by federal, state, and local governments should be disclosed. Especially important is identification of governmental bodies that will be responsible for implementation of planning recommendations.

Special *development* factors need to be examined. Questions concerning land availability need to be answered. This is especially important for government or private sector decisions on development. Financial, labor, managerial, and tourism expertise need to be examined to determine whether regional or outside resources will be able to complete plans.

Step 3. *Synthesis-conclusions.* This step is one frequently omitted or minimized in planning studies. Too often planners proceed directly from the research phase to their recommendations. Instead, there is need for careful and studied synthesis, the bringing together of the many fragments of information collected during research. This step requires thoughtful application of logic and reasoning to derive meaning from the many facts. It ends with descriptive *conclusions* that are general enough to be comprehensive but specific enough to lay the foundation for conceptualizing solutions and making recommendations.

Step 4. *Concepts.* The fourth step is that in which both creativity and ideation have full sway. Since tourism planning is neither scientific nor mathematically predictable, it needs heavy inputs of conceptual thinking. It is at this step that both lay citizens and professionals study the research data and the conclusions, and then look forward to resolution of problems and improvement of tourism. The final outcome depends upon how well older and tested practices are challenged by new ideas and the logic of the research findings. But before these ideas are formulated into recommendations, they are tested with feedback to local geographic areas, the existing socio-environmental situation and local decision-making inputs.

Step 5. *Recommendations.* The final step includes recommendations for action on two fronts: physical development and program. The recommendations are for a relatively short time frame, perhaps one to five years because the plan will need to be regularly updated with new inputs. The recommendations must speak to all the problems and issues within the several factors listed in Figure 10-3. Issues of social, economic, and environmental impact must be addressed. The degree of growth, or no growth in some instances, would be explained. The setting of priorities and sequence of staging would be part of the recommendations. Basically, the recommendations lay the regional context base for local identification of planning opportunities and responsibilities for tourism. Perhaps the most valuable output, based on the market-product match principle, will be the geographical identification of potential destination zones—areas that can then follow up with their own tourism potential and plans.

## Central Texas Example

Perhaps the regional strategic planning process for tourism can be explained best

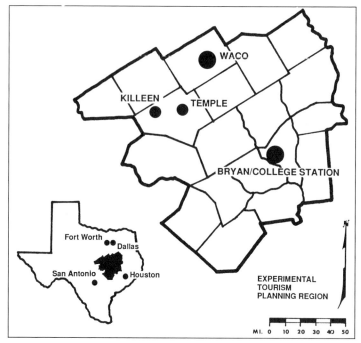

**Figure 10-4.** Location of Tourism Planning Experiment. A study of tourism potential of this group of 19 counties in central Texas was made in 1982. Principal cities included Waco, Temple, Killeen, Bryan, and College Station. (Gunn: 1982).

by use of an example. In 1982 Gunn experimented with such a process in a 19-county region (16,079 square miles, or 41,805 square kilometers) located in central Texas, an area about the size of the Netherlands (See Figure 10-4). In that year travel expenditures within the area were approximately $335 million, generating a payroll of $64 million in connection with 7,853 jobs and providing state tax revenues of $8.4 million and local taxes of $3.6 million. The total 1982 population was 780,554, containing the major cities of Waco, Temple, Killeen, Bryan, and College Station.

The objectives of the project were to: (1) evaluate present tourism development, (2) identify zones with the greatest tourism potential, and (3) identify kinds of potential for tourism expansion.

One part of the next step was a review of market data for three prime market sources. (1) Within the region there are four prime market areas—Waco, Killeen-Temple, Bastrop, and Bryan-College Station—with a total population of 580,240. An additional 200,308 people live in small towns and rural areas within the region. (2) Within a radius of 100 miles are three major population concentrations with a total population of over 7 million (Dallas-Fort Worth, Houston-Galveston, San Antonio-Austin). Other markets in the surrounding area include a population of 409,000. (3) Beyond a radius of 100 miles lie many additional markets—in Texas, surrounding states, and prime tourist markets in California, New York, Illinois, and some foreign countries, particularly Canada and European nations. Traveler interests and purpose factors were reviewed, identifying both business and personal activity preferences.

A second part of this step was research of present supply and physical factors. *Water* resources were found to be plentiful—six floatable rivers of good quality, 15 lakes (66,000 acres) with beaches, fishing, and boating potential. The *vegetation* varies from woodlands to prairies. The overstory of hardwoods (oak, mesquite) and an understory of grasses and wildflowers occur in bands separated by "blackland" prairies, well suited to livestock and agriculture. These offer not only an esthetically pleasing landscape but good habitat for *wildlife*— deer, quail, rabbits, squirrels, migratory waterfowl, and an abundance of birds. The climate and *weather* of the region is humid subtropical with cool winters and hot summers. Prevailing winds are southerly, mean annual precipitation ranges from 19 to 40 inches, and the region receives 62 percent of the total available sunshine annually. Severe storms are infrequent. *Topography* of the area ranges from hilly and rolling to level with major land relief along the Balcones escarpment in the northwest portion. Several river corridors provide an interesting variety of land relief.

The rich *historical* background begins with Cherokee occupations and continues through periods of Spanish control, settlement, the Texas Revolution, formation of the Republic of Texas, and statehood. "Washington-on-the-Brazos" and Independence were important sites in the formation of the Republic of Texas. Many elaborate Victorian homes and historic buildings can be found throughout the area. The region has an abundance of legends and myths and is enriched by a varied ethnic heritage. Whereas the *esthetics* of the region may not be as spectacular as exotic beach and mountain destinations, the landscape is competitive with other regions equally available to the dominant markets. The waters, trees, wildflowers, land relief, and wildlife offer all-year interest for visitors. Some cities have developed attractive settings; others are cluttered and unattractive. Existing *attractions* are already diverse and reasonably abundant. Many parks, campgrounds, beaches, historic sites, educational institutions, manufacturing plants, and festivals already are available.

The four major *urban complexes* have sound infrastructure—water supply, waste removal, fire control, police. In addition, they have historic sites, entertainment, and educational institutions, as well as lodging and food services of importance to travelers. The county seat towns have a limited supply of quality lodging, food services, and attractions. Even smaller communities are distributed throughout the region but have even more limited development appropriate for travelers. The highway *transportation* network is especially good. Interstate 35, located to the west of the region, supports an average of 32,000 vehicles per day where it penetrates the region; I-45 skirts the east side of the area. Interstate, U.S., and state highways provide good linkage between the region and Texas markets. Commuter airlines connect the major cities with the Houston Intercontinental Airport and the Dallas-Fort Worth Airport.

Computer graphic technology was enlisted for aggregating these physical factors. Because development may be slightly different between touring circuit and longer-stay tourism, two map sets were prepared. For each, expert opinion was employed to give a different comparative weight to each factor. Table 10-1 shows the resulting weightings, providing an "index"—the maximum that can be credited to each factor.

### TABLE 10-1
### WEIGHTED TOURISM DEVELOPMENT FACTORS

Weighted index is based upon comparative importance of the several factors for the development of two types of tourism—touring circuits and longer-stay.

| FACTORS | WEIGHTED INDEX FOR TOURING CIRCUITS | WEIGHTED INDEX FOR LONGER-STAY |
|---|---|---|
| Water, waterlife | 6 | 21 |
| Vegetative cover, wildlife | 6 | 11 |
| Climate, atmosphere | 4 | 8 |
| Topography, soils, geology | 8 | 5 |
| History, ethnicity, legends | 14 | 8 |
| Esthetics | 15 | 8 |
| Institutions, attractions | 13 | 12 |
| Service centers | 14 | 12 |
| Transportation | 20 | 15 |

Based upon the literature search, observation, and study of the region, zone maps for each factor were prepared. This subjective evaluation gave scores to the several zones with "very weak" to "very strong" support offered by that factor. Figure 10-5 illustrates a typical computer map for each factor, in this case for touring circuits.

Research was also made (but not mapped) for factors such as existing promotion, information, socioeconomic status, governmental policies, and financial capability.

The next step—synthesis-conclusions—consisted of two parts. First, the several factor maps were aggregated by computer, producing a sum distribution of all factors of touring circuit and longer-stay potential, as shown in Figure 10-6 and 10-7, respectively. The darker zones have the greatest amount of support for tourism development; the light zones have the least.

The second part of this step consisted of reviewing all earlier descriptive data and reviewing maps to draw conclusions about tourism potential in the region. Those conclusions were:

1. Two of the most developable and promotable assets of the region are its reputed image of hospitality and its four-season potential.

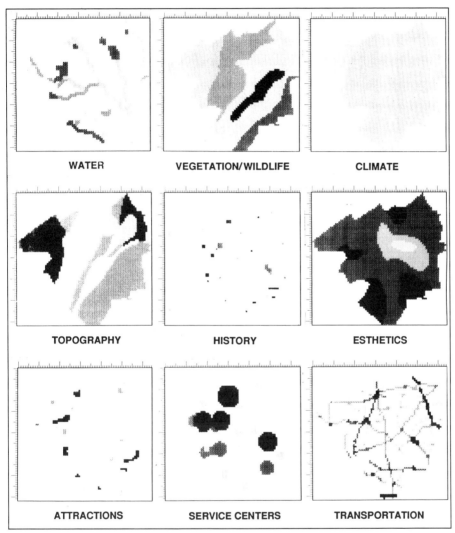

**Figure 10-5.** Assessment of Nine Factors for Tourism. Maps of nine physical factors, showing relative distribution of support for future tourism development (dark=strong; light=weak).

2. The region has an adequate yet underdeveloped physical resource base for recreation and tourism, which, with careful planning, stimulation, and leadership, could absorb extensive expansion while still protecting the environmentally sensitive sites such as waterfronts, forests, wildlife, and historic areas.

3. Based on the region's apparent ability to attract primarily internal (within the region) and nearby markets, it has greater potential than adjacent areas to the northwest and can compete well with surrounding Texas regions.

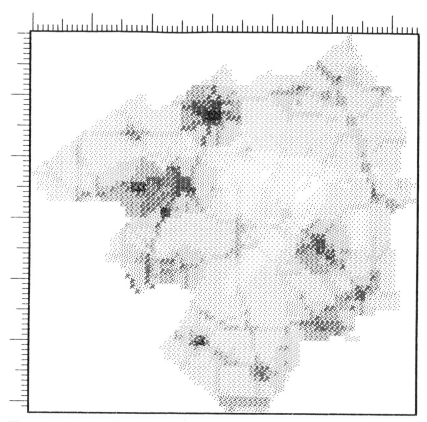

**Figure 10-6.** Touring Circuit Potential. Map of nine computer-aggregated factors indicating areas where factors offer strongest support for future touring circuit tourism development (dark=strong; light=weak).

4. The greatest potential for development for these markets exists in the historical, religious, medical, educational, ethnic, military, water recreation, organized sports, and natural resource activities.

5. Areas of greatest resource strength and potential are clustered in the western and southeastern sections of the region where they have better access and support from service centers and transportation. The center of the region is weakest in potential due to lack of a significant mix of resource assets, services, and access.

6. The region is well served by several transportation modes, but highway signage and directional information are not always readily and clearly understood by the traveler.

7. At present the community services throughout the region are dominantly directed toward local ranching, oil and industry rather than to the recreation and tourism trade. Whereas the region has pockets of strong tourism-recreation leadership, it is fragmented and suffers from a lack of collaboration and central coordination and organization.

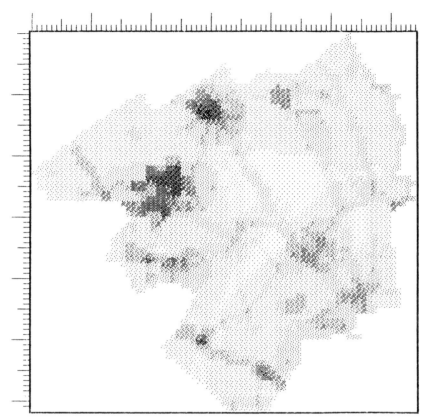

**Figure 10-7.** Longer-Stay Potential. Map of nine computer-aggregated factors indicating those areas where all factors offer strongest support for future longer-stay tourism development (dark=strong; light=weak).

8. Although lacking a major regional theme or attraction, the study area has many small destinations and attractions that already serve as both "touring circuits" and "longer-stay" destinations.

9. It appears that there is an abundance of publications describing the travel and recreation opportunities of the region, but these are not adequately promoted or readily available to those seeking them.

10. The resource base does not suggest strong appeal to out-of-state or foreign markets, but further research may reveal opportunities for such specialized development.

From the map and descriptive results of analyzing the area and drawing conclusions regarding strengths and weaknesses, concepts for development were created. These are the prefeasibility recommendations requiring in-depth feasibility as a next step by developers, both public and private.

Eight *touring circuits* were recommended, provided that attraction complexes based on the special resources of the region were developed (illustrated in Figure

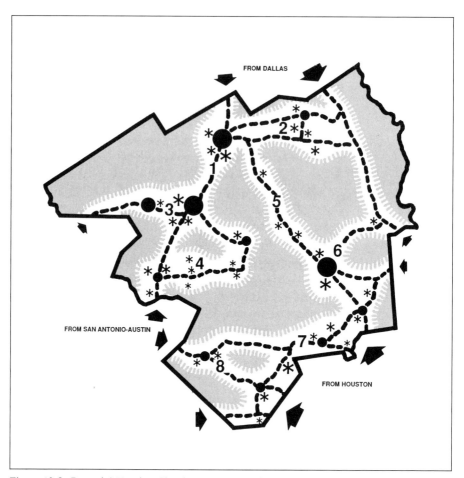

**Figure 10-8.** Potential Touring Circuits. A concept of touring circuit potential based on future development of attractions and other tourism expansion in most favorable zones shown in Figure 10-6.

10-8). Each one has linkage with prime access routes from market sources. For example, Route 7 has potential for the development of many new or enlarged attraction complexes—historic buildings, ethnic festivals, crafts, foods, Washington-on-the-Brazos State Park (signing of the Texas Declaration of Independence), festivals, events at Round Top and Winedale, spring flower festival, and sightseeing around Lake Somerville.

For *longer-stay* tourism (resorts, conventions, conferences, festivals, vacation homes, organization camps), four destination zones were identified as having the greatest potential (Figure 10-9). For example, Zone B has resource potential for longer-stay visits in two subzones—one centered on Killeen-Temple and one on Georgetown-Round Rock. Two major reservoirs, Lake Belton and Stillhouse Hollow Lake, dominate the natural resource recreation and resort opportunities. The Scottish ethnic resource, historic homes, the historic village of

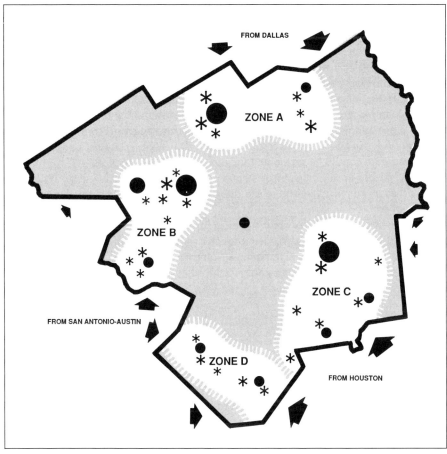

**Figure 10-9.** Potential Longer-Stay Development. A concept of longer-stay destination zones based on future development of attractions and other tourism expansion in most favorable zones shown in Figure 10-7.

Salado, Old Summers Mill, and the well-developed service centers are major assets for expanded vacation and business tourism. The University of Georgetown, Lake Georgetown, Inner Space Caverns, the slave cemetery and grave of outlaw Sam Bass, restored homes and shops provide the foundations for longer-stay development. The zone is easily accessible for millions of market potential from Austin and the Fort Worth-Dallas area.

The study resulted in several overall recommendations for development and management:

> For touring circuits, historic homes and sites need restoration, interpretation, and ancillary services; industrial, agricultural, and ranch areas have potential for interpretive tours; tours to parks and rural areas for scenic appreciation and photography could provide visitor satisfactions if developed.

For longer stay, the several zones need greatly improved meeting and service facilities for the medium to small-size conferences; a great deal of improvement is needed in entertainment if longer-stay visitors are to be attracted; the abundance of water-based recreation has hardly been touched with major resorts; and much organizational clustering of attractions is needed.

Information should be organized for greater understanding and interpretation; guide books are needed; better information distribution and centers are needed.

When more attractions are truly available, much greater promotional effort should be initiated for advertising, publicity, public relations, and incentives; promotion should be segmented for the several target markets.

Regarding social and environmental issues, no major constituencies were found, but all tourism expansion plans should have heavy public involvement and support.

The major obstacle for greater tourism appears to be apathy. No organizations, public or private, have identified tourism as a major growth opportunity, not only for economic reasons but for the social value of cultural exchange and well-being of the communities and surrounding areas.

## Conclusions

What can one conclude from this discussion of three suggested levels of tourism? Certainly not that these are the only planning approaches. These conceptual processes are conclusions unto themselves—deductions from past planning work and ideas for further experimentation in the very complicated phenomenon of tourism and its need for better planning. Therefore, the following are not hard and fast conclusions from scientific testing but rather represent some common threads of tourism planning.

*A mix of planning processes is most effective.* Tourism is such an abstract, complicated, and comprehensive a phenomenon that it defies simple planning processes. Projected here are three levels of processes—continuous, regional strategic, and community. Whereas the fundamental functions requiring planning for best goal-seeking are similar everywhere, different processes may be required for different political and traditional settings. Goals and objectives are universal. Implementation must be adapted to specific situations. Planning processes need further study and experimentation.

*Barriers to planning should be removed.* Over many years, all nations have developed complex organizational structures for governance. Each nation has it own agencies, policies, practices, and traditions for economic and social development. However, very few nations have overtly incorporated tourism objectives and responsibilities within the many segments of the body politic. In most nations turf protection, lack of commitment to tourism, and bureaucratic mandates have made integrated tourism planning either impossible or very difficult. These barriers must be removed if tourism is to progress.

*Market-product match is fundamental.* No matter the model or the political ideology, tourism planning requires market-product match. Instead of over-emphasis on promotion, as has been true of the past, today's tourism calls for equal consideration of the product side and how it meets market preferences. Such a conclusion means also that models for physical land analysis are driven as much by markets as by resources. For this reason planning for market-product mix must be flexible because of the dynamics of markets.

*Fact-gathering is a prerequisite to tourism planning.* Professional planners have long known the essential need for base data for planning. However, tourism's preoccupation with promotion has not favored financial support for research. Too often research is seen as esoteric and of little practical use. Or only proprietary research is seen as valuable to the firm. Research is time-consuming, requires trained expertise, and therefore can be of considerable cost. But until agencies and private sector groups are willing to support market and supply research, tourism will continue to be based on guesswork rather than sound information for concerted planning.

*Planning is conceptual.* At the same time that planning is based on sound data, it is equally dependent on creative ideas. Research about land resources, business success, and tourism impacts are not determinants—they are influential factors. To them must be added visions of a future in which new developments are blended into the existing structure. Creative thinking must be utilized to experiment with several concepts that will be directed toward goals and objectives. Scientific research can produce foundation information, but new opportunities will be unleashed only when new concepts—planning, design, managerial, operational, policy—are added.

*Not all lands have equal tourism potential.* Because of the great emphasis by governments and the private sector on promotion, there is an attitude that all places can obtain tourism if promoted. This half truth is misleading. Some areas have more favorable factors—physical, political, program, traditional—than do other areas. New research and computer cartographic techniques make rapid analysis and delineation of potential possible today. Both public and private sectors can obtain greater social and economic rewards when moneys are invested where the potential is greatest.

# Bibliography

Blank, Uel, Clare A. Gunn, and Johnson, Johnson & Roy, Inc. *Guidelines for Tourism-Recreation in Michigan's Upper Peninsula.* Cooperative Extension Service. East Lansing: Michigan State Univ., 1965.

*Comprehensive Outdoor Recreation Plan: State of Hawaii.* Donald Wolbrink & Associates (C.A. Gunn, contributor) for Department of Planning and Economic Development, Honolulu, 1970.

Friedmann, John. "A Conceptual Model for the Analysis of Planning Behavior." In: *A Reader in Planning Theory,* Andreas Faludi, ed. New York: Pergamon Press, 1973.

Gunn, Clare A. *Tourism Development—Assessment of Potential in Texas,* Bulletin MP-1416. Texas Agricultural Experiment Station. College Station: Texas A&M Univ., 1979.

Gunn, Clare A. "Tourism Planning for East Texas," Report 73-1, no. 5, Recreation and Parks Department. College Station: Texas A&M Univ., 1972.

Gunn, Clare A. "Tourism Potential in Central Texas," unpublished paper. Recreation and Parks Department. College Station: Texas A&M Univ., 1982.

Hall, Peter. *Urban and Regional Planning.* New York: John Wiley & Sons, 1975.

Lang, Reg. "Planning for Integrated Development," paper presented at conference on Integrated Development Beyond the City. Mount Allison University, Sackville, New Brunswick, Canada, June 11-14, 1986.

Mutunayagam, N. Brito and Ali Bahrami. *Cartography and Site Analysis with Microcomputers.* New York: Van Nostrand Reinhold, 1987.

Saltus, Richard. "Drought Forces Catalina Island to Cut Water Use," *The Eagle,* Bryan-College Station, Texas, May 29, 1977, p. 1G.

*Tourism Development in Ontario A Framework for Opportunity.* Balmar & Crapo and Clare A. Gunn for Tourism Development Branch, Ministry of Industry and Tourism. Toronto: Tourism Development Branch, 1976.

Trist, Eric. "Referent Organizations and the Development of Inter-Organizational Domains," *Human Relations,* 36 (3), 1983, 269-284.

White, Judy. *A Review of Tourism in Structure Plans in England.* Centre for Urban and Regional Studies. Birmingham, England: Univ. of Birmingham, 1981.

# Chapter 11

# Community Tourism Planning

Throughout the world, decision making for tourism planning and development is the most critical at the local level and especially in urban areas. Yet most communities have had little experience with tourism and are ill-prepared for it. All travel is linked with communities no matter how urban or remote the purpose may be. This inescapable fact can be a blessing or a curse depending upon how well a community accepts its tourism role and maintains a balance between traveler and resident development and management. The misconception that all tourism is escapism from the city persists. Facts today demonstrate that urban tourism dominates and continues to increase. Tourism can be of great benefit but also of great stress. Whereas all the systems, policies, and planning principles for the regional scale include the city, elaboration at the urban scale provides special insight into urban tourism development issues. The following discussion contains suggestions that if implemented could avoid many of the detrimental aspects of community development for tourism.

Cities are far more important for tourism than popularly thought. The image of vacation travel tends to overemphasize areas beyond the city. Blank and Petkovich (1987, 166) list five factors that endorse the urban areas as being important to tourism:

- Cities are population concentrations; therefore, the target of "visiting friends and relatives," a large segment of tourism.
- Cities are transportation nodes.
- Cities are focal points for commerce, industry, and finance—the cause for much travel.
- Cities contain "people services"—health care, education, government and headquarters for religious, ethnic, industrial, and interest groups.
- Cities contain concentrations of cultural, artistic, and recreational opportunities.

Unfortunately, most cities do not have the legal and organizational structures to plan for tourism's growth and development. In spite of this, they regularly _ublic and private decisions that greatly influence tourism. If greater considerations were made from a travel perspective the entire community and tourism could gain greatly. Rovelstad and Logar (1981, 13) offer six succinct reasons for community planning for tourism:

- Can provide better understanding of interdependence between attractors and service businesses.
- Promises greater community harmony by avoiding problems.
- Reduces business failures by assuring sound growth.
- Fosters community acceptance of tourism.
- Assists in obtaining needed human and financial resources.

The following discussion about community tourism planning includes impacts on the community, planning concepts, a process for planning, and organization.

## Impacts on the Community

The naive community attitude that tourism offers a quick fix to economic doldrums is gradually being replaced by a more enlightened and realistic view. Such a view recognizes that tourism development may be slow, costly, and disruptive of past living patterns as well as providing new economic growth. Every community contemplating tourism development should recognize that there will be social, economic, and environmental impacts.

### Social Impacts

Perhaps tourism exacts greater social impact on a community than any other form of economic development because it depends on invasion by outsiders, both as visitors and frequently as developers. When the number of visitors exceeds the number of residents, there is bound to be some social response, negative and positive.

Planning before installing massive tourism expansion can alleviate many of the negative impacts. Based on her work in British Columbia, Cooke (1982) believes that communities can preserve their identity and protect their lifestyles in the face of tourism development. She offers several guidelines:

1. At the local level, tourism planning should be based on overall development goals and priorities identified by residents.

2. The promotion of local attractions should be subject to resident endorsement.
3. The involvement of native people (as the Inuits of British Columbia) in the tourism industry should proceed only when the band considers that the integrity of their traditions and life-styles will be respected.
4. Opportunities should be provided to obtain broad-based community participation in tourist events and activities.
5. Attempts to mitigate general growth problems identified in a given community should precede the introduction of tourism or any increase in existing levels of tourist activity.

Studies of rural development and tourism in Hawaii (Chow: 1980, 600) revealed that tourism could be a positive influence in rural areas but that it required integrated planning for best results. Chow suggests that smaller communities could be the production, distribution, and processing centers for agricultural, horticultural, aquacultural, silvicultural, and marine products for sale in shops and hotels to tourists. Stronger planning policies that integrate regional and local development could stimulate more and better quality tourism. Chow cites several variables necessary for a better mix between visitors and suppliers for greater retention: longer-stay segments, repeaters, near- rather than long-distance origins, local owners and suppliers, and group rather than independent travelers.

Whereas some authors dwell on the increased negative social impact on communities as tourism grows (Doxey: 1975; Butler: 1980), others (Cohen: 1979) indicate that adaptation is more likely the rule. Murphy (1983, 10) found that three groups locally had quite different perspectives on community tourism development. When he researched several tourist centers in England, the administrative group—political, professional, planning, and official—was most positive, believing that greater development would enhance employment and improve local facilities. Whereas the business sector was positively inclined, it was somewhat skeptical that all the benefits from tourism would materialize as promised. Perhaps this was influenced by concern over new competition and allegiance to local resident markets. Opinions of the third sector, the residents, ranged from very favorably inclined to negative. The author concludes that if leaders favoring tourism wish support from all three sectors, greater accountability and better understanding of tourism are required at an early stage.

Perspectives of small town rural land residents often vary greatly from those of developers. From the developer's perspective in a free market economy, the resources of rural lands are seen as potential wealth and open to development whenever economically feasible. Because of their need for greater know-how and financial backing, developers frequently come from outside local rural areas. Public land owners in the U.S.A. and other countries generally exercise policies laid down at the bureaucratic centers outside local areas. More than 1.4 million visitor use days take place at U.S. sites administered by the National Wildlife Refuge System; the U.S. Forest Service supports over 230 million recreation visitor days per year; the Army Corps of Engineers, 480 million recreation days, just to name a few (Clawson and Van Doren: 1984). Because policies vary greatly, the local situation is often confusing, contradictory, and beyond local control except in the long range by an alert electorate.

Another group, generally beyond local control and yet greatly influencing tourist use of rural lands, is made up of outside industries and other nonrecreation activities. The use of sites for radioactive and toxic waste dumps and the damming of rivers for power and flood control are most often the result of decisions from outside local jurisdictions. Major mining and extractive industries can greatly alter the environmental resources of areas that might have had potential for tourism.

Swinnerton (1982) in his research of the social impact of recreation in Canada found that, not surprisingly, local farmers held strong proprietary attitudes toward rural lands. This meant that travelers and recreationists were seen as invaders of their private environments. Furthermore, land owners who often bear the costs of allowing tourist use of their lands do not feel they are adequately compensated. Pizam (1978) also identified several negative concerns by local people toward tourism development on rural lands near Boston: traffic, litter, noise, vandalism, high prices, drugs, and alcoholism. Legal issues are also of concern. Many landowners fear lawsuit from users of rural lands who may be injured during their travel or recreation.

However, many localities make remarkable adaptation to invasions of tourists even though these outsiders are known to disrupt usual community life. Rothman (1978) found that resort cities realized that visitors required extra local effort to cope during the peak season: church schedules were changed, residents tended to avoid popular visitor places, and the pace of activity and congestion increased.

From a planning perspective, all attitudes and opinions should be brought to full awareness before major development takes place.


## Economic Impacts

There is little question that tourism has economic costs and benefits as does any other community economic development. As has been stated often, the positive impacts come most directly from traveler expenditures in service businesses used by tourists—mostly lodging, food service, entertainment, and shopping. These, in turn, add to local revenues, employment, and local taxes paid.

But not all communities are alike concerning tourism economic development. For example, Blank and Petkovich (1987, 166) reported that travel purpose can make quite a difference. Some cities are known for their shopping, others for their entertainment as shown in Table 11-1.

**TABLE 11-1**
TOURIST USE OF CITIES BY PURPOSE

The purpose of travel to cities varies between cities. Listed is percentage of person-trips, from 1977 U.S. Census of Travel.

| Purpose | Orlando, FL | Indianapolis, IN | Wilmington, NC | Portland, OR |
|---|---|---|---|---|
| Visit friends and relatives | 18 | 38 | 27 | 29 |
| Business/convention | 12 | 30 | 10 | 30 |
| Outdoor recreation | 6 | 3 | 47 | 3 |
| Entertainment/sightsee | 53 | 8 | 10 | 13 |
| Personal | 5 | 13 | 2 | 17 |
| Shopping | 1 | 0 | 0 | 3 |
| Other | 5 | 8 | 4 | 5 |
| | 100 | 100 | 100 | 100 |

(Source: Blank and Petkovitch: 1987, 166)

Even so, it must be emphasized that most all trips to cities are multipurpose. Most travelers engage in more than one activity while in the city (Blank and Petkovich: 1980). This means that economic impact may be spread differently by different groups of travelers and for different cities.

Tourism development should be viewed slightly differently within large versus small cities and rural areas. For the large city it will take a great amount of new tourism to make an impact, whereas the same increment may have a substantial impact in smaller cities. For example, a study of New London County, Connecticut, compared tourism impact on four segments of the county—urban-industrial, coastal suburban, inland suburban, and rural (Kottke: 1986). When tourism sales were compared to total retail sales, the greatest relative impact occurred in the coastal suburban towns (16 percent) and rural areas (14 percent). Whereas the total dollar volume of tourism was much greater in the urban-industrial cities, the ratio to overall sales was much less (3 percent).

As already stated, tourism has economic costs. Again, these will vary from community to community and with different kinds of markets. If a city already has a surplus of infrastructure, the incremental cost for added tourism may be minimal. However, for major tourism destination development, new expenditures on streets, water supply, sewage disposal, and control may add a new burden to public agencies. Additional private capital will be required for new construction and development.

For example, for Walt Disney World to have adequate access, the state of Florida had to invest $5 million in new highway development. Towns near major winter sports development, such as Ketchum and Sun Valley, Idaho, were for a time under a moratorium of development until improved effluent standards of sewage disposal could be attained. Community public health, safety, and welfare costs may escalate with tourism development (Weaver, 1978, I/11).

In the past, most communities have equated economic success with growth and often view tourism as a driver of growth. Now, the single factor of greater size is being questioned and greater emphasis is being placed on quality of community economic and social life. In the 1970s Molotch (1976, 328) questioned land development as a prime political goal for communities. Growth in numbers and land development exacts costs of environmental degradation, social problems, increased costs of infrastructure, and public taxes and may be perceived as benefitting only a few. Tourism can create new jobs for local residents and not necessarily require imports of new people to the community. Especially in communities where other economic foundations have weakened, tourism may take their place. But political leaders need to be constantly aware of the community axiom of economic diversity rather than growth.

## Environmental Impacts

Community environments often are enhanced by addition of new amenities resulting from tourism incentives. New museums, historic restorations, parks, convention centers, theaters, and improved craft centers have often been added because tourism not only provided the motivation but new revenues to pay for them. It must be emphasized that virtually all of these are improvements in the quality of life for residents.

On the negative side, calling for better planning and management are some detrimental impacts of tourism. When these are anticipated, they can be ameliorated or avoided entirely. Probably congestion is the most frequently cited environmental issue. However, more frequently than not this is a physical traffic and circulation problem that has not received adequate attention by traffic engineers and planners. When urban attractions are planned for pedestrian use and vehicular traffic is routed over systems with adequate design capacity, the issue disappears. Additional litter and site erosion are problems that are solved with greater design and management controls. Standards of mass use established by the Disney enterprises decades ago are proof that these problems can be solved.

Planning on the urban and surrounding scale can enhance preservation of resources at the same time that tourism is fostered. The number of restored historic buildings can be increased by providing most with economic tourist functions inside. In this way, not all become a burden on the public sector, which would result if all were made into museums. Mahadin's plan (1984) for redevelopment of the ancient city of Al-Karak, located southeast of the Dead Sea, provides an illustration of protection and tourist use. Historic sites, such as Al-Quala (the castle), would be protected while new business growth would take place outside this zone. Because of the dramatic topography, several historic sites could also be used for overlook views toward the valley and towns of Palestine. However, without reinforcing the plan with restrictions, the traditionally crafted stonework of the ancient buildings will be esthetically eroded by further invasion of modern concrete buildings. Many cities throughout the world require the specialized inputs of architects, landscape architects, and historians for compatibility between conservation and tourism development.

## Planning Concepts

When one speaks of community tourism planning, the image varies greatly from a city of more than a million population to those of 20,000 or 2,000. It is likely that the larger city already has a history of tourism because it already contains a critical mass of attractions, services, and professional staffs. The large city has usually recognized tourism and given it resources for development and promotion, sometimes on a very large scale. But more and more, small and medium-size cities are recognized as having tourism potential, especially where other economic foundations, such as for mining, forestry, and agriculture, have dwindled.

For most cities the focus for both residents and visitors has been the downtown area, at least for many years. The decades of suburban growth shifted the emphasis away from downtown. Suburbia became the preferred locale for hotels, motels, food services, shopping, and even amenities such as parks and museums. Now for many cities, downtown is experiencing a renaissance, a renewal of cultural and economic emphasis. Old buildings are being restored and reused. New shops are being added. New convention centers and other visitor amenities are again going downtown. And for a few cities, new housing is bringing residents back downtown creating new supporters of parks and other amenities.

But equally significant for tourism (and often ignored by tourism developers) is the area surrounding cities. Jurisdictional boundaries and traditional rivalries often preclude cooperation and collaboration between surrounding areas and cities even though both can gain thereby. The city remains the logical location for greatest economic impact through the service businesses—hotels, food service, entertainment, retail sales. But these services can be greatly enhanced by adding tourist development in nearby areas. Most cities contain many undeveloped assets within a relatively short radius. Natural resource assets such as forests, rivers, hills, mountains, and wildlife nearby often lend themselves to new park and recreation areas in demand by visitors as well as residents. Often historic sites, trails, and locales of legends can be found nearby.

Even small towns and rural areas within a greater radius—50 to 100 miles—may have potential for tourism. Some market segments prefer these to the more congested downtown areas of larger cities. With increased use of motorcoach tours, once fallow lands become increasingly valuable for tourism development. Regarding demand, Table 11-2 lists some of the more popular tourist activities that take place in rural areas and small towns (Gunn: 1986, 2). These activities attract a wide diversity of markets.

**TABLE 11-2**
TOURIST ACTIVITIES IN RURAL AREAS

Rural and small towns often contain resources that support a variety of tourist activities.

| | |
|---|---|
| Picnicking | Canoeing |
| Camping | Crosscountry skiing |
| Hiking | Swimming |
| Horseback riding | ORV use |
| Bicycling | Resorting |
| Hunting | Retirement residences |
| Fishing | Historic touring |
| Boating | Scenic touring |
| Waterskiing | Festivals, events |

(Source: Adapted from Swinnerton, 1982.)

Through many writings about nature and the values of outdoor recreation, many people have been stimulated to seek experiences in the rural areas. As evidence, in 1983 (Mills: 1984) the economic impact of outdoor recreational trip expenditures in Texas was $9,280,841,000, a substantial amount for a state with a population of 15 million.

Figure 11-1 diagrams the typical relationship between a major central city's destination zone and secondary destination zones with their special attractions in a surrounding rural and small town area. Even though the majority of travelers may return to the larger city for services and other amenities, others will prefer the slower pace and special features of small towns.

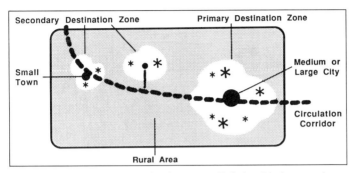

**Figure 11-1.** Rural-urban destination zones. Relationship between larger city primary destination zone and surrounding rural area containing smaller city destination zones. Each will complement the other if planning and management take their relationships into account (Gunn: 1986, 10).

This geographical relationship between cities and rural areas is well exemplified by Adirondack Park (6 million acres) and surrounding areas in northern New York State. Cities surrounding the park such as Plattsburgh, Saratoga Springs, Utica, and Watertown benefit from the many attractions of Adirondack Park—canoe trails, cross-country ski trails, hiking trails, camping areas, and wilderness interest. Within the park there are many hamlets, once important villages for the timber harvest (Trancik: 1986). These small towns now hold promise of providing new tourism services. One plan, shown in Figure 11-2, calls for containment of service development in the hamlets and cities rather than dispersed in the resource areas. New information centers at gateway communities and nearby recreational development are planned, leaving the majority of the park in wild land and remote recreational uses.

## Planning Process

Increasingly, guidelines for community development are being issued. *Tourism Research for Non-Researchers* (1985) is an example of a simplified outline of steps prepared for communities in western Australia. *Top Secret* (1984), prepared for community tourism stimulation in Canada, offers many constructive ideas for action. It identifies five keys to development: (1) clustering—the need for cooperation between communities, (2) no duplication—build on special qualities to avoid competition, (3) key players—early involvement of those able and willing to cooperate, (4) attitude and awareness—establish program to gain cooperation from entire community, and (5) money—gain basic support for first projects before promoting more tourists. In addition, it offers 20 tips for successful tourism development. The province of Alberta has prepared its own *Community Tourism Action Plan* (1987) containing four popularly written

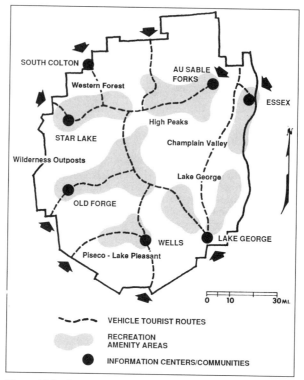

**Figure 11-2.** Concept of tourist zones. Conceptual map showing areas best suited to tourism development in Adirondack Park based on service centers, attractions, and access. When so planned, the remainder of the park area can be managed as a protected natural resource region (Trancik: 1986, 9).

booklets and worksheets. Book 1, Introduction, describes tourism and its advantages and disadvantages. Book 2, Organization, outlines policies and organizational roles of public and private sectors. Book 3, Process, describes the specific steps a community should take to attract more tourists and minimize negative impacts. Book 4, Appendices, contains a sample community tourism action plan and sources of technical assistance and financial aid.

Based on review of these and other sources, five steps are concluded as action needed for community development of tourism—establish leadership, evaluate assets and liabilities, identify opportunities, establish an action program, and conduct a postdevelopment evaluation.

Step 1. *Establish leadership.* A first step is to identify individuals with sufficient motivation, reputation, and commitment who can lead a program of tourism development in a community. The key leader may be a mayor, planner, architect, landscape architect, service business leader, community organization leader, or other individual capable of directing planning and implementation. If there is already a tourism promotional agency, this group may have a logical leader. However, total *development* will require many other skills and efforts, suggesting that a collaborative program in addition to promotion will be needed. Not only will individual leadership be necessary, but usually a new organization will be required. Such an organization should guide what *should* be done to enhance tourism development and *how* to do it. A nonprofit organization, corporation, or trust that stands beside the business sector and government and that can foster specific project development may be needed.

Step 2. *Evaluate assets and liabilities.* This step is one of objective study of those aspects of the community and surrounding region that are most important to tourism. An Australian agency calls these "resource analysis" and "market analysis" (Tourism Research: 1985, 14). An expansion of this approach suggests the following eight topics that a community should investigate if it expects to develop tourism.

1. Evaluate the market situation.

Larger cities may be able to afford professional market studies, but smaller towns can do much to understand markets on their own. The Australian bulletin, *Tourism Research for Non-Researchers* (1985), contains many pointers for self-market study.

The first recommendation is to obtain market data from the national, provincial, or regional tourism organization. Often such data are collected but not generously distributed to the local level. These studies often provide generalized data on origins, expenditures, preferred activities, and demographic profiles. Sometimes sector data are available from hotel, restaurant, or attraction associations. Often transportation authorities have data on highway and air traffic flows.

Beyond these general data, communities can gain further insight on their travel markets by means of "intelligent observation." Listening to traveler comments, observing competitor business, taking note of queues at popular places, and watching visitor response to new attractions can provide much information about travelers. Hotels, motels, and RV parks could make simple surveys of their guests, noting party size, age range, and whether travelers are on business or pleasure trips. Attendance records at attractions, such as parks, museums, art galleries, and information centers can provide clues to the nature of present markets. Whenever possible these data should be kept on a daily, weekly, and annual basis to determine fluctuations and trends.

Every effort should be made to identify the several market segments because different development and promotion may be needed for each. One way is to divide total travelers by purpose (Rovelstad and Logar: 1981, 150):

Business
Convention
Visiting friends and relatives

Outdoor recreation
Sightseeing
Entertainment
Other (shopping, medical, and so on)

For many communities, employment data are a matter of public record and can be reviewed for trends. Suggestion boxes can provide sources of constructive criticism. Several places within the community lend themselves to casual interviewing of visitors, such as the bus depot, parks, manufacturing plants, and other attractions. Some of the questions to be asked are:

- Are you here on business or personal reasons?
- Do you come here often?
- What are the things you like about this community?
- What do you think would make this community better for visitors?

The purpose of these investigations is to gain not only an understanding of present travel markets but also clues to potential markets if only more were done to attract and promote them.

2. Evaluate the attraction potential.

One should take stock of existing and potential attractions within and around a community. Simply taking local citizens on a tour of their own community with the objective of observing opportunities and problems can be very revealing. Some items to be considered are:

- Is the downtown area a fun place to visit?
- Is it attractive with shade, plazas, and night lighting?
- Are the shops and businesses oriented to visitors?
- Are there interesting historic buildings and sites to be visited downtown?
- Are there evening entertainment and fine food downtown?
- Are some older buildings worthy of restoration and reuse?
- Could miniparks replace old dilapidated buildings?
- Is downtown pleasantly walkable?
- Is the remainder of the city pleasant and attractive?
- Are recreation areas, parks, zoos, and other attractions available to visitors?
- Are there ample facilities for meetings, seminars, and conferences?
- Could manufacturing and processing plants be given visitor tours?
- Are handicapped visitors welcome?
- What are the natural and historical resource assets in the surrounding area?

The intent of such a local investigation is to become aware of resources that may have been taken for granted or not viewed from a visitor's perspective.

Another valuable technique is for a select group of local citizens to visit successful tourism communities. Observation and discussion with local leaders can reveal how they have met obstacles and utilized their assets in developing tourism.

In addition to identifying potential attraction places, one must be cautious about premature promotion. Sites with potential will not become attractions until owners and managers develop the properties explicitly for visitors. A scenic area or archeological site could be ruined if publics were allowed to enter before interpretation, parking, services, and control were secured. This early investigation can be valuable to the community decision-makers regarding whether more effort should be placed on tourism development.

3. Evaluate transportation.

Key to all tourism development is access. Highways and airports are costly and not changed readily. A community should obtain from transportation authorities the extent to which changes in transportation are contemplated. Communities not now readily available by visitors may have great difficulty in expanding tourism because no improved transportation is planned. In some cases, relatively small changes, particularly intermodal, can give a community better access.

Some questions to be asked locally are:

- Is downtown readily accessible from main highways, airport?
- Is there ample parking near attractions at all seasons?
- Does the city have a clean and reliable public transportation system?
- Do major highways cross the city causing congestion and safety hazards?
- Are entranceways to the city attractive and well maintained?

Sometimes the result of this local investigation will reveal issues that can be solved with new collaboration between local and major transportation authorities. No community can anticipate tourism growth if it does not have ample and high-quality access.

4. Evaluate the tourist-oriented businesses.

An essential part of examining the community supply side is identifying the quality and quantity of visitor services. For many small and medium-size cities, the established service businesses are likely to be oriented to a previous economy of forestry, agriculture, or mining and only to residential needs. Visitors may be of a dramatically different socioeconomic status requiring new and different services. This new perspective requires changes in attitudes and business practices and may even require outside investors and managers more experienced in satisfying today's travel markets.

Some of the questions to be asked when analyzing the lodging, food, car service, travel, gift shop, and other supply businesses are:

- Are lodging accommodations suited to all markets?
- Are food services adequately oriented to travelers?
- Are traveler goods and services available in shops?
- Can competent and reliable automobile service be obtained readily?
- Are all businesses needed by travelers easily located?
- Are health and community services available to visitors?
- Are service business employees hospitable and competent?
- Can foreign language translation be obtained locally?

This investigation can become a sensitive local issue especially among established lodging enterprises who fear overbuilding and new competition. However, market intelligence may indicate that some travel segments are not being satisfied and that some new facilities of a different kind may not be competitive after all.

5. Evaluate the information.

As travelers become more knowledgeable and sophisticated, the need for a greater amount and better quality of information becomes more and more important. Information of guidance to visitors is very different from advertising. Increased travel for intellectual purposes requires more information from a variety of sources locally.

Some of the points to include in an evaluation of information are:

- Is descriptive information (maps, booklets, exhibits) readily available?
- Is it prepared from a visitor's perspective, providing accurate and clearly presented information?
- Does it provide clearly understood information on location, price, and open hours of attractions and services?
- Are information and interpretation centers open when needed the most?
- Is information prepared in the several languages for key markets?

6. Evaluate promotion.

Probably no other aspect of tourism is practiced as poorly as promotion in spite of its receiving the greatest amount of monetary support. Promotion generally includes advertising, publicity, public relations, and incentives. Promotion can be very effective in bringing tourism products to the attention of potential markets. This is especially important today as the field of tourism becomes more proliferated and competitive.

However, overzealous or poorly targeted promotion can be misleading. Because technology can be used by promoters to produce colorful illustrations and artistic creations, advertising sometimes creates images that cannot be realized. Promotion developed prematurely can bring masses of tourists to sites not yet prepared to handle them.

Some of the questions to be asked about promotion are:

- What advertising, publicity, public relations, and incentive programs are now in place?
- Are they adequate and effective?
- What media have been used and how effectively?
- Are federal, state, and city promotional roles clearly defined?
- How competitive are present public and private promotional programs?

7. Evaluate the infrastructure.

Often forgotten in evaluation for tourism is the adequacy of community infrastructure for tourism expansion. By infrastructure is meant water supply, waste disposal, fire protection, police, street systems, lighting, power, and communications. For most communities these are public utilities provided by or controlled by government. Because they are public sector items, they require support by local taxpayers or outside public sources. Major tourism development in a community may require public support for new infrastructure that will not be paid by the investor.

Some questions to be asked are:

- Are water and waste systems already under capacity stress?
- Would the needed infrastructural changes to improve tourism be equally justified for local citizen use?
- Can new tourism improvements be adapted to present infrastructure?

8. Evaluate the regulation policy.

Nearly everywhere in the world, a multitude of laws and regulations have been established to guide and control many aspects of local community life. Some countries are more restrictive than others. Because these have accumulated over time and because they were directed toward only the local population, they may be overrestrictive or not applicable for tourism development.

Any community contemplating tourism expansion should make a careful review of regulations and policies. Some may inadvertently restrict the building of new attractions or service businesses. Building codes and zoning ordinances established many years ago may no longer be suitable.

A thorough investigation should answer:

- Are present regulations overrestrictive and inhibiting tourism development?
- Are new regulations needed to enhance and control tourism?
- Has bureaucratic overlapping and control become so complicated that tourism is restricted by red tape?

Nearly all of the evaluations suggested here can be done by existing local citizens on a volunteer basis. For larger cities, the volume of investigative work may require professional project help. Although this cataloging of assets and liabilities may seen a formidable task, experience has shown the need for all points to be considered.

Step 3. *Identify opportunities.* The purpose of all the previous investigation has been to provide insight into opportunities for development. The Australian research booklet (Tourism Research: 1985, 14) calls this step "gap analysis." A comparison between market demand (especially when segmented) and the existing supply and resource base should reveal those gaps that may be filled with new supply development. New opportunities for attractions, service businesses, information, and promotion should become apparent.

Special emphasis should be placed on attractions and attractiveness. It is likely that the need for new service businesses will not become apparent until greater volumes of visitors are attracted to the community. Those performing the evaluations should now summarize the opportunities for development and improvement according to the several topics:

Markets                      Information
Attractions                  Promotion
Transportation               Infrastructure
Services                     Regulation/Policy

At this stage, the input by a joint venture consulting team of planners, landscape architects, and economists may be appropriate. Such creative professionals can conceive of several alternative solutions to development problems. They can assist in creating concepts for filling those gaps between market demand and supply based on the special resources of the community. They can act as catalysts to bring the several actors together for solution.

Step 4. *Establish an action program.* In addition to local leadership needed for community development of tourism, an action program should be established with roles clearly presented for all sectors.

The public sector has a strong responsibility in carrying out tourism development. In the U.S.A. this would involve the city (sometimes township), county, state, and federal governments. From the standing of a small community, this sometimes presents a formidable specter—the many governmental agencies having either land use or other regulations and control. However, the better a program is thought out and presented, the better are the chances for the public sector to make the necessary inputs for tourism development.

For example, if the opportunities include the delineation of a scenic road, several public actions may be needed. The highway department (city, county, state) may need to make site and engineering modifications, such as turnouts, overlooks, and parking areas. The adjacent jurisdiction may need to enact zoning regulations to protect the scenic qualities that make the landscape worthy of visitor appreciation. Existing signs, billboards, or derelict structures may need to be removed. Perhaps state legislation may be needed (as was done in Wisconsin) to give local jurisdictions the authority to designate scenic roads according to specific criteria.

Other public roles may include changes in airport facilities, water, waste, fire control, police, parks, and many other features as well as regulations and controls.

The private enterprise commercial sector in a free market economy usually responds to demonstrated or speculated market need. Therefore, until additional attractions and attractiveness create increments of new visitors, there may not be market evidence for new travel businesses. On the other hand, some entrepreneurs may see opportunities for new shops, entertainment, or historic restoration that would stimulate markets to come to the community. Certainly, the private sector organizations should indicate how they can contribute to increased tourism development. Also, they need to identify barriers to their expanded role so that measures can be taken to remove them.

The nonprofit sector in many countries, and especially in the U.S.A. and Canada, is greatly increasing its role in tourism development. Many health, religious, recreation, historic, ethnic, archeological, and youth organizations are capable of developing lands and carrying on programs that, in turn, are valuable for tourism.

Knechtel (1985) believes that this "third sector" (voluntary, informal occupation, family) holds great promise for tourism expansion, especially in developing countries. Rather than inviting the large multinational firm to invest outside capital and even labor, local talent can be harnessed for many indigenous types of tourism services. Because the goal is less for profits than for community well-being, the development is more likely to strengthen the community.

Knechtel (1985, 5) cites two tourism examples—Isodores Restaurant in Granville Island, Vancouver, and Many Hands Catering in Ottawa. The co-op restaurant sold shares to the public who were offered lower meal prices. Bed-and-breakfast, farm, and tourist rooms in resort areas provide intimate service appreciated by visitors and enjoyed by the providers. Kimberley, British Columbia, with more than 130 voluntary groups out of a population of 8,000 people, has established a plan to coordinate its own tourism development. The first $100,000 investment for upgrading downtown came from the residents themselves. Now over $2 million in private investment has transformed a decaying mining town into a viable tourist community.

Certainly all three sectors should cooperate in the action plan needed for developing the community's tourism. This statement is contrary to popular belief in many communities that tourism is solely the responsibility of the

Chamber of Commerce or the tourist agency. Although these groups may assert some leadership, tourism must be a total community affair.

Also at this stage, all concepts for development need to be evaluated for economic, social, and environmental impact. This prefeasibility stage should result in identifying specific project proposals for physical and program development.

One way of organizing project opportunities and concepts is to divide them into two main groups: those that require physical development and those that are action programs.

1. Development.

There is hardly a community that will not require new physical development and improvement of the present physical plant if tourism is to grow. Some of the *downtown* projects that may be viable for tourism development are:

| | |
|---|---|
| Sidewalk cafes | Restored historic district |
| New craft, gift shops | Historic reuse restaurants |
| New nightclubs | Miniparks |
| Remodeled museum | Landscaped parking |
| Convention center | Courthouse/city hall restoration |
| Landscaped plazas | Visitor/interpretation center |
| Traffic/pedestrian improvement | Specialty restaurants |

Some of the community *environs* projects that may be suited to tourism expansion are:

| | |
|---|---|
| Renovated parks | Bed/breakfast inns |
| Rodeo arenas | Landscaped industrial parks |
| Sports arenas | Recreation areas |
| Historic home restoration | Outdoor theaters |
| Waterfront parks | Waterfront resorts |
| Golf courses | Fair/exhibition grounds |
| Historic inns | |

But some of the greatest opportunities may lie in the *surrounding region.* Depending upon market demand and resources, the following may be project opportunities in outlying areas:

| | |
|---|---|
| Natural phenomena parks | Lake resorts |
| River recreation | Hunting, fishing areas |
| Ranch resorts | Campgrounds |
| Mining interpretation | Organization camps |
| Ski resorts | Vacation homes |

| ORV areas | Mountain climbing sites |
| Agricultural interpretive centers | Historic sites |

It must be emphasized that this prefeasibility stage may show potential for some of these projects, but only when developers make detailed studies of specific sites and projects will true feasibility be established.

2. Programs.

In addition to physical development, a great amount of tourism enhancement can be accomplished without major capital investment. Four kinds of programs should be considered by communities seeking to improve tourism.

For most communities, new *activities* can stimulate travel interest. Frequently, large and small downtown areas could be enlivened greatly by adding evening entertainment, indoor or out. Most communities, through local school or adult groups, have talented amateur performers that could initiate new programs. Retired professional entertainers might volunteer their talent if asked.

Rural communities might find opportunities for horse shows and rodeos. As travelers increasingly come from urban areas, nostalgia toward rural roots becomes an attraction. Rural areas with wildlife, forests, and water resources may find opportunities for guided scenic tours, hunting and fishing trips, or nature interpretation hikes. Rural tours may include demonstrations of life and crafts of an earlier era. If archeological sites have been established and protected and exhibits and interpretation have been established, new tours to these places could provide interesting trips for visitors.

Probably the fastest growing form of visitor activity is festivals and events. Many communities have gained greater visitor interest in festivals that originally were held only for residents. One guide for organizing and holding festivals, *The Urban Fair* (1981), describes how valuable events can originate from a social as well as an economic perspective. Not only do these events bring in new dollars, but they stimulate local cooperation, civic pride, and spinoff effects of civic improvements.

"Nostalgiafest" (Urban Fair: 1981, 9), a week-long festival in Petersburg, Virginia, including over 20 concerts by local and state performers, attracts more than 75,000 people with significant economic impact. However, organizers cite other results as even more important—downtown revitalization, enhanced historic district, underground power lines, park beautification, scholarships, and local charities.

"Winterlude," a winter festival in Ottawa, Canada, attracted some 576,000 visitors in 1985 (Final Report: 1985, ii), producing a direct local impact of $15.5 million and a national contribution of $24 million within one week. But more than this, the event has improved the image of the capital city, stimulated visits to landmarks of the national capital region, enhanced winter spirit, and brought about public-private cooperation that did not exist before.

The holding of festivals and local events requires considerable planning and organizing. Such events provide not only direct input from tourism but general civic improvement as well.

In addition to activities, a community seeking tourism will also need to be involved in improved *information, promotion,* and *research.* These activity functions need not entail large outlays of money. Frequently, volunteer effort can start making improvements with new brochures, exhibits, maps, and publicity. As development grows, more sophisticated promotion and market research will be needed, requiring input by paid professionals.

All three actor groups—commercial enterprise, governments, and nonprofit organizations—should be alerted to the opportunities to see which ones can best carry out the project development. At this point, estimates of economic costs and returns can be made. Anticipated employment and taxes paid because of new traveler volumes can be estimated. Depending on the size of the community and the extent of development anticipated, there may be need for a study by a competent tourism economist.

For social impact, several approaches including public hearings, workshops and nominal group technique are available. Public hearings in which publics are presented plans for development often provide only token public involvement. Preferred is input earlier in the process. Workshop meetings, where a large meeting group is broken down into smaller discussion cells to discuss one topic at a time, have proven effective. This allows everyone to participate on an equal basis.

Recently, the nominal group technique (NGT) has been used effectively to obtain general input of ideas and opinion regarding specific issues. Ritchie (1987) used NGT in Alberta with the public and private sector groups—Travel Alberta (the provincial public agency) and the Tourism Industry Association of Alberta (TIAALTA) (the provincial private sector). The NGT approach is usually carried out in several stages and provides two main results. "First, it provides a list of ideas relevant to the topic in question. Second, the technique provides quantified individual and aggregate measures of the relative desirability of the ideas raised in the session" (Ritchie: 1987, 440). In his study, Ritchie sought to bring out key tourism issues and actions to deal with them through three phases:

Definition of priority issues and problems facing tourism in the province.
Identification of initiatives, actions, and programs to deal with priority issues and problems facing Alberta tourism.
Monitoring of recommendations concerning initiatives, actions, and programs for tourism development to ensure timely and effective implementation.

The application of NGT resulted in identifying 15 themes of highest priority and 22 themes of second level priority. The same approach could be used at the community level to gain public involvement in the analysis of research findings and project recommendations for improved tourism. Pearce (1981, 44) indicates that characteristics of the host society need to be taken into account—population, size, demographic composition, and vitality, structure of ethnicity, and religion. A small and weak community may feel greater social impact than a larger and stronger one.

Assessing the environmental impact anticipated from tourism growth may require input from specialists in natural resource and historic resource fields. Pearce (1981) has provided Table 11-3 as a framework for the environmental stress possibilities from tourism. When the sites and resources that may undergo greatest stress are known, the use of good planning, design, and management principles can avoid problems and yet allow some utilization of the resource asset.

Step 5. *Conduct a postdevelopment evaluation.* Often neglected by communities is evaluation of project success after projects have been started. Evaluation requires regular data collection so that comparisons can be made. Before a community takes any specific steps toward tourism development, base data should be collected. This effort should include inventories of present attractions, services, and businesses as well as some measure of travel market activity in the community.

Then, after downtown is revitalized, after new festivals and new attractions are established, surveys of both market and supply can provide information on the successes or problems these changes produced. If the changes did not bring new visitors as anticipated, what went wrong? Were the attractions and businesses poorly managed or was there no market for what was developed? Every community, right at the start, needs to set up machinery for evaluating the outcome of all tourism development.

## Organization

It is difficult in a general publication to provide fundamentals on community organization for tourism because communities vary so widely. So much depends upon the local fiscal stability, leadership, tradition, and legal mandates. But each community, if it wants to develop tourism, must develop its own organizational structure to get the job done.

Experience shows some functional responsibilities that must be carried out no matter the organizational structure. At the community level these could be divided into three groups: development, program, and operation. Further subdivisions may be needed depending upon the size of the city and existing

**TABLE 11-3**

**ENVIRONMENTAL STRESS FROM TOURISM**

A framework for the study of tourism and environmental stress (after DELD).

| STRESSOR ACTIVITIES | STRESS | PRIMARY RESPONSE ENVIRONMENTAL | SECONDARY RESPONSE (REACTION) HUMAN |
|---|---|---|---|
| 1. *Permanent environmental restructuring*<br>(a) Major construction activity<br>urban expansion<br>transport network<br>tourist facilities<br>marinas, ski-lifts, sea walls<br>(b) Change in land use<br>expansion of recreational lands | Restructuring of local environments<br>expansion of built environments<br>land taken out of primary production | Change in habitat<br>Change in population of biological species<br>Change in health and welfare of man<br>Change in visual quality | *Individual*—impact on aesthetic values<br>*Collective measures*<br>expenditure on environmental improvements<br>expenditure on management of conservation<br>designation of wildlife conservation and national parks<br>controls on access to recreational lands |
| 2. *Generation of waste residuals*<br>urbanization<br>transportation | Pollution loadings<br>emissions<br>effluent discharges<br>solid waste disposal<br>noise (traffic, aircraft) | Change in quality of environmental media<br>air<br>water<br>soil<br>Health of biological organisms<br>Health of humans | *Individual defensive measures*<br>Locals<br>air conditioning<br>recycling of waste materials<br>protests and attitude change<br>Tourists<br>change of attitude toward the environment<br>decline in tourist revenues<br>*Collective defensive measures*<br>expenditure of pollution abatement by tourist related industries<br>clean-up of rivers, beaches |
| 3. *Tourist activities*<br>skiing<br>walking<br>hunting<br>trail bike riding<br>collecting | Trampling of vegetation and soils<br>Destruction of species | Change in habitat<br>Change in population of biological species | *Collective defensive measures*<br>expenditure on management of conservation<br>designation of wildlife conservation and national parks<br>controls on access to recreational lands |
| 4. *Effect on population dynamics*<br>Population growth | Population density (seasonal) | Congestion<br>Demand for natural resources<br>land and water<br>energy | *Individual*—Attitudes to overcrowding and the environment<br>*Collective* Growth in support services, e.g., water supply, electricity |

(Source: Pearce, 1981, 47)

bureaucratic and private organizational structure. The reponsible organization may be governmental, business, or nonprofit.

## Development

One part of a community organization needs to accept the responsibility for stimulating and guiding physical development. It should be formed early enough in the planning process to understand the foundations and analysis performed. This group would first publicize the findings of the evaluations of markets and resources in order to stimulate interest and support from the three action sectors.

A next step would be counseling with leaders of the three sector groups to provide further information regarding development. This counseling would assist investors and developers to develop individual project feasibilities. It would act as a catalyst between the three sectors to remove barriers and facilitate development. This agency would coordinate the three geographic areas for special attractions—downtown, environs, and surrounding area. Increased awareness of existing regulations as well as the need for modifying controls would also be its function.

In some communities, this organization could be a nonprofit or commercial corporation sanctioned by government but operating independently. The advantage of this kind of organization is its relative independence to buy and sell land, select locations most desirable for major development, and work with all sectors to get the job done. In some countries, this corporation could receive grants or concessions from government for start-up funding.

## Program

Another part of the community organization for tourism development would be responsible for stimulating and guiding four kinds of programs. Based on findings and recommendations, this group would encourage the establishment of new activities in and around the community. New entertainment, tours, pageants, performing arts, festivals, parades, conferences, meetings, celebrations, and sports events would be encouraged. A catalytic role of finding the right sponsors for carrying out these events would be central to this effort.

A second role would be to contact potential sources for the creation of new and better information—guide books, brochures, maps, slide talks, signs, exhibits, kiosks, communications, and information centers. Coordinating these projects so that all information is prepared in a clear, concise and accurate manner would be an important function.

A third role would be promotion. This may require a separate staff trained in the several modes—advertising, publicity, public relations, and incentives. Guiding the several sectors regarding jurisdictional boundaries—who will promote what and for whom—would be an important function.

The fourth function centers on research, ranging from simple observation in a small community to paid professional research projects where needed, especially in a larger community. Research of both supply and market would be needed.

## Operation

A staff arm of the community tourism organization would be needed to actually perform the necessary tasks of running the organization and taking action. Responsibilities would encompass three functions: (1) supervising projects, (2) obtaining finance, and (3) government liaison.

## Conclusions

Development at the community level dominates all tourism planning. Yet most communities are ill-prepared regarding the many facets of tourism that are essential to its success. Communities may be able to benefit from expanded tourism, but such rewards depend on many factors. Several important planning conclusions can be derived from this discussion of community tourism development.

*Planning for community tourism must make use of all local amenities.* Most local residents are much surprised at their own potential when alerted to their many assets for tourism expansion. Everyday life tends to dull our perspective of our own community, making it difficult to see it from a visitor's point of view. Therefore, special awareness efforts are needed to spark attention to projects. Local tours, educational seminars, and outside help aid greatly in improving local understanding and attitudes toward tourism.

*Planning must include the surrounding area as well as the city.* The city is the focal point for the greatest economic benefits from tourism. Revenues from hotels, food services, other tourist service businesses, and shopping provide for employment, income, and taxes paid. Because these businesses depend greatly on local as well as traveler markets, the urban location for both markets is most logical. However, all these tourist businesses depend on the attractions of the surrounding area as well as the city. The outlying natural and cultural resources often provide the foundation for much travel and tourism to the community. City governments and organizations must collaborate with other organizations in the surrounding area if tourism is to be of the greatest benefit for both.

*Community tourism development demands commitment and leadership.* The belief that tourism expansion is possible with only casual effort by a hotel association or Chamber of Commerce is a myth. Total community commitment is required because tourism involves nearly every facet of the community. Many volunteers and nonprofit groups can play an important role in guiding as well as actually developing tourism. Governments at all levels must have the support of publics for many public actions. The business sector must be in accord regarding the specific actions that only they can provide. The several public factions—religious, ethnic, educational, cultural, environmental—must play an important role. But, above all, strong and competent leadership will be required to pull together the several complexities of tourism within a community.

*Communities must approach expansion of tourism in an orderly strategy.* Haphazard, sporadic, and uncoordinated efforts are not likely to yield expected results from community tourism development. Planning ahead to offset negative impacts and to gain positive results from tourism requires an orderly step-by-step process. Among the several steps, the very first is an understanding of whether a community really has sufficient market potential and resource foundation for even considering tourism expansion. Then a clear view of the specific projects and programs will have to be delineated. Finally, public and private support will be needed to make the kinds of changes required for any benefits to be realized.

*Community tourism development can and should enhance total community life.* If it is found that a community has the potential markets and the resources necessary for expanded tourism, the entire community will be the beneficiary—not just travel businesses. Better tourism demands a better community with quality amenities. These also improve the quality of life for residents. Tourism cannot be given a cosmetic approach, but rather requires a deep communitywide integration of every facet.

# Bibliography

Blank, Uel and Michael D. Petkovich. "The Metropolitan Area: A Multifaceted Destination Complex." In: *Tourism Planning and Development Issues,* D.E. Hawkins et al. eds. Washington, DC: George Washington Univ., 1980, 393–405.

Blank, Uel and Michael D. Petkovich. "Research on Urban Tourism Destinations," Chap. 14. In: *Travel Tourism and Hospitality Research*, J.R. Brent Ritchie and Charles R. Goeldner, eds. New York: John Wiley & Sons, 1987, 165–177.

Butler, R.W. "The Concept of a Tourism Area Cycle of Evolution: Implications for Management of Resources," *The Canadian Geographer,* 24, 1980, 5–12.

Chow, Willard T. "Integrating Tourism With Rural Development," *Annals of Tourism Research,* 7 (4), 1980, 584–607.

Clawson, Marion and Carlton S. Van Doren, eds. *Statistics on Outdoor Recreation.* Washington, DC: Resources for the Future, 1984.

Cohen, Eric. "Rethinking the Sociology of Tourism," *Annals of Tourism Research,* 6 (1), 1979, 28.

*Community Tourism Action Plan.* Alberta, Canada: Alberta Tourism, 1987.

Cooke, Karen. "Guidelines for Socially Appropriate Tourism Development in British Columbia," *Journal of Travel Research,* 21 (1), Summer 1982, 22–28.

Doxey, G.V. "A Causation Theory of Visitor-Resident Irritants: Methodology and Research Inferences." *Proceedings of the Sixth Annual Conference,* 1975, Univ. of Utah, Travel Research Association.

*Final Report of the Winterlude 1985 Visitor Study,* prepared by Ekos Research Associates. Ottawa, Canada: National Capital Commission, 1985.

Gunn, Clare A. "Small Town and Rural Tourism Planning," presentation at conference on Integrated Development Beyond the City, Mount Allison Univ., New Brunswick, Canada, June 11–14, 1986.

Knechtel, Karl. "The Role of the 'Third Sector' in Tourism Development," unpublished paper. Ottawa, Canada, November 1985.

Kottke, Marvin W. *Comparison of Economic Impacts of Tourism on Rural and Urban Towns in New London County,* Research Report 80, April 1986. Storrs: Univ. of Connecticut.

Mahadin, Kamel O. "Urban Landscaping of Al-Karak," unpublished thesis. School of Landscape Architecture. Baton Rouge, Louisiana: Louisiana State Univ., 1984.

Mills, Allan S. *1983 Outdoor Recreation Trips Expenditures in Texas,* Parks Division. Austin: Comprehensive Planning Branch, Texas Parks and Wildlife Department, 1984.

Molotch, Harvey. "The City as a Growth Machine: Toward a Political Economy of Place," *American Journal of Sociology,* 82 (2), September 1976.

Murphy, Peter E. "Perceptions and Attitudes of Decisionmaking Groups in Tourism Centers," *Journal of Travel Research,* 21 (3), Winter 1983, 8–12.

Pearce, Douglas. *Tourist Development.* London: Longman, 1981.

Pizam, Abraham. "Tourism's Impacts: The Social Costs to the Destination Community as Perceived by Its Residents," *Journal of Travel Research,* 16 (4), 1978, 8–12.

Ritchie, J.R. Brent. "The Nominal Group Technique—Application in Tourism Research," Chap. 37. In: *Travel, Tourism and Hospitality Research,* Ritchie and Goeldner, eds. New York: John Wiley & Sons, 1987, pp. 439–448.

Rothman, Robert A. "Residents and Transients: Community Reaction to Seasonal Visitors," *Journal of Travel Research,* 16 (3), 1978, 8–13.

Rovelstad, James M. and Cyril M. Logar. *Creating Economic Growth and Jobs Through Travel and Tourism,* for U.S. Department of Commerce. U.S. Department of Labor,

Small Business Administration, W.Va. Governor's Office of Economic and Community Development. Washington, DC: U.S. PO, 1981.

Swinnerton, Guy A. *Recreation on Agricultural Lands in Alberta.* Edmonton, Alberta: Environment Council of Alberta, 1982.

*Top Secret—Why Tourism is So Important.* Ottawa: Tourism Canada, 1984.

*Tourism Research for Non-Researchers,* Perth: Western Australian Tourism Commission, 1985.

Trancik, Roger. "Hamlets of the Adirondacks: Regional Strategies for Recreation and Tourism," presentation at ASLA Annual Meeting, section of "Landscape/Land Use Planning" proceedings, San Francisco, CA. Washington, DC: American Society of Landscape Architects, 1986.

*The Urban Fair: How Cities Celebrate Themselves.* HUD-PA-661, prepared by Porter, Novell & Associates. Washington, DC: U.S. Department of Housing and Urban Development, 1981.

Weaver, Glen D. "Appraising Tourism Potential," Vol. 1. In: *Tourism USA.* Washington, DC: U.S. Travel Service, 1978.

# Chapter 12

# Conclusions and Principles

The objectives of this book are to describe the concept of tourism planning and explain why its implementation today is important for a better tourism future. Preceding chapters focus on specific aspects directed toward these objectives. These chapters stand on their own and readers should develop their own conclusions.

However, it also seems desirable in this chapter to highlight some conclusions and provide some interpretation that might lead to basic principles. The following 18 statements are offered not as hard and fast rules of tourism planning, but as general principles that have evolved during the study for this book. Presented here are broad statements that are supported in greater detail throughout the text and in other documents.

1. *Tourism policy must go beyond platitudes and marketing.*

Whereas more nations and their subdivisions have developed statements of tourism policy, these policies are still dominated by older lofty goals with token strategy for accomplishment beyond marketing schemes. Perhaps the fact of issuing policy for the first time, no matter how loosely structured, is a logical first step toward greater national recognition.

In policies and plans one frequently finds generalized statements of how tourism fosters international understanding and enhances the quality of life. Universally, statements about economic windfalls from the development of tourism—jobs, incomes, taxes generated—dominate policy. Tourism is touted as a smokeless industry with no impact on the environment. Recent emphasis upon health, fitness and personal derivatives from travel is appearing in policy statements.

Perhaps no other theme is more dominant in federal tourism policies than that of marketing. It receives greatest attention, huge funding, and largest staff support throughout the world. The working definition of marketing in most tourism plans is selling. It is as if the entire supply side is ready—made and just waiting for a tourism bonanza if adequate promotion were initiated. Although nations, provinces, and states continue to give great support to promotion in order to obtain and maintain market share, some new policies contain greater breadth.

Gradually the supply side, especially physical development, is gaining greater attention and policy emphasis. Gradually the many problems and issues that have arisen from tourism are seen as cause for new policy focus. Newly admitted are economic and social costs. Admitted also are environmental impacts—nearly all of which can be reduced or avoided through planning.

Gradually the political side of tourism is being recognized. Seen previously as primarily the role of the private commercial sector, the implications of political action are being felt. Political decisions regarding a multiplicity of bureaus and agencies have considerable influence on tourism.

Tourism policy must be the result of deliberate and conscientious discussion, continuing planning processes, and actions by many agencies and organizations. Tourism policy formulation cannot be left to a few bureaucrats or big business monopolies. The combined impact of many constituencies including commercial enterprise, nonprofit organizations, and governmental agencies of all sizes and levels is essential to policy development for tourism.

*2. Policy sets goals and administrative framework.*

A private sector perspective might deny the need for federal, state, or provincial tourism policy. Policy suggests government involvement, which is the antithesis of private enterprise freedom. But this is a misunderstanding of ideal tourism policy as a foundation for planning and fostering business success.

The multiplicity of decision-makers in many sectors of tourism demands a complexity of policy formation. The private sector does not have a monopoly on tourism development. A federal policy must be an amalgam of commercial, nonprofit, and governmental development policy. But seldom has the commercial sector declared policy, other than for individual businesses. This is the role of associations—hotel, restaurant, airline, motor coach tour, and others. An example in the U.S.A. is the National Tour Association with broad policy mandates including support of education through scholarships.

Tourism has become so large a contributor to economic growth worldwide that it demands policy recognition and guidance at the highest possible level. Tourism, once considered merely the handling of visitors by airlines and hotels, is far more complicated. When tourism is allowed to drift, expand, and blanket societies, many negative results appear. It is a racehorse run wild. Quipped an observer, "Tourism is like fire; it can cook your food or burn your house down."

Policy is needed to declare goals. Review of the many facets of tourism now suggests at least four goals that are worthy of declaration by governments at all levels:

To stimulate and guide the economic rewards from the development, management, and promotion of tourism services and facilities.

To assure high-quality visitor satisfactions from the experiences of travel.

To utilize natural and cultural resources in ways that can perpetuate and not destroy the quality of resource assets.

To enhance the community quality of life by means of integrating tourism with all other social and economic activity.

3. *Balanced social, environmental, and economic goals are essential.*

Over the years, many nations have developed ideologies of conservation, recreation, and tourism in ways that appear to conflict. Examples of disagreement on tourism projects are abundant. The shift from an age of industrialization with its environmental neglect to an age of communication and enlightenment has been painful but productive.

Now a more straightforward view of tourism development requires policy and planning that builds on the symbiosis of rather than conflict between these ideologies. The concept of recreation, considering personal and social values from active and passive recreations, benefits from and contributes greatly to the success of tourism. The dominance of market preference for activities associated with the physical environment demand policies and laws that conserve and protect resources.

Planning tourism must enlist the cooperation, assistance, and input from many constituencies that in the past have been antagonistic to one another. With greater enlightenment, conservationists and environmental protectionists will be seen as allies of tourism development. With greater education and communication, exponents of recreation will understand the great contribution and role they have in tourism. With greater breadth of understanding and forward-looking planning, tourism leadership and constituencies will obtain the special input of recreation supporters and specialists.

Throughout all planning for tourism, and at all levels and by all sectors, social values of tourism must be given priority equal to that of the economy and environment. Tourism can be culturally broadening, a catalyst for world understanding and peace and a rich local exchange between hosts and guests. But these goals will not be reached unless the barriers are broken down through integrated planning.

4. *Federal-to-local governmental roles need clarification.*

This book is not a treatise on preferred governmental ideologies. Each nation has arrived at its present pattern through its own process and over many years. But within these several ideologies, there is need for all governments to clarify roles and policies on tourism planning and development among the several levels.

Report after report describes the lack of coordination between the several levels for tourism policy and planning no matter the political faith. But some general patterns appear to be emerging.

For greater advantages from international travel, nationwide promotion appears to be an accepted federal role, especially in countries where national boundaries define a tourism destination image. For command economy countries, federal intervention through the development of services appears to be accepted as a federal role. In market economy countries, there is a strong trend toward delegating planning, development, and promotion to the subdivisions—states, provinces, prefectures. The identification of tourism as an important element of national, social, and economic activity is increasingly being recognized in federal policy.

The trend toward more localized delegation of tourism planning is a reflection of the true nature of the functioning tourism system. The complexity of tourism is difficult to comprehend at the national level. The interdependency of the several elements of destinations is more visible and more effectively planned at the state and local levels.

No matter the pattern used in a nation, the cause of tourism is fostered by clarifying the governmental roles at the several levels. For example, the dominance of governmental provision of infrastructure—water, waste removal, police, fire protection, streets, roads—suggests strong integration of bureaus and agencies for best tourism planning and development.

5. *Integrative, not segregative, planning is needed for tourism.*

The "go-it-alone" policies of many tourism sectors of the past are giving way to stronger cooperation and collaboration, but not for idealistic reasons. The reasons are stemming from practical operational success. The site-scale perspective, such as for hotels and resorts, is being augmented by the many externalities contributing equally to tourism success.

Better understanding of the tourism "product"—a satisfying traveler experience—is fostering integrative planning. No one business or governmental establishment can operate in isolation. Everyone contributes to and is affected by the many other developments.

So, when tourism planning is being considered, the many advantages of a continuing and integrative planning process become clear. Segregative planning is narrow, sporadic, reactionary, and lacks continuity, so important to travelers. Integrated planning is both short and long range, reflects interdependencies, fosters continuity, is realistic in phasing that is related to budgets, and continues to reflect yearly changes in many tourism factors.

These desirable objectives of integrative planning are not easily accomplished. They are fraught with many bureaucratic and sectoral barriers. But as each tourism developmental element gains understanding of dependency upon outside influence, greater cooperation and even collaboration on tourism planning will take place.

6. *Market-plant match is essential.*

Within tourism organizations, public and private, there has been a tendency to work with markets and supply separately. Much political, governmental, and private attention has been given to market promotion. Many federal policies and plans are focused on marketing. In market economy countries the supply side has been expected to develop its own response to individual market need.

However, several factors now suggest a more sophisticated understanding of how markets and supply are related. Whereas plans of the past have been divided into marketing and supply separately, there is a strong trend toward matching markets with supply and vice versa. Planning should not be driven by each separately but by discovering the gaps between what is desired and what is available.

Market-supply match fosters the individuality of destinations. Instead of all destinations attempting to satisfy all markets, the special qualities of place can be emphasized. Year-round warm beach development in the North may not be physically feasible, although some market segments may prefer it.

A better market-supply match demands better information of both market demand and resource potential. The guesswork of experimentation in development that hopefully might succeed is being replaced by more detailed and technical study and analysis. Planning for future tourism, therefore, must include better and greater data bases upon which better plans may be conceived.

*7. Geographic heterogeniety is a tourism planning postulate.*

The shallow simplicity of past approaches to the development of tourism is giving way to the recognition that not every place has the same tourism potential. Some places have many more assets than others. The geography of tourism, at only an elementary stage of study, has profound planning implications.

Two realms of geographic difference important to tourism are physical and cultural (or programmatic). Physical resource assets—natural resources, cultural resources, transportation, cities—are not equally distributed. Their individual location and reference to other parts of tourism vary across the land. With costs of development at such a high level, it is far more economic to invest wherever assets are greatest. Some areas, therefore, may have much greater tourism potential than others because of better physical assets.

Equally significant are the differences among areas regarding their programs and policies. Tourism may have little potential in a community that dislikes visitors, that sees only disruption of social values, or that cannot possibly expand budgets to develop tourism. Locations where governing bodies are committed to tourism and where publics desire tourism have much greater potential regardless of resource assets.

No, all areas are not alike. The strength of tourism lies in how well the special assets of an area are planned and developed.

*8. Slow-paced, indigenous tourism is best.*

If the experience of past tourism development has taught anything at all, it is the importance of pacing and local fit. Many of the tourism ills of the past can be traced to too much too fast and inappropriateness of development.

Any drastic change, especially in small communities, can be traumatic, and tourism is no exception. The lack of planning for adjustment to crowds of foreigners, to outside investors and power holders, to outside management and work forces, and to outside erosion of local environments has too often caused many problems. But these are not so much problems inherent in tourism as the shock of massive and too-rapid change.

Communities are organic organizations, each with its own degree of adaptability. When tourism growth is planned in palatable increments, internal adjustments can be made more easily. Any conspicuous errors can be headed off before they become major problems. When local residents and governments are constantly involved in slow-growth tourism, there is likely to be less damage to the social and physical environment. Whereas local communities may need the financial and technical assistance of outsiders to develop tourism, it should be kept at the level of assistance, not dominance.

9. *Continuous and sporadic planning need to be balanced.*

As nations and areas seek better planning for tourism, the trend has been to engage a tourism consultant. In this way, management plans, marketing plans, and development plans have been created. The advantages of this approach are several. It provides for a more objective perspective than might be produced locally. It allows for input of specialists in the technical and professional aspects of tourism. It may be the first time that anyone has researched the many aspects of tourism foundation so deeply. There is little doubt that this sporadic plan approach will continue to be employed in the future because it does have much merit.

However, too often such plans were so detailed and so complicated that they were difficult to digest by the decision-makers and actor segments. In an effort to do a thorough job, much documentation is accumulated and many facets of tourism are given consideration. For the uninitiated, these planning documents can present such a formidable list of needed projects to cause the opposite reaction from that intended. Instead of stimulating better development, the area or region then saw tourism as too huge and too complicated a task and gave up.

Actually, the most feasible tourism planning is a mix of continuous planning by agencies and organizations combined with specific development, management, and marketing plans by consultants. When agencies contract for consultant plans, popularly written summaries can be requested and printed for widespread distribution. Another requirement should be staging—descriptions of how overall tourism development can be staged over time. When government agencies and private organizations are engaged in tourism planning on a regular day-to-day basis, the separate consultant plans can be much more meaningful.

10. *Urban-rural tourism demands a regional approach.*

There is a tendency among the tourist service businesses to think exclusively about the urban setting. There is a tendency among motor coach tour agencies to consider touring circuits between and among cities. There is a bias among outdoor recreation specialists to be concerned over only nonurban travel. Of course, each is correct, within each one's perspective. But since all depend on many of the same physical properties and programs, they need to be planned in concert.

Whereas most of the tourism economic impact occurs within cities because that is best for services, these services depend greatly upon the attractions in the surrounding region as well as within the city. Major cities are major travel destinations. But satellite cities and rural regions may make up a large portion of the attractions important to the overall destination.

Unfortunately, there are traditional, political, and attitudinal barriers to integrated planning of the whole region. Often a rural constituency has been accustomed to certain traditional patterns of land use and economic development, and it resists change. Jurisdictional political boundaries make cooperation and collaboration difficult. And the social acceptance of tourism may be quite different between the city and surrounding rural areas.

When both the hinterland and the urban areas open up communication on tourism development issues, the benefits of cooperation are readily obtained. Tourism can benefit both, but only if coordinated planning is initiated. In many parts of the world more mechanisms are being devised to allow greater integration of all forms of development on a regional basis. City-county and regional levels of government are fostering more widespread coordination for many common issues such as transportation, water supply, waste disposal, education, electrical energy, and welfare. Tourism should become a major agenda item for these new approaches to community and rural problem-solving.

11. *Clustering is superior to dispersal.*

There are many advantages to clustering. With the dominance of the present expressway mode of travel, the traveler finds it more convenient not to make many brief stops along the way but to keep traveling between major focal points, either for enjoyment or for services. Air travel, destined to airport locations, also favors clustered development within a radius of these airports and airport cities. Experience has demonstrated that even facilities and services (motels, restaurants, service stations) gain by clustering. Not only has this principle proven to be good marketing, it is efficient engineering as it reduces the per-unit cost of infrastructure (water, waste, power).

Clustering is contrary to dispersal, once favored by park designers. Clustering favors more efficient management. Instead of scattering swimmers and beach users over many miles of waterfront, clustering not only favors the gregariousness enjoyed by most beach users, but policing and other operational activities are much more efficient. The California coastal plan "encourages the concentration of new development in those places that are already developed, so as to minimize resource damage, economize on public facilities and make nonautomobile transportation possible" (Healy: 1976, 221). The corollary of clustering is leaving other areas open for low-density use or conservation protection (Gunn: 1972, 133).

12. *Services succeed only because of attractions.*

There is a temptation for inexperienced developers to favor locations of services-facilities at attraction sites. Certain minimal functions, such as snack bars at fishing sites, marina services at harbors, and restroom facilities, are needed at attraction sites. But unless they are of great magnitude (such as a Disney World), major clusters of services and facilities thrive better at the nearest community, especially when commuter linkage is provided. It is at such a community that there is support also from local trade. Furthermore, it is at these communities that other services are available to travelers: medical aid, communications, money exchange, legal service, and frequently nearness to friends and relatives. Finally, incremental expansion of infrastructure is more easily accomplished at such communities.

As planning hindsight has demonstrated, the bulk of service location is best at communities. For example, Gatlinburg has proven to be a better relative location for a service facility center to serve Great Smoky Mountains National Park than has Yosemite Village, buried within the attraction complex of Yosemite National Park.

13. *Tourism planning requires understanding of the natural and cultural resources.*

For tourist attraction development, the natural and cultural resources of an area provide much of the foundation. In other words, the forces that protect the special qualities and quantities of natural and cultural resource assets directly support attraction creation for tourists.

Whereas this may appear to present conflict, the issue is largely resolved by design and management practices. The protection of the natural resources—waterfalls, mountains, wildlife, vegetative cover, special natural features—provides much of the magnetism for tourists. By design and management, masses of visitors can be provided enjoyment from these assets without employing resource-destructive methods. An aerial cable car transport system would provide many more visitors access from the rim to the bottom of the Grand Canyon and would be far less destructive of the natural resources, physically and esthetically, than the present donkey trail. In other instances, interpretive programs in visitor centers, hotels, and shopping areas can often provide suitable experiences for masses of people without having them physically come in contact with fragile sites.

Because of the popularity of educational, physical, and generally enriching experiences within many natural and cultural settings, such places offer potential for development of tourism. But it cannot be said that resource assets determine developmental success. Resource advantages are helpful but cannot determine success of development or visitor experiences.

Generally, researchers and planners cite these three types of capacity concerns: *physical* (just not enough room for the visitors), *biological* (too much walking or driving on fragile sites), and *managerial* (inability to cope because of staff, practices, budget, or mandate limitations). Solutions to physical limits can be found in building more facilities, increasing size of buildings and services, installing more equipment, and "hardening" site development in appropriate locations. Solutions to biological-ecological constraints imposed by fragile sites include utilizing harder sites for services and reducing erosiveness by increasing interpretive programs at the edges of fragile sites. Management solutions may require increased appropriations, changes in practices (such as stronger people orientation), changes in legislation, or turning some functions over to private enterprise.

Generally, the principle of capacity is very elastic provided that proper design and management practices are implemented.

14. *Tourism transportation requires special planning.*

A fundamental of tourism planning is access. Although this seems a very elementary statement, it deserves deep study and application. In planning, the means of getting to and from attractions, services, and facilities are very important. Because most major transportation systems are multipurpose, they are located where they can serve the greatest numbers. Therefore, service from the existing network, especially highway and air, becomes a basic fundamental for tourism development.

Even so, this need not be taken too literally. The pattern of strip development, typical of the prejet and pre-expressway era, related the attractions, facilities, and services tightly to transportation routes, with little regard for other factors. Today the principle of clustering at locations that can be linked to travel nodes (expressway interchanges leading to cities, airports) is more important than ribbons along the transportation route.

A good case in point is coastal development. It is far better to keep major transportation routes back from the water's edge so that all coastal regions can retain their esthetic appeal and yet be developed without being sliced by transportation systems. Modern tourism transportation systems require much more than application of engineering technology. Visitor and resident images and behavior suggest great emphasis be placed on how such systems will function for people.

15. *Touring circuits and long-stay destinations require different planning.*

Even though tourism is very comprehensive and contains great diversity, development and use varies between two major subsystems: touring circuits and long-stay destinations.

The *touring* subsystem includes the attractions, transport, services, and information/promotion functions for those who utilize a tour of several locations. Attractions are most closely associated with the travelway, even at nodes, and they are usually visited only once by the same party. The activities are slightly more passive, and time constraints become very important because of a fixed touring schedule. The geographic distribution is a circuit rather than a point.

The *longer-stay* destination subsystem, however, is more tightly self-contained geographically. All activities are utilized at the destination point, which must be designed to accept repetition. The travelway between the destination and origin of residence is merely a spacer and its design is of little significance except to move people to destinations.

Where these two subsystems overlap, the physical development must be planned to accommodate both. The same highway and community service center may be used by tourists traveling through as well as by those who are depending on them for access and service for a nearby resort, convention center, vacation home complex, or youth camp destination.

The two subsystems of tourism, touring and longer-stay, need to be recognized in the tourism planning process because of the differences in resources utilized, and the difference in markets.

16. *Research, training, and education for planning are needed.*

Much progress has been made in training at the level of the tourism firm. Managers and employers of hotels, restaurants, and other services related to tourism now have many opportunities for training and education. Education for the engineering and technical sides of highways, airports, hotels, and other tourism elements is available. But very few schools—at any level—teach tourism policy and planning.

The complexity of tourism demands a multidisciplinary approach to tourism planning. Urban and regional planning programs would need to be expanded to include information on tourism market analysis; social, environmental, and economic aspects of tourism; and certainly the geography of tourism. Political science departments need to add tourism planning concepts if future political leaders are to carry out their duties effectively. Recreation and park departments need greater inputs regarding tourism planning so that future managers understand how to plan and manage for tourists visiting parks. Transportation educational programs need much greater understanding of tourism markets and how tourist movement can be given better planning.

When administrators of universities, colleges, and technical institutions become aware of the great importance of tourism throughout the world, there may be stronger acceptance of new programs, particularly in tourism policy, planning, and development.

17. *Creativity and innovation are essential to planning.*

Whereas all tourism planning must be grounded in more and better information, the future will be a mere repeat of yesterday unless innovation and creativity are added. The vision of a better tourism future must occur before it has a chance of becoming reality. The dynamism of markets and changes in people's interests and attitudes demand equally dynamic and innovative solutions. New design and planning of traditional resources will be called for.

So, there is danger in too-prescribed and too-restricted planning. There is danger in elitist planning because no one individual can conceive of all aspects of planning at once or even in a lifetime. Tourism planning must tap the free-running imagination of specialized tourism designers and planners but also must be tempered by input from many others. It must be dynamic and not static. Policy documents and legal statutes can provide the foundation and permission for tourism planning, but they will not be the producers of imaginative, appropriate, and sparkling plans that meet new needs.

Creativity is usually associated with individuals in science, invention, and the arts. But others, such as tourism developers, investors, governmental agencies, and citizens of affected areas, although not possessing the training and experience of professional planners, may have very helpful creative concepts and recommendations. The trick is for leaders of tourism policy planning to tap all productive sources of creativity.

Creativity has been defined as "the ability to formulate new combinations from two or more concepts already in the mind" (Haefele: 1962, 5). Applied to the processes of tourism planning described in this book, every step demands creative input. Even at the very first step of setting objectives, those involved in tourism planning will have begun to formulate a few ideas regarding tourism's future. Certainly the step of research in which greater understanding of the several factors are obtained evokes ideas about what is good or poor about the present and how the future could be better. Even more directly, the conclusions about these factors can suggest many ideas for the future. So concepts do not come as a bolt of lightning out of the blue sky, but only after intensive study of the region. It should not be difficult for the planning actors to have "two or more concepts already in the mind." But this only brings us to the brink of new concepts.

Implicit in tourism development in the future is change. In recent years, with greater emphasis on nongrowth and environmentalism, change is looked upon with less and less favor. Perpetuation of the same—the same theme park design, the same fast-food businesses, the same motels—seems to be a strong rule of tourism development. Much of this is fostered by the world of finance that demands evidence that a business type has established records of success over time. It is further supported by government park and recreation agencies that routinely use the same policies and planning concepts year after year.

However, the lessons from tourism history well show that markets change, transportation changes, and innovations are regularly introduced in services and attractions. Tourism is dynamic. If the future of tourism is to be served, it will depend as much on creative new expressions as upon conformity with past patterns.

Along with acceptance of change is what Storr (1975, 113) calls "divine discontent." This is not the same as chronic complaint. Chronic complaint is negative whereas visualizing a newer, greater, and better future is positive. In other words, no region should enter into tourism policymaking and planning unless it is willing to fantasize a better future. A tourism region is the way it is because of collective policies and decisions throughout both urban and rural areas. A city now recognized as "beautiful" or "a great place to visit" was first fantasized as such by its leaders and citizens. Without fantasy it is doubtful if desired goals and objectives will be reached. Our fantasy "leaves us with, as it were, a divine discontent, which is constantly urging us on to achieve a better match between our subjectivity and objectivity of the world in which we find ourselves" (Storr: 1975, 113).

If creativity is to be exercised, certain characteristics of public and private planners must prevail. Attitudes of unimaginative conformity with every other place are not the best qualifications. To the contrary, favorable attitudes toward risk-taking and toward fresh approaches are to be sought. Haefele (1962, 135) has outlined several traits desired of creative people:

Independence of judgment, particularly under pressure
Assertiveness, boldness, and courage
High level of resourcefulness and adaptability
Capacity to be puzzled
Openness to new experiences
Enthusiasm

In their search for creative answers to regional tourism development, planners are likely to encounter many barriers in the process. Adams (1974, 13–29) has identified these "perceptual blocks" as: (1) difficulty of isolating the problem, (2) tendency to delimit the problem too closely, (3) inability to see the problem from various viewpoints, (4) seeing what you expect to see—stereotyping, (5) saturation, and (6) failure to utilize all sensory inputs.

For example, is the local attraction problem a lack of numbers of attraction features or is there a sufficient number of parks and natural areas that are not known because of poor information? Do the owners of an excellent attraction view their low patronage as caused by poor signage on the highway or is the problem that of not being a part of package tours to attractions? Is the apparent lack of tourist business in a small city with high traveler volume due not to the lack of billboards but to the lack of parking? Although cause and effect are not readily determined, it is easy to define the problem too loosely or incorrectly.

But problems can be defined too narrowly as well. A tourism promotional group may see the problem of increased tourism as a lack of advertising dollars. Actually the difficulty may be more complicated, perhaps needing a greater number and a greater variety of types of attractions to bring in visitors. The broader problem may be more difficult to solve, but may be the right one. Or the lodging industry may view their major problem as minimum wage laws, whereas the broader problem may be the lack of a new convention center. Problem definition can be one of the greatest catalytic accomplishments of public and private planners at all levels.

Because of the complexity of tourism, there are many different viewpoints. But policymakers and planners at the regional scale cannot afford to exercise such bias. Their approach must be broad and objective. For example, a planner with experience only in the economics of tourism may fail to recognize the importance of encouraging camping and recreation vehicle segments of visitors even though they make somewhat less economic impact per capita as compared to motel travelers. Or a preservation government agency's pressure on the planners may minimize the "people" dimensions of tourist resource development.

Stereotyping is an easy trap for tourism planners. Present patterns of resorts, theme parks, public parks, and motels are so strong that it is difficult to visualize new types. But a creative approach stimulates innovation. For example, beginning in the 1930s and for over twenty years, tourist courts and hotels were the two stereotypes of overnight lodging. Gradually, this has been broken with the creation of several other lodging types.

At the regional level, planners should be creative enough to break from stereotypes and conceive of new ideas for development. But this takes input from open-minded representatives of many segments of tourism. A major problem of integrating the many fragmented parts of tourism is the stereotyping of the parts: highway design as engineering instead of people-movement; hotels and motels as bed sales rather than service to travelers; parks as protection areas rather than fun places to visit; and state tourism agency roles limited to promotion rather than including research and planning.

Saturation refers to those situations in which sensitivity to creativity is dulled because of repetition and adaptation. The unattractive approach to a city along its entering highways is saturated with its existence. We adapt to our surroundings, often without questioning them because our senses are so saturated we do not see the details. Local citizens' senses often become so saturated with their pleasant vistas that they no longer see these either. They assume they have all the details in their minds until questioning reveals that they do not. The special project emphasis that tourism planning can offer could break this saturation, suggesting potential for improvement and new development.

All-sensory approaches are seldom made to tourism development. Instead, usually only formal written research and problem-solving takes place. A part of the reconnaissance of a region should include travel and inspection at sites long enough to grasp the sounds (birds, surf, wind, storms, traffic), odors (fragrant flowers, livestock feed lots, new-mown hay, oil fields, petrochemical plants), touch (sand, rock, grass, mud, sunburn), kinesthesis (hiking, climbing, swimming, biking), and sight (distant vistas, skylines, people, buildings, farmlands, mountains). The planning process should allow all constituencies ample opportunity to express their impressions and concepts of change that relate to all the senses. Experimentation in graphics, even when not artistically done, can help solve development problems. Even brainstorming sessions that put ideas (words, diagrams, stick figures) on the blackboard or chart paper can assist in the development of new concepts.

Not only do these blocks to creativity get in the way of the search for new solutions, but other blocks can obstruct acceptance of creative solutions. Even when creative solutions to tourism development are offered, others may resent them only because they do not conform with the past and present. This psychological set, preparedness based upon past experience ("this is the way we have always done it"), can block the acceptance of new concepts for tourism development.

> A part of set is necessarily made up by the established rules one lives by, those things which are axiomatic, indisputable, not to be questioned. Therefore, a given set, or expectancy, may reject as unfitting pattern a solution which is correct. The offered solution is rejected as not matching the temporary specifications, or as violating one's axiomatic truth (Haefele: 1962, 228).

There may need to be some "selling" of concepts even though founded in very logical conclusions about the region's characteristics for tourism development. Just because existing property values, zoning rules, and existing development may suggest otherwise, the concept may be the "right" way to develop. When leaders can anticipate these barriers to new concepts, they can prepare the way for them by publicity and communication and by close cooperation and collaboration with those who prefer the status quo.

These few reflections on creativity are insufficient for either training or utilizing creativity. They are meant only to emphasize that creativity is an essential ingredient in tourism planning. Further reading of how creativity can be stimulated is recommended. But in the actual process of tourism planning, the key to a brilliant and viable tourism future is one that adds creativity to all the valuable experience learned from the past.

18. *Tourism must not be overplanned.*

To say that tourism must not be overplanned may appear to be antithetical to the entire purpose of this book. Instead, it is a final statement of principle, a caveat.

It has been said that planning is prediction—if certain proposed action is taken, the results can be predicted. For tourism, the many precepts delineated throughout this book are directed toward predicting a better tourism world if the many facets of tourism are given more sophisticated and more innovative planning effort.

However, planning must not be so definitive that it proscribes the very personal freedom that can be obtained only through travel. So much of life denies individual expression—work, duty, law, social responsibility—that the greatest emancipation may come only through travel.

Tourism planning must be accomplished with such restraint that travelers can be free to obtain the enriching rewards of discovery, adventure, and achievement. Planned physical settings, planned programs, planned political action, planned management decisions, and planned promotion can and should foster, not inhibit, the individual originality and personal satisfaction that can be derived only from travel. If domestic travel is to break down the barriers of parochialism and if international travel is to stimulate world understanding and peace, all policies and plans must be so dynamic and flexible that these objectives can be reached by travelers. Policies and plans are tools, not ends.

# Bibliography

Adams, James L. *Conceptual Blockbusting.* San Francisco: San Francisco Book Co., 1976.

Gunn, Clare A. "Concentrated Dispersal and Dispersed Concentrations: A Pattern for Saving Scarce Resources," *Landscape Architecture,* 62 (2), January 1972, 133–134.

Haefele, John W. *Creativity and Innovation.* New York: Reinhold, 1962.

Healy, Robert G. *Land Use and the States.* Baltimore: Johns Hopkins Univ. Press, 1976.

Storr, Charles A. "The Dynamics of Creativity." In: *The Creative Process in Science and Medicine,* Hans A. Krebs and Julian H. Shelley, eds. Amsterdam: Excerpta Medica, 1975.

# Appendix

Included in this appendix are selected examples of tourism policy statements with planning implications, intended as a supplement to the text. These statements are edited and condensed versions without commentary. Annual statistical reports and marketing plans are omitted to provide a focus on development. However, it will be noted that specific tourism development plans are not well represented as a result of government policies not to release such documents to the public. Whereas no one example provides a model plan, these summaries are filled with concepts and elements of planning, policy, and process.

The following areas are included:

United Kingdom of Great Britain and Northern Ireland
   British Tourist Authority
   English Tourist Board
   Thames and Chilterns Tourist Board
Japan
   Department of Tourism
   Other Organizations and Programs
United States of America
   Tourism Legislation
   Hawaii
   New York State
Australia
   Western Australian Tourism Commission
   Tourism Development Plans
Ireland
Canada
   Saskatchewan
   Alberta
Zanzibar
Taiwan
Israel

## United Kingdom of Great Britain and Northern Ireland

Following is a condensation of policies, organizations, and functions of levels of tourism development and management in Great Britain and Northern Ireland.

### British Tourist Authority

Because Great Britain is an amalgam of four countries (England, Northern Ireland, Scotland, and Wales), one canopy tourism organization, the British Tourist Authority (BTA), oversees tourism matters in that region. It was created under the Development of Tourism Act of 1969 and then assumed the functions of its predecessor, the British Travel Association (Strategic Plan: 1981–1985).

The Tourist Authority and Tourist Board Act of 1969 created four bodies: the British Tourist Authority, the English Tourist Board (ETB), the Scottish Tourist Board (STB) and the Wales Tourist Board (WTB). The British Tourist Authority consists of a chairman, five members appointed by the Board of Trade, and the chairman of ETB, STB, and WTB. The act identifies the purposes of the functions of the BTA (Annual Report: 1986):

- To promote or undertake publicity in any form.
- To provide advisory and information services.
- To promote or undertake research.
- To establish committees to advise them in the performance of their functions.
- To contribute to or reimburse expenditure incurred by any other person or organization in carrying on any activity over which the Board has power.

Only the BTA has the authority to carry on activities (promotion, marketing) outside the United Kingdom, but the several tourist boards can do so on behalf of the Authority. The BTA is the primary source of financial assistance. Its main responsibilities are to:

- Promote tourism to Britain from overseas.
- Advise the government on tourism matters affecting Britain as a whole.
- Encourage the provision and improvement of tourist amenities and facilities in Britain.

More specifically, the objectives are to:

- Take all possible steps to spread overseas visitors to destinations throughout Britain, thus emphasizing the attractions of Scotland, Wales, and the

English regions, whilst continuing to identify London as Britain's principal gateway and its major visitor attraction; in this regard, promotion will aim to benefit areas in the country that could attract more tourists and where unemployment is high.

- Increase the proportion of traffic in off-peak periods, which relieves congestion and improves prosperity of tourism services.
- Work in partnership with the trade and other tourist interests and to encourage support from the trade for BTA's promotional work overseas.
- Advise government on tourism matters affecting Britain as a whole, to help government maximize the widespread national benefits from successful tourism by creating favourable conditions for the prosperity of the tourist interest.
- Encourage the provision and improvement of visitor amenities and facilities within Britain.
- Complement the work of the national tourist boards within Britain by encouraging the tourist interests in developing their services to meet international market needs, thus enabling them to compete successfully with other countries.

A major role of the BTA is tourism promotion overseas. Promotional emphasis has been on Britain's heritage appeal, extension of the season, new marketing guides, and joint marketing schemes with local authorities and tourist boards.

Because business travel accounts for one-fifth of the British overseas visitors, cooperation is extended to convention tours by means of joint promotion, assistance in establishing and holding trade fairs, familiarization tours, incentive travel, and sales management training.

Overseas public relations include articles, press releases, and newsletters. Press, film, and television representatives are hosted. Local press coverage in 1985/86 emphasized reassurance of travel to Great Britain in response to the acts and threats of terrorism.

Fulfilling requests for travel information is a significant function and has been computerized. Sharing of facilities with ETB has improved efficiency.

The research function concentrated on marketing. Visitor surveys, evaluation of marketing programs, economic impact studies, and potential growth studies are representative of the research effort.

Promotional travel offices for BTA are located in Toronto, Chicago, New York, Atlanta, Dallas, Los Angeles, Mexico City, São Paulo, Buenos Aires, Johannesburg, Sidney, Singapore, Hong Kong, Tokyo, Madrid, Rome, Zurich, Paris, Frankfurt, Brussels, Dublin, Amsterdam, Copenhagen, Oslo, and Stockholm.

Consultation work of BTA is carried out through five major committees: market, infrastructure, development, hotels and restaurants, and the British Heritage Committee.

The BTA *Strategic Plan-1981–1985* sets forth prospects for growth, implications of growth, conditions for success, and policies. Of particular interest to planning are the latter two topics.

Included in the conditions for tourism success are six topics. First, emphasis is placed on the *marketing* function as very important to future growth. The marketing strategy is based on meeting objectives, demand for the products already available or easily created, responsiveness to promotional activity, potential size, and economic or political restraints. Collaboration with local and regional authorities and all commercial interests is important for success. A second condition is *product development.* Improved information and guidance, particularly for smaller operations, and better linkage between promotion and product are part of the strategy. Greater emphasis on new uses of history, sport, culture, education, business, recreation, and health is required. Levels of *standards* comprise a third condition for success but will not be achieved by statutes alone. Quality and price relationships are important for maintaining competition. A fourth condition is *local authority planning.* Better liaison between government, local authorities, and tourist organizations must take place for better product development. Another condition for success is *investment.* The private sector requires financial inducements or it will develop elsewhere. Public investment in infrastructure may be required for some tourism development. And, finally, changes in *legislation* may be needed to allow tourism development equal access to certain governmental assistance now available to other industries. Other sources of funds need to be sought.

The BTA policies include seven key topics in order to achieve the conditions for success outlined above.

The *marketing* policy needs constant review and adaptation to change. In addition to established markets of North America and Europe, priority of marketing should be placed on the European Economic Commission, particularly Germany. A second priority should be placed on Southeast Asia, Latin America, West Africa, and the Middle East (pending greater political stabilization).

Greater sensitivity to *market segments* is needed. Expanding marketing of tourism areas for the visiting-friends-and-relatives segment could stimulate the segment's greater use of hotels and other services. The senior citizen, independent traveler, business conference, trade fair, car tour packages, farm and guest house, low-cost package tours, and resort and spa segments need greater attention.

Several *longer-term developments* require emphasis:

- High-standard self-catering
- Activities with self-catering
- Semiserviced accommodations

- Tennis centers
- Equestrian centers
- Other sporting facilities
- Boating, cruising, and sailing opportunities
- Spa/health treatment centers
- New hotels in specific locations
- Rented cottages and chalets
- Summer camps

Regarding *investment,* BTA seeks and encourages links with new partners overseas. Further, it is encouraging greater private sector and joint private-public ventures.

New *economic and fiscal measures* need to be introduced to stimulate investment. Hotel building allowances need to be increased, allowance needs to be extended to all forms of building, increased support for historic restoration, and extension of public sector grants to all parts of Britain are key issues.

Regarding *standards,* some strategic measures are recommended such as better market research on accommodations, hospitality training, and awards and incentives for outstanding improvements in quality.

A final strategy is to promote the *uniqueness* of Britain's attractions as high value for the money. This long-term strategy includes emphasis on its clean countryside, historical and cultural traditions, high standards of educational, sporting, medical, and health services, orderly conduct of its life, and the friendliness and welcome of its people.

## English Tourist Board

The statutory duty of the English Tourist Board (ETB) based on the Development of Tourism Act of 1969 is to advise government and public bodies on matters relating to tourism promotion and development in England. More specifically, the ETB has as its objectives (Annual Report: 1986, 40) to:

- Maximize tourism's contributions to the economy through the creation of wealth and jobs.
- Stimulate growth in the overall level of tourism through the development of a highly competitive domestic tourism industry serving the needs of both domestic and overseas visitors.
- Encourage the development of projects that will strengthen England's competitive position in the international tourism marketplace by enhancing the image of tourism as an area for investment and by providing strategic advice and financial support.

- Enhance the image of England as a tourism destination with products offering value for money to both domestic and overseas visitors.
- Advise government on tourism matters with a view to spreading the benefits of tourism and increasing prosperity and enterprise throughout England as a whole.
- Work in partnership with BTA, other national and regional tourist boards, local authorities, and the trade to encourage support for marketing programs both at home and overseas.
- Work in partnership with the regional tourist boards, local authorities, and commercial interests in establishing Tourism Development Action Programs.
- Provide information in conjunction with the trade on tourism attractions, accommodations, and other facilities in order to maximize the use of resources throughout the whole country during 12 months of the year.
- Research growth markets of tourism and the most effective means of promoting England's resources and the performance of the industry.
- Encourage the application of new technology to the tourism industry.
- Enhance the status of tourism as an attractive sector for employment through the provision and stimulation of education and training programs.
- Protect and enhance England's unique character and heritage and to use tourism in a positive way to improve the quality of the environment.

ETB works closely with the Department of Employment and has responded to its recommendations for changes in licensing and other constraints on employment. ETB cooperates with the Department of Transport, the Manpower Services Commission, Ministry of Agriculture, Fisheries, and Food, and the Department of Environment. It liaises closely with the Scottish, Wales, and Northern Ireland Tourist Boards.

The ETB sponsors 12 regional tourist boards responsible for promoting and developing tourism. It also has liaison relationships with the Sports Council, Nature Conservancy Council, English Heritage, Arts Council, Countryside Commission, and the water authorities. It works closely with the trade and professional associations.

ETB maintains close ties with tourism developers, operators, and investors. Its development strategy includes:

- To maximize tourism's contribution to the economy through creation of wealth and jobs.
- To stimulate the development of a highly competitive tourism industry.
- To encourage projects that will strengthen England's competitive position in the international tourism marketplace.

Its role is essentially one of catalyst, adviser, researcher, coordinator, promoter, and financier.

In 1985–1986, ETB assisted Sporthuis Centrum Recreatie NV by £1.5 million for the development of a holiday village. It contains 600 units surrounding a complex of sporting activities. ETB grant-aided the Sandpipers, a residential holiday center for severely disabled people in Southport. It has fostered new budget inns and hotels. In York, ETB grant-aided the Jorrik Viking Centre, a shopping center with strong tourism features. ETB helped finance the feasibility study for cultural and historic redevelopment of Portsmouth Royal Dockyard. It also fosters private financial backing of tourism development, primarily through changing attitudes of investors in this sector.

ETB has a multidisciplinary Business Development Team capable of advising on a wide range of tourism development topics. It is particularly active where specialist knowledge and expertise are in scarce supply. Initially conceived for the city of Bristol, ETB is fostering tourism development through its Tourism Development Action Programme. The emphasis is on action, involving a coordinated and corporate approach from local authorities, tourist boards, and the commercial sector. It identifies the strengths and weaknesses of a particular area. Through its Resorts 2000 competition, 38 district councils entered presentations for seaside resorts, many communities have been given technical assistance, and two winners, Bridlington and Torbay, received special grants. The development grant programs (by the third year, 1986) generated investments of more than £300 million, creating 7,220 jobs. In 1985, 542 projects were offered grants out of 678 applications—a total of £7.9 million.

The research arm of ETB studies effectiveness of promotion, visitor surveys, tourism potential, regional data on tourism activity, and evaluation on development strategies and plans.

ETB has also been involved in special tourism programs. In 1985 the Crown classification system, a new national, voluntary means of classifying accommodation services, was introduced. Through consultation with the private sector organizations, the scheme will assist visitors' selection of services. Standards of buildings and services must be met to be approved. New signage guidance and direction were initiated in 1985. Tourism Information Points and several leisure drives have been established. Sign posts are permitted on motorways if an attraction has more than 150,000 visitors.

In the field of education and training, ETB has several programs. It publishes books on employment opportunities—*The Handbook of Careers in Tourism and Leisure* and *The Directory of Courses in Tourism and Leisure.* Assistance has been provided for the development of two first-degree courses of the Dorset Institute of Higher Education and at Newcastle Polytechnic in conjunction with New College, Durham. Further assistance is given in the development of new courses, job placement, pilot training program for small hotel operators, marketing short courses, and linkage with educational organizations.

In the fields of marketing and promotion, the ETB worked closely with the several regional tourist boards, emphasizing travel to the nation as a whole. It works with the travel trade through sales of commissionable products; development of sales manuals, window displays, and holding trade fairs. National advertising, promotional events, an amateur talent contest, liaison with Travel Information Centres, press releases, a marketing advisory service, and development of electronic information systems round out the marketing and promotional functions of ETB.

## Thames and Chilterns Tourist Board

England has divided tourism organizationally into service regions. One of these is the Thames and Chilterns region, encompassing five counties located just west of London: Oxfordshire, Buckinghamshire, Bedfordshire, Hertfordshire, and Berkshire. The initial tourism strategy was published in 1979.

Because of several changes since then, a new set of guidelines, *A Tourism Strategy for the Thames and Chilterns,* was published in 1984. Key changes to stimulate revisions were a changing need for emphasizing tourism in certain areas, better utilization of idle resources, greater emphasis on overseas markets, a need for more domestic destination development, and capitalizing on growth factors.

The overall objective is "to manage growth to the benefit of the existing tourist industry and the national, regional, and local economies. The strategy is to encourage investment in the existing resources and one of selective and managed growth" (Tourism Strategy: 1984, 8).

This objective is to be carried out by means of a *marketing* and a *development* strategy. The marketing strategy includes:

- Helping existing businesses maintain and improve their viability through upgrading, packaging, discounting, and marketing consortia.
- Stimulating travel out of London.
- Ensuring expectations of travelers are met.
- Improving directional signage.
- Special emphasis on certain markets: business travelers, overseas visitors, British tourists, cruise rental.

The development strategy includes:

- New investment in improvements, attractions, accommodations, and infrastructures, especially for business and conference travelers.

- Modernizations of older facilities and addition of budget facilities.
- New self-catering cottages, conversion of farm buildings.
- New holiday villages.
- New caravaning (RV) and tour camping services.

Because all counties of the region have navigable waters, including the River Thames, special policy emphasis is placed on tourism development of these resources. Among the opportunities listed are the extension of the Bedforshire Ouse, restoration of the Kennet and Avon Canal, Grand Union Canal links with the Oxford Canal through to London, improved water supply from the reservoir at Banbury, and extension of attraction value to canal-side locations in addition to boating. The constraints are identified as the problem of maintaining diverse use without restricting hire (rental boat) cruising, the planning of adequate moorings, and adjustment to markets (hire launches make 10 times as many passages through locks as privately registered boats). The policy emphasizes the importance of catering to the hire cruising market, which spends more than other holiday makers and offers great overseas potential. Recognition is given to the need for waterway corridor development—linear parks, historic sites, visitor interpretation centers, information centers, picnic and campsites.

The strategy report provides more detailed description for each county. For example, following, with some editing, is the entire section for Berkshire county.

Berkshire is of considerable significance to the region's tourism as the county includes Windsor, the River Thames, Newbury with its racecourse, its museum, and an attractive canalside area, an Area of Outstanding Natural Beauty, the Kennet and Avon Canal, and an important center of business and population in the Reading/Wokingham area. Berkshire County Council has expressed commitment to the development of the tourism and travel industry within the county. In particular, the industry has the potential to provide jobs, especially for the less qualified or less skilled. Planning policies are being reconsidered in the review of structure plans. Within these policies it is proposed to give encouragement to a wide range of leisure and tourist facilities, while taking into account the need to protect the county's environment.

The strategy for tourism in Berkshire should reflect the main aims of the "Beautiful Berkshire" promotional campaign, initiated by the County Council, but funded jointly by the County Council, the District Councils of Newbury, Reading and Windsor and Maidenhead, and the local tourist industry. The aims of the campaign, which commenced in 1982/1983, are to:

Promote the county as a location for short breaks.
Promote Berkshire as an excellent location for conferences.
Promote the increased use of tourist facilities.
Promote specific local areas, especially in the west and center of the county.

Therefore, the strategy should be:

To make adequate provision for current and future business tourism, including taking advantage of the opportunities inherent in conference tourism.

To encourage weekend and short stay packages in order to improve the utilization of existing accommodation facilities.

To encourage tourism opportunities on the Thames compatible with the need to protect its special qualities.

To develop tourism on the Kennet and Avon Canal as restoration progresses over the length of the waterway from the Thames to the Avon. It is anticipated that the Berkshire part of the canal will be restored within the next few years. This will provide a major new tourist facility with potential for associated developments, e.g., marinas, which would be particularly appropriate within the Lower Kennet Water Park. This will also result in promotional opportunities for Reading, where it stands at the confluence of the Thames and Kennet. The County Council's planning policies include the development of water recreation in some gravel extraction areas, some of which could provide facilities of more than local interest, notably in the Lower Kennet Water Park and the Colne Valley.

To encourage visitor attractions that have spare capacity in their role of providing days out for tourists, day visitors, and residents.

Windsor: In the spring of 1981 Windsor and Maidenhead published a Strategy for Tourism, which set out positive measures to change the pattern of tourist behavior in Windsor. It called for the development of an infrastructure to attract and more ably cater for longer-stay tourists on the one hand, whilst on the other hand liaising more positively with the travel trade and local tourist-related businesses in marketing the range of tourist attractions present in the town.

By 1983 progress had been made with a scheme for redeveloping Clewer Mead for a Leisure Pool Complex and 47 self-catering flats. Rather less progress had been achieved on the proposed new coach park and Tourist Reception Centre at Goswell Road, whilst the proposed hotel development on the River Street Car Park site had been amended to a combined redevelopment scheme to include the Jennings Yard site to the east of the car park.

A traffic management scheme to segregate coaches, visitors, and local traffic into Windsor was discussed with the County Council in the autumn of 1982 and is now being worked up for incorporation into the Local Plan for Windsor. During 1983 Windsor and Maidenhead considered once again its fundamental approach to the borough's tourism industry. As a result the Council reaffirmed its commitment to the management of tourism within the borough in a more positive way and also agreed to take over the operation of the Windsor Tourist Information Centre.

In Berkshire in 1979, maximum assumptions for investment in serviced accommodations were for 1,200 new rooms by 1985. This figure has already been exceeded by 179 rooms. Despite this, there is still an outstanding demand for serviced accommodations, in particular in the medium price and budget price categories. (Budget accommodation is good quality but probably small rooms with en suite bathroom priced at up to £25 for a double room and up to £15 for a single.) New hotel accommodations will be based on business demand, although there is still unrealized potential for accommodating holiday visitors at weekends.

The consultant's maximum assumptions for facilities for camping and caravaning in Berkshire was for up to 500 pitches by 1985. However, no new permanent sites have been established so far, although there is an outstanding planning permission for 100 pitches near Newbury at Sandleford Farm. It is likely that pressure for development and increased activity in the industry will become apparent on the completion of the M25 around London, in particular where the new motorway joins into the M4 and along the M4 corridor. Any new development will have to conform to the relevant planning policies in that area.

The report for the Thames and Chilterns concludes with a discussion of *implementation* and *monitoring.* Because of the complexity of tourism development, the report calls upon cooperative action by five major groups: the Thames and Chilterns Tourist Board, local authorities, other public authorities (such as regional recreation and river authorities), the English Tourist Board, and the commercial sector. Also called for is regular monitoring of tourism, including measures of demand, supply, plans/policies, forecasts and conservation, and preservation.

A final statement charges the Thames and Chilterns Tourist Board with a review of these policies and strategies every other year.

# Japan

The national tourism organization of Japan dates back to 1930 when the Board of Tourist Industry was established by the Ministry of Transport. After several reorganizational changes, especially in 1984, the Department of Tourism now functions under the Bureau of International Transport and Tourism, Ministry of Transport (Tourism in Japan: 1986, 48).

## Department of Tourism

The department has three divisions: (1) planning, (2) travel agency, and (3) development.

The key functions and responsibilities within the Planning Division are:

1. Overall coordination and planning of tourism administration.
2. Supervision of the Japan National Tourist Organization (JNTO).
3. Improvement of reception services for foreign visitors.
4. Research and study on tourism.
5. Subsidies to the tourist industry.
6. Matters relating to the acquisition of stocks by foreign investors in the field of tourism.
7. Collection and compilation of tourism-related documents.
8. Handling of general affairs for the Council for Tourism Policy.

Within the Planning Division, an International Affairs Office was established in 1978 to improve international relations of tourism. It is responsible for:

1. Liaison, cooperation, and exchange of information with tourism administration authorities in foreign countries and international tourism organizations.
2. Research and study on tourism policies and situations in foreign countries.
3. Planning and guidance concerning international tourism publicity.
4. Collection and compilation of international tourism documents.

The Travel Agency Division has the responsibility for: (1) supervision of the travel agency business, (2) supervision of travel agents associations, and (3) supervision of the guide-interpreter business.

The Development Division contains several subdivisions related to development, recreation, youth, and convention activity.

Responsibilities for development include:

1. Financial affairs and a taxation system relating to the tourist industry.
2. Registration of hotels and *ryokan,* and supervision of the registered hotels and *ryokan.*
3. Improvement of tourist souvenirs in quality.
4. Promotion of tourist morality.
5. Planning and guidance concerning tourism publicity.
6. Planning for the development of tourist facilities.
7. Guidance for the improvement of tourist facilities in Japan including the Youth Hostel Center.

The Tourism and Recreation Planning Office was established in 1972 and is responsible for:

1. Investigation and improvement of tourism resorts.
2. Affairs relating to tourism promotion in the comprehensive national land development plans.
3. Investigation, preservation, and utilization of tourist resources.
4. Planning and coordination for the improvement of the tourism and recreation areas and the Youth Travel Villages.
5. Planning of systems to be formulated for the consolidation of the tourism and recreation areas.
6. Financial aid for the development of the tourism and recreation areas.
7. Investigation and research on tourism and recreational activities.

The Youth Hostel Center provides guidance for operation of youth hostels in Japan. A Senior Planning Officer for Tourist Industries was established in 1984 and oversees work with the tourist industry. The Convention Counseling Office, established in 1986, assists local governments on holding conventions, exhibitions, and events.

## Other Organizations and Programs

In 1963 the *Council for Tourism Policy* was established, composed of 30 nonofficial civilians with learning and experience from private and academic areas. The purpose is to provide tourism administration with opinions and information, either requested by administration or volunteered from the Council. Examples are "Theory and Method of Forming Desirable Domestic Tourism," developed in 1982, and "For the Future Development of International Tourism in Japan," completed in 1984.

In order to interrelate with several other ministries and agencies that have interest in tourism, the *Inter-Ministerial Liaison Council on Tourism* has been formed. It includes 21 director-generals of related ministries and agencies.

*Local tourism* development is shared with the Department of Tourism through the 47 prefectural governments. At this level, plans are prepared for regional tourism development, promotion programs, tourist facilities improvement, and park and cultural protection. Also at the local level, nine district bureaus of the Ministry of Transport cooperate on travel agency management, and liaison with prefectural governments, and local organizations.

Several *private and quasi-private organizations* have active roles in tourism development and management. The *Japan Tourist Association* was established in 1964 to promote domestic tourism. It holds seminars and programs to stimulate better local understanding of tourism. The *Japan Travel Bureau*, a nonprofit foundation, carries on tourism research, provides consulting service for tourism development, and offers personnel training. The *Japan Guide Association* offers training and coordination for all government-licensed guides. The *Japan National Trust* is a foundation that fosters protection of special resources. *Japan Tourism Development Foundation* was established in 1971 to promote wholesome tourism and recreation and is also responsible for the Youth Hostel Center in Otsu City. Additional nonprofit organizations include the Japan Hotel Association, Japan Ryokan Association, Japan Tourist Hotel Assocation, Japan Business Hotel Association, Japan Minshuki Association, Japan Economy Foundation, Japan Association of Travel Agents, and Japan Association of Domestic Travel Agents.

*The Japan National Tourist Organization* (JNTO), also nonprofit, was established in 1959 (reorganized in 1979 and 1985) to promote inbound travel to Japan and to provide information to Japanese travelers on traveling safely overseas. It is supervised by an administrative council of 30 members selected from among people of learning and experience in tourism, appointed by the president. Its work is carried out through six departments:

General Affairs
Finance
Planning and Research
Overseas Promotion
Tourist Assistance
International Cooperation (including the Japan Convention Bureau, Overseas Offices, Representatives, and Tourist Information Centers).

*Home Visit System* is an unusual program designed to increase the host-guest relationship. It is conducted free of charge to further international understanding and friendship. As of 1985, almost 1,000 Japanese in 14 areas have registered to receive over 2,700 guests.

*Guide-interpreters* (language) are licensed by the government following a national examination. The exam includes tests of foreign language and speaking ability, personality, and general knowledge.

Perhaps the most innovative program and related to planning is the *New Sites of Discovery* project, set up in 1984. The aim is to assist lesser-known areas in developing their tourism potential. The Ministry of Transport has established standards for the selection. The area must:

1. Have great appeal for overseas tourists.
2. Have a well-developed transportation system.
3. Have readily available medical facilities.
4. Offer Western-style accommodations.
5. Have "i"-centers, or local tourist information centers for overseas visitors.

6. Contain a population that shows a strong interest in attracting foreign tourists and improving reception facilities for overseas visitors.
7. Need strong support from the central government.

On March 24, 1986, the Ministry of Transport officially designated 15 areas as New Sites of Discovery, covering 15 prefectures and a total of 153 cities, towns, and villages.

# United States of America

## Tourism Legislation

The National Tourism Policy Act was passed on October 16, 1981. This discussion builds on the background described in Chapter 2 and places the act in context of overall U.S. policy. The act reinforces federal support of expanding export receipts from tourism, but it also includes several other goals (Edgell: 1982, 121):

1. Optimize the contribution of the tourism and recreation industries to economic prosperity, full employment, and the international balance of payments of the U.S.A.
2. Make the opportunity for area benefits of tourism and recreation universally accessible to residents of the U.S.A. and foreign countries, and ensure that present and future generations are afforded adequate tourism and recreation resources.
3. Contribute to personal growth, health, education, and intercultural appreciation of the geography, history, and ethnicity of the U.S.A.
4. Encourage the free and welcome entry of individuals traveling to the U.S.A. to enhance international understanding and goodwill, consistent with immigration laws, the laws protecting the public health, and laws governing the importation of goods into the U.S.A.
5. Eliminate unnecessary trade barriers to the U.S. tourism industry operating throughout the world.
6. Encourage competition in the tourism industry and maximum consumer choice through the continued viability of the retail travel agent industry and the independent tour operator industry.
7. Promote the continued development and availability of alternative personal payment mechanisms that facilitate national and international travel.
8. Promote quality, integrity, and reliability in all tourism and tourism-related services offered to visitors to the U.S.A.

9. Preserve the historical and cultural foundations of the nation as a living part of community life and development, and ensure for future generations an opportunity to appreciate and enjoy the rich heritage of the nation.

10. Ensure the compatibility of tourism and recreation with other national interests in energy development and conservation, environmental protection, and the judicious use of natural resources.

11. Assist in the collection, analysis, and dissemination of data that accurately measure the economic and social impact of tourism to and within the U.S.A. to facilitate planning in the public and private sectors.

12. Harmonize, to the maximum extent possible, all federal activities in support of tourism and recreation with the needs of the general public and the states, territories, local governments, and the tourism and recreation industry, and to give leadership to all concerned with tourism, recreation, and national heritage preservation in the U.S.A.

It must be emphasized that the implementation of these objectives, desirable as they may have been by the supporters of the legislation, has been dominated by one primary effort—promotion of travelers to the U.S.A. This is not surprising when one places tourism in an American historical and ideological context.

As Richter's research (1985, 832) points out, there are many ways in which public and private policy and actions are carried out for tourism in the U.S.A. U.S. tourism is dominantly a domestic matter and not as dependent upon foreign tourists as are most other nations. As such, states, not the federal government, are very active in support of tourism development and by both private and public sectors. Promotion is the major public role at the state level.

In such a strong market economy country, one finds action driven as much by market forces and voluntary cooperation as by legislation. Richter (1985, 835) found that out of all 50 states, 40 tourism offices cooperated with others on joint promotions. Forty-two state offices regularly provide city and county governments with advice on financial and legal matters, convention planning, land use planning, tourism literature, or assistance in development of local celebrations. Matching fund programs with local governments, private associations, and nonprofit organizations occur in 34 states. At least some planning is identified as a policy function by 38 states; 15 states reported that they perform regular impact studies of proposed tourism development. In contrast to countries overseas, the states consistently *avoid* licensing and other private sector monitoring.

In spite of the overwhelming success (especially economically) of the present balance and extent of public policy in tourism in the U.S.A., tourism leaders and scholars believe the public responsibility is greater than now carried out at the several levels, federal to local. Certainly a more comprehensive role was identified in the National Tourism Policy Act. Many states recently are implying that environmental and social issues must be dealt with at the policy level.

One expression of this new attitude toward tourism policy was a Model State Tourism Policy draft prepared by David Edgell of the Travel and Tourism Administration (1986). It is written in the form of a bill that could be introduced by a state legislature. This model policy bill is divided into six sections.

Section I, "Findings," includes general statements regarding tourism fundamentals, such as the advantages that can be derived from tourism and the need for state policy. Section II, "Policy," presents 19 statements that identify a state role of "encouraging, fostering, providing, facilitating, and ensuring" actions and programs. For example, item 13 states: ". . . facilitate tourism to and within the state by developing an essential tourism infrastructure; providing investment incentives to tourism businesses; and encouraging municipal and county officials to plan for tourist needs and capitalize on local tourism resources."

Section III, "Duties and Responsibilities of the Governor," identifies the governor's office as the key implementing agent. Furthermore, to assist the governor's role, a state travel division is recommended. Several specific duties are assigned to this state agency, including market research, stimulation of development, education, cooperation between state agencies, creation of informational materials, and resource protection as well as tourism promotion. This state agency would be given a catalytic role to act between the many other agencies to foster environmental and social as well as economic enhancement of tourism.

Section IV, "Tourism Policy Council," asks for an interagency coordinating body at the state level. Its functions include acting as a review panel to oversee potential impacts of acts by the travel divisions, implementing aspects of the tourism policy, and indentifying legislative needs. Section V, "Tourism Advisory Board," calls for a board made up of industry representatives to advise on programs. Section VI is a final statement regarding the date of approval of the state tourism policy.

Another approach to a model state tourism policy was presented at a seminar of state tourism travel directors (Gunn: 1984). This proposal suggests that the traditional role of promotion be broadened at the state level to include responsibilities not now accepted by any organizations or agencies. The details of organization are not defined. Five roles in addition to promotion are described:

1. *Guidelines for development:* In order to assist the private and public sectors in establishing new and improving older destinations, this program would provide timely technical information for better decision making. Following research, three popularly written documents would be prepared and disseminated widely on an annual basis. The first would provide detailed and up-to-date market information: characteristics, preferences, origins, trends. The second

would summarize resource assets and physical factors important to the development of tourism. The third report would combine market-product match information derived from the other two reports to identify potential.

2. *State incentives.* This role would initiate new incentive mechanisms to stimulate development where most needed and most appropriate. Financing methods, matching funds, improvement contests, tax incentives, changes in land use regulations, and other forms of incentives would form part of the state policy.

3. *State development.* Needed is a monitoring function at the state level that can assist agencies in better integration with tourism. This may mean changes in enabling legislation to allow broader application to tourism. It may mean only new regulations and practices that ask each agency to consider tourism implications in all decisions. The intent is to provide better coordination of the many agencies that control resources and develop lands and programs related to tourism but have no stated policy for integration.

4. *Policies/regulations.* Needed is regular monitoring of regulations and policies that impinge upon tourism development, especially for the private sector. Many laws and regulations were prepared for an earlier period and for different purposes than tourism. Some are overly restrictive and obsolete. At the same time that health and safety standards are assured, revision of older and enactment of new regulations more appropriate to tourism are needed.

5. *Research/education.* A responsible role of research and education needs to be broadened. Tourism education is needed at all levels for learning how to travel and about the many career opportunities. Few economic areas are so poorly researched as tourism. This role calls for studies that are specific to tasks as well as on a time series basis. The topics are abundant in both the market and supply sides.

6. *Advertising/promotion.* A continuing but sharpened role at the state level is advertising and promotion. New techniques need to be reviewed for applicability. Better accountability, better impact and better integration with all forms of promotion within the state should result from an improved policy for promotion.

## Hawaii

Not necessarily influenced by the state tourism models, the state of Hawaii has for several years been engaged in planning for tourism. This section briefly describes Hawaii's status of tourism planning.

Background: The foundation for Hawaii's tourism plan is the enactment of the Hawaii State Plan in 1978, the first U.S. state to do so. It was in response to a felt need, more conspicuous on the islands, to coordinate and integrate major governmental and private sector actions. It followed several earlier expressions of concern over how the islands could protect and maintain their special resources and accept growth and development at the same time. The first land use law was enacted in Hawaii in 1961. A shoreline setback law was passed in 1970, and the Coastal Zone Management law was established in 1975 (Hawaii: 1986).

Because of its small size, limited resources, and relative isolation, Hawaii has been engaged in tourism planning for many years. The burgeoning growth of tourism since the 1960s caused much of the planning emphasis to be placed on overall economic and physical growth. For example, in 1972, a two-volume tourism impact study was completed by the Department of Planning and Economic Development for the western region of the island of Hawaii (West Hawaii: 1972). Volume one summarizes public and private investment in facilities, physical guidelines for resort area planning and design, a socioeconomic case study of tourism's impact in the North Kahala community, and statewide findings and recommendations. Volume two is a planning study of West Hawaii, relating tourism development to the overall economic and physical development of the island, especially the west side. Called for is overall community and regional growth, not just resort hotels. Specific proposals and recommendations are:

- A college campus as part of the state university system.
- Shoreline trail linear park along the ocean.
- A study of potential use of ancient Hawaiian fish ponds.
- Consideration of new interisland ferry impact.
- New urban planning, historic protection, alternative growth patterns, interrelationships between public and private lands, and interrelationships between *mauka* and *makai* areas.
- A property tax study to equalize assessments.
- Sewage and recreational facilities.
- New policies for state lands.
- Protection of scarce water resources (potable, brackish).

The Department of Planning and Economic Development in 1974 issued a major document, *State of Hawaii Growth Policies Plan: 1974–1984* (1974). On the heels of the energy crisis, it called attention to the need for new policies and directions regarding social, environmental, and economic consequences of unplanned growth. This report made two major recommendations for tourism: state actions to slow growth in tourism and direct it to the neighbor islands, and state actions to promote the economic health of the visitor industry.

The Hawaii State Plan of 1978, not a law, calls for preparation of 12 functional plans: agriculture, conservation lands, education, energy, health, higher education, historic preservation, housing, recreation, tourism, transportation, and water resources development. These plans were not directives but guidelines (Fact Sheet: 1982).

By action of the Twelfth State Legislature, all but "agriculture" and "education" functional plans including tourism were adopted. Considered by many a landmark document, the *State Tourism Functional Plan* (1984) is in response to the great growth of tourism in Hawaii in only a few decades. Approximately 4.2 million persons visited the state in 1982, compared to only 362,000 in 1962. The assets of resort development, beautiful mountains and seacoast, ideal climate, and hospitable people have made it an attractive tourist destination.

The State Tourism Functional Plan was developed by an advisory committee representing the general public, the visitor industry, the state, labor, and the counties. This plan, as is true of all other functional plans, is a set of guidelines as an expression of legislative policy, but it is not a law or a statutory mandate. Four major concerns were addressed, offering objectives, policies, and implementing actions.

For *tourism promotion,* the policy states, "Assist in overseas promotion of Hawaii's vacation attractions." Implementing action emphasized maintaining market share by investigating new markets and encouraging more conventions. This policy was to be implemented by the Hawaii Visitors Bureau in cooperation with the tourist business community.

The *physical development* recommendations included several policies:

Ensure that visitor industry activities are in keeping with the economic and physical needs and aspirations of Hawaii's people.

Improve the quality of existing visitor destination areas.

Encourage greater cooperation between the public and private sectors in developing and maintaining well-designed adequately serviced visitor industry and related developments.

Ensure that visitor facilities and destination areas are carefully planned and sensitive to existing neighboring communities and activities.

For all these policies, specific implementing action was stated. Most action was to be carried out by county planning and land use agencies with cooperation from private sector developers. Budgeting and specific zoning recommendations are made.

For *employment and career development,* two key policies were stated: (1) develop the industry in a manner that will provide the greatest number of primary jobs and steady employment for Hawaii's people, and (2) provide opportunities for Hawaii's people to obtain job training and education that will allow for upward mobility within the visitor industry. Action steps were to be taken by the HVB, governmental agencies, educational institutions, and unions.

The fourth major concern, *community relations,* includes three policies: (1) ensure that visitor industry activities are in keeping with the social needs and aspirations of Hawaii's people, (2) foster a recognition of the contribution of the visitor industry to Hawaii's economy and the need to perpetuate the Aloha Spirit, and (3) foster an understanding by visitors of the Aloha Spirit and of the unique and sensitive character of Hawaii's cultures and values. Implementation of these policies was placed in the hands of the State Ombudsman, Office of Consumer Protection, Chamber of Commerce of Hawaii, State Foundation on Culture and the Arts, Bishop Museum, Hawaiian Historical Society, and several government agencies.

As this functional plan was being formulated, a survey of the residents of Hawaii was made in order to obtain their perception of tourism (Liu and Var: 1984). The survey of 3,000 residents covered 150 items pertaining to a wide range of topics, including economic, social-cultural, and physical-environmental effects. Several interesting conclusions were drawn from this study.

First, it confirmed the generally understood economic importance of tourism. Cultural benefits were also cited. Second, surprisingly, the residents did not agree with the media and literature that blamed tourism for crime, congestion, and environmental degradation. Third, there seemed to be a contradiction in the findings. On the one hand, residents regarded protection of the environment as more important than the economic benefits from tourism, but, on the other hand, they were not willing to lower their higher standard of living to achieve this goal.

As a forerunner to the development of the state tourism plan for Hawaii and as an expression of opinion about tourism, a Travel Industry Congress had been held in 1970. Realizing the success of this congress in resolving some controversial issues, particularly between the public and private sectors, a new meeting, the *Governor's Tourism Congress* was held December 10 and 11, 1984 (Proceedings: 1985). It was called for by Hawaii Senate and House resolutions to convene representatives of state and county government (Hawaii has joint city-county governments), visitor industry, labor, and other interested organizations. The Congress was organized around three issue areas: (1) economic, (2) physical and environmental, and (3) social. Four task forces were appointed for each of the first two issues and five task forces for the social topics. More than 50 speakers made presentations, discussions followed, and a questionnaire was prepared for opinion on economic, land use, and environmental and social issues. Results of the questionnaire were based on an average of 380 ballots returned for each of four time periods.

Majority agreement on several economic issues was shown. Tourism promotion should continue; no limits should be placed on tourism growth; more trade fairs and exhibitions should be held; state funds for promotion should be increased; emphasis should be placed on longer-stay, higher-spending tourists;

economic base should be diversified; growth on neighbor islands should be proportionately greater than on Oahu; and international, educational, and cultural Hawaiian attractions should be increased.

Regarding planning issues, highlights of majority agreement included: retaining the Waikiki Special Design District; improvement of the physical infrastructure and police control of Waikiki; enhancement of appearance of private property in Waikiki; resolve the complicated control of Waikiki Beach maintenance to one agency; regulation of commercial facilities at Waikiki should be assigned to one agency. Also agreed upon were: a major convention center should be built, preferably at the Fort DeRussy site; state and county governments should preserve special natural and cultural areas; certain recreational activities including helicopter access should be restricted in some areas; waterfall stream flows should be assured; an environmental "sense of place" should be maintained; resort area carrying capacity should be determined; public access to beaches should be maintained; local publics should be better informed regarding planned resort development, and public disclosure should reveal resort payments to public agencies.

Among the social issues, there was majority agreement on topics relating to social impacts, racial tension, Hawaiians and tourism, and employment. It stated that Hawaii's tourist industry should be more alert to social impacts; nontourist industry should be increased to reduce major dependency on tourism; and the HVB should study ways of reducing the social impact of high-spending tourists. The majority approved broadening activities of the Hawaii Business Roundtable to improve local attitudes toward tourists and tourism; visitor educational programs should be initiated to enhance visitor respect for local attitudes and cultures; and cultural festivals and other means should be used to minimize misunderstandings between the industry and the Hawaiian community. Regarding employment, there was strong support to encourage stronger "partnership" between the visitor industry and community college systems; career opportunities should be publicized; state support should be given to heighten community awareness of tourism; and the School of Travel Industry Management at the University of Hawaii should be given greater research support.

In his final statement, the chairman of the polling procedure, Aaron Levine (Proceedings, 1985, 210) stated: "Underlying all the comments made today is the hope that we can preserve the rare quality that is Hawaii. This is the assignment that each of you has to carry out, each in your own way, whether you are in government, whether you are in the industry, whether you are members of the general public."

In 1986, the Department of Planning and Economic Development prepared a *progress report* on the State Tourism Functional Plan (Progress Report: 1986).

Some progress had been made on all items, but only a few objectives had been reached by that date.

Regarding promotion, HVB had the largest budget ever—$6.5 million. New emphasis was placed on promotion in Asia, Far Western promotion, for Molokai as a destination, for sports promotion, and funds for a visitors' survey. A transient accommodations act was passed, which adds $60 million for the visitor industry.

Within physical development, several public and private investments in improvements have been made. Over $12 million were invested in new infrastructure. Street improvements, new water supply, and a new fire station have been funded. Land use changes to support new resort destinations have been made. All county plans have been modified to reflect the needs of the Tourism Functional Plan. Act 159 authorized the establishment of a Tourism Impact Management System within DPED.

For employment and career development, a career guide was revised and new in-service training workshops were held. Two new arms of the School of TIM, the Hawaii International Hotel Institute, and a Tourism Research and Training Unit were established. High school and community college expansion now offers more programs related to tourism.

In the field of community relations, Tourism Week and Aloha Week were expanded for greater public awareness. A new educational film was produced. A new "Hawaiian Ecosystems: A Living Heritage" exhibit was prepared. Proposed legislation for strengthening historic preservation laws failed.

Public and private sector groups continue to strive for meeting the several recommendations and policies set forth in the Tourism Functional Plan. It is not only a land use plan but a comprehensive form of democratized planning for guided growth (Keith: 1985, 30).

After a two-year comprehensive review of the Hawaii State Plan, revisions were recommended and approved. This step demonstrated that this plan is a flexible process that is constantly being updated to meet current needs. While most of the public surveys show continuing concern over the environment, priority has shifted to crime, housing, education, and jobs (Hawaii State Plan: 1986, 11). Tourism continues to grow in significance. In 1963 it represented 9 percent of the gross state product and by 1986 it represented over one-third of all civilian jobs and total personal income. The major amendments and revisions to the plan included updating of need and emphasizing that it provided guidelines for democratic action and not a centralized command.

## New York State

The Office of Marketing and Tourism, an arm of the state Department of Commerce (DOC) of New York, contains four functional offices: Deputy

Commissioner for Advertising, Deputy Commissioner for Tourism, Director for Media Services, and Director for Business Marketing. The Division of Tourism (DOT) administers several programs:

Matching funds program
Regional tourism marketing liaison
Hospitality training
Familiarization tours
Trade shows
International promotion
Meetings and conventions
Collateral distribution

Outside New York City the state is divided into 11 regions for tourism promotion. Each region has its own coordinator and is made up of the several county tourism organizations.

In 1985 the DOC was directed by the legislature to develop a master plan for tourism. Although the "I Love New York" promotional campaign of 1977 dramatically increased the number of visitors for a few years, symptoms of several tourism problems, together with loss of market share in the early 1980s, stimulated the study, *Proposed Tourism Master Plan,* completed in December 1986 (Price Waterhouse and Gunn: 1986). Key elements of the study include: the study process, objectives, nature of tourism of New York, issues, findings, and recommended strategy.

Study process: The consultants employed six steps in performing the study:

1. Data collection. Collecting and analyzing secondary data on the tourism economy, industry makeup, agencies involved, and market data.
2. Interviews. In order to obtain information on key organizations and their functions, interviews were held with 150 representatives throughout the state.
3. Workshops. For the purpose of obtaining local experience and opinion on new state policy and management, eight workshops were held throughout the state.
4. Regional Economic Development Strategy (REDS) meetings. All REDS conferences of 1986 were attended to learn how tourism was viewed by the economic development agencies.
5. Other data. Specific information on tourism programs was collected from 11 states and provinces outside New York.
6. Analysis and recommendations. From the findings of the several study steps, a tourism strategy was developed and implementation plans were recommended.

Objectives: The purpose of the master plan was to give direction to the public and private sectors of New York State to ensure continued expansion of travel and tourism in the State. More specifically to:

Create a *strategy* that defines tourism policy, strategic goals and objectives, and encompasses a combination of specific organizations, responsibilities, programs, and activities.

Describe the *implications* of this strategy—time frames, resources, and program impacts.

Describe *background, issues,* and *opportunities* for the state, such as the tourism economy of New York and its regions, assessment of promotional and development programs, and organizations and identification of issues and opportunities.

Nature of tourism in New York: The tourist industry of New York includes many state and local government operations, development by commercial enterprise, and nonprofit installations. Included are 52,413 businesses involved in tourism that in turn contributed $8.4 billion to total wages in 1985. Between 1975 and 1985 employment from tourism increased 24.7 percent, and wages, 119 percent. Tourism has provided 13.9 percent of new nonagricultural jobs in New York over the last 10 years.

Issues: A tourism plan was needed because of several changing factors in recent years. Competition from other states had increased greatly. Markets had become more demanding. Destinations required updating. Some areas suffered from undercapitalization and inefficient management. New attractions were needed to stimulate expansion of facilities and services. Tourism, as a separate and significant "industry," had not been recognized by many public and private segments of the state. Better assessment of economic development was needed. The state legislature and the DOC agreed that a management plan was necessary in order to define and clarify public and private roles, identify tourism opportunities, clarify the importance of tourism, and identify the needed new programs.

Findings: 1. Major weaknesses were found in the marketing, promotional, and information programs. Advertising was directed to markets already saturated. Little was known about the effectiveness of advertising. Scheduling of advertising often missed target dates. Matching promotional funds have been utilized, but a county focus may not present the best image. Information centers lack coordination. No uniform signage policy is in effect.

2. It was found that business assistance and regulation needed improvement. Tourism is explicitly excluded from Job Development Authority (JDA) funds. Loan rejections were often due to undercapitalization and insufficient management. There are several technical and business assistance programs offered, but

lack of communication reduces their effectiveness. Several regulations restrict tourism development. Liability insurance is a major problem.

3. The need for changes in organization and coordination was identified. The narrow focus of promotion limits the DOC's ability to provide adequate assistance to the supply side. Clarity of DOC roles and better coordination with other state agencies were needed. A need was cited for better communication and coordination between DOC and the private tourism sector.

4. Better data definition, collection, and evaluation was needed. The tourist was not clearly defined and existing data were not readily available. Tourism-related data collected by separate agencies and organizations were not routinely shared. Little data were available on the supply of tourism development.

5. Resources for tourism development are abundant but not evenly distributed. Each region has a distinct set of foundations for attractions, service centers, and relationship to markets and access. Both business (conventions) and personal travel (vacations) have potential in many regions.

Recommended strategy: The consultants recommended a specific objective of increasing employment related to tourism by 10 to 15 percent by 1992, to increase sales tax revenues by 5 percent by 1992, and increase average wages in tourism by 8 percent by 1992. These specific objectives were presented within the goal framework of increasing employment, increasing personal and business income, increasing tax revenues, and diversifying the state's economy.

Detailed recommendations for the government and private sectors were identified and included the following major points of strategy:

- Strengthen the management of tourism responsibilities at the state government level by: strengthening the *organizational capability,* improving the *information system,* expanding *research,* and developing *communication* between the public and private sectors.
- Simplify the target information to potential travelers by *targeting* advertising and promotion, and *simplifying* the process of information.
- Strengthen the tourism industry as an integral part of regional economic development by *education, training,* and *technical assistance.*
- Encourage tourism growth as an integral part of regional economic development by utilizing the *special resources* of each region for balanced total economic growth, and *incorporating tourism* into regional economic development plans.

The final conclusion was that for New York State to maintain its competitive edge, the strategy set forth in the master plan should be implemented, including: (1) redefining roles and responsibilities of the public and private sector organizations, (2) implementing specific programs for action, and (3) providing adequate funding for the strategy and to evaluate its success.

# Australia

At the federal level in Australia, a minister of tourism oversees the governmental roles in tourism. This office has supervision over three bodies (Western Australian: 1986?, 3). The *Department of Sport Recreation and Tourism* (DSRT) has as its goal the creation of "an effective environment conducive to private enterprise development" and stimulation of an effective tourism industry. It advises the ministry on tourism-related policy. It monitors industry trends and conditions and maintains close liaison with the industry, the states, and the territories of Australia.

The *Australian Tourist Commission* is a statutory body whose policies are determined by a board of commissions responsible to the Minister for Sport, Recreation, and Tourism. Its primary function is overseas promotion of travel to Australia through its offices in Auckland, Frankfort, London, Los Angeles, New York, Tokyo, Singapore, Sydney, and Melbourne. Major activities include market research, promotional and publicity campaigns, productions of publications and films, product education, familiarization tours, and involvement in trade shows and exhibitions.

A *Tourist Ministers Council* is composed of the ministers of tourism from the six states and the Australian Capital Territory (Canberra). Its role is coordination of tourism activities between the several states.

Each of the states of Australia, including the Capital Territory and Tasmania, exercises its own land use planning guidance and controls. Current interest, involvement, actions, plans, and policies for tourism and recreation for 30 agencies and organizations in Australia are summarized in *Recreation and Tourism Land Use Planning in Australia: An Information Manual* (1987). This manual has stimulated greater interest among many agencies to strengthen their tourism roles, but more especially the opportunities for better planning integration.

## Western Australian Tourism Commission

A description of one of the state tourism bodies, the Western Australian Tourism Commission (WATC), provides insight into the typical functions of government and the private sector at the state level (Western Australian: 1986?). This state includes the entire western one-third of Australia (westward from 129° longitude), an area of 975,920 square miles (2,527,621 square kilometers). Its main aims and objectives are to:

Market Western Australia as a tourist destination for international, interstate, and intrastate visitors.

Increase the length of tourist visits and the use of tourist facilities.

Assist in the development of the state's tourist infrastructure, support and coordinate the provision of new tourist facilities, and provide for more effective and efficient investment in the state's tourism industry.

In order to accomplish these goals, the WATC is able to:

Acquire, lease, and deal in property.

Construct, establish, maintain, and operate tourist facilities by itself or through arrangement with others.

Coordinate the development of new or existing tourist ventures in liaison with interested individuals or groups.

Enter into agreements to enable it to participate in tourist promotions in Western Australia and elsewhere.

Make loans or grants to help research or investigation into setting up, maintaining, extending, or developing tourist facilities.

The WATC includes five major divisions and several other functional bodies. The *Marketing Division* includes advertising and promotional functions as well as overseeing the Holiday WA Centers and Communications. The *Investment and Regional Development Commission* is primarily concerned with development of public infrastructure and that marketing and visitor services are carried out efficiently. A *Research and Planning Division* oversees the collection and analysis of comprehensive tourism-related information to assist planning and decision making by the WATC, government, and private enterprise. The *Finance and Administrative Division* provides internal management and fiscal control. The *Human Resources Division* is responsible for the training of tourism operators and the personnel and human resource functions of the commission.

In addition, the America's Cup Unit promotes the America's Cup Yacht Race in Perth, and the Perth Convention Bureau and the Western Australian Tourism Industry Training Committee are closely allied to WATC.

This state has been divided into nine regions (some with subdivisions), each of which is represented by a regional travel association. These are primarily marketing and promotional bodies made up of members from Tourist Bureaux and Tourist Information Centres.

## Tourism Development Plans

The state of Western Australia is the only one in the country to have initiated, through its Research and Planning Division, a program of development plans.

These regional Tourism Development Plans (TDPs) are not statutory documents, but they contain actionable tourism projects for implementation by both governmental and private sectors. They are prepared by consultants in close cooperation with local community groups. They are designed to provide comprehensive and integrated strategies for tourism development in order to:

Enhance the tourism sector's contribution to the economic and social development of the community.

Ensure that visitors are provided with appropriate facilities and services so that the potential for an enjoyable stay is maximised (Tourism Development Plans: n.d.).

Four steps are used in the development of the plans:

1. Resource analysis—inventory and analysis of the current tourism infrastructure and market performance.
2. Market analysis—assessment of potential demand for tourism product in the region and identifying future target markets.
3. Gap analysis—recommend strategies to resolve gaps between supply and demand of tourism product. Key development opportunities are taken to prefeasibility stage and presented as project profiles.
4. Action plan—presentation of regional development strategies (Tourism Statistics: 1986, 9).

By 1986, 11 TDPs had been commissioned, completed, and were in various stages of implementation. These plans are not designed to be ends in themselves but rather only one stage in a continuing tourism development process at the regional level. The WATC recommends a model for implementation consisting of five steps:

1. Presentation of the TDP for public comment.
2. Collection of tourist industry, community, public, and private sector feedback.
3. Formation of regional task force.
4. Assessment of submissions.
5. Presentation of task force's preferred development strategy.

One example of the TDPs was prepared by consultants for the *South West Region,* (1986) located in the southwestern extremity of the continent. It comprises 15 local government authorities and includes an area of 10,640 square miles (26,601 square kilometers). Its population is 110,590 (as of June 1984).

This TDP resulted in identifying six major issues:

1. There is a wide diversity of product, but as yet the various product components are uncoordinated.
2. There are deficiencies in the basic tourism infrastructure.
3. Local government is tending to react to, rather than plan for, the development of tourism.
4. Financial support for tourist associations from the business sector is, in most cases, minimal.
5. There is a lack of awareness in some communities of the benefits of tourism, and little effort had been made by the industry to rectify the problem.
6. Marketing is uncoordinated, generally lacks professionalism, and there is considerable duplication of effort.

It was observed that in the past, too much emphasis had been placed on promotion rather than product development. The TDP includes many recommendations for both promotion and development, but especially for linking them together.

The TDP for the South West Region includes several recommendations for improved functions, such as adoption of tourism policy documents by local authorities, improved policies and practices by the several tourist bureaus, better organizational structure, increased funding grants, enlistment of better community support, new tourism awareness educational programs, and recommendations for improved operation of tourist information centers.

The South West Region includes four subregions or tourist zones. For the zone of Vasse, the following were recommended (Tourism Development Plan, SW: 1986, 10):

Busselton be recognized as a node for development of upgrading the foreshore; add holiday unit and chalet rather than more caravan parks, deny the construction of a proposed road from Sugarloaf Rock to Yallingup.

Margaret River be identified as a node to include: a winery, convention center at Margaret River, a "Club Mediterranee"-type resort at the mouth of Margaret River, a hot air balloon base, a new visitor center at Lakes Cave, and the establishment of a wine festival.

Beyond the specific recommendations for each subzone, the following general actions were recommended:

An integrated system of designated scenic routes be established.
Upgrading signposting for scenic routes.
The Margaret River signpost project be used as a pilot.

Establishment of events: "Man of the Forest" at Dwellingup, "Wine Festival" at Margaret River, "Karri Forest Festival" at Pemberton.

Marketing of new packages: farmhouse/guest referral system, budget holidays, and adventure weekends/holidays.

Improvement in the range, format, and distribution of descriptive information for visitors (Tourism Development Plan, SW: 1986, 12).

One of the recommendations in the TDP for the development of the South West Region was creation of new coastal resort development around Margaret River. For this project, a planning design firm was commissioned to prepare a *Leeuwin Estate Tourism Development Plan* (1986). The 1829-hectare (4,573-acre) property is located about 173 miles (280 kilometers) south of Perth and contains vineyards and a winery, dairy and beef cattle, sheep grazing, and small woodlots.

The study objectives included assessment of resource potential, market potential, integration with surrounding tourism potential, maintenance of privacy for owners, and continued economic land use for forestry and farming. This study represents the integration of planning from the state to the region and local level.

Research included analysis of physical features, cultural factors, attractiveness, study of surrounding tourism, and reviews of detailed market studies. The research for this project revealed both needs and opportunities. Among the needs identified:

Insufficient development for up-scale and down markets.

Lack of integration with surroundings.

Poor access diversity.

Too few events.

Insufficient promotion.

Poor information.

For tourism development opportunities, the study research built upon earlier studies and particular emphasis was placed on market segments. A marketing study had identified six of the most important market segments for Australia:

1. New enthusiasts
2. Big spenders
3. Antitourists
4. Stay-at-house tourists
5. New indulgers
6. Dedicated Aussies

Study of the resource base and these market segments revealed opportunities for matching tourism development with potential market groups as listed in Table A-1.

### TABLE A-1
### MARKET-PRODUCT MATCH FOR LEEUWIN ESTATE

A comparison between proposed product opportunities and potential market groups for future development of the Leeuwin Estate for tourism.

| TOURISM OPPORTUNITY | POTENTIAL MARKET GROUPS |
|---|---|
| An exclusive resort and convention centre with a range of excellent recreation facilities, time share and short-term accommodations, and guided tours | • New indulgers (individuals)<br>• Big spenders (individuals and corporations) |
| A top class equestrian centre and riding school | • New indulgers<br>• Big spenders |
| A culinary arts centre, where people can learn to cook and appreciate the best in wine and food | • New indulgers<br>• New enthusiasts |
| A lifestyle adjustment and health resort | • New enthusiasts<br>• Antitourists |
| Small to medium self-contained conference chalets business workshops | • New indulgers (business and other established professional groups) |
| A low key rural retreat with rustic chalets or cabins in a peaceful setting | • New enthusiasts<br>• Antitourists<br>• New indulgers |
| A steam railway with stations along the Bussell Highway and at key attractions connected with branch lines | • New enthusiasts<br>• New indulgers<br>• Stay at homes<br>• Dedicated Aussies |
| Hot air balloon flights | • Wide appeal to all groups |
| A high-quality art gallery | • Wide appeal to most groups |

(Source: Leeuwin Estate: 1986, 20)

Three alternative concepts were prepared for tourism development: option 1, minimal development, option 2, moderate development, option 3, major development. The concept plan for option 3 is illustrated in Figure A-1.

This study concludes that there is ample opportunity for tourism development here, and with planning, the ongoing farming, forestry, and private estate can be protected at the same time. It recommended even deeper market study,

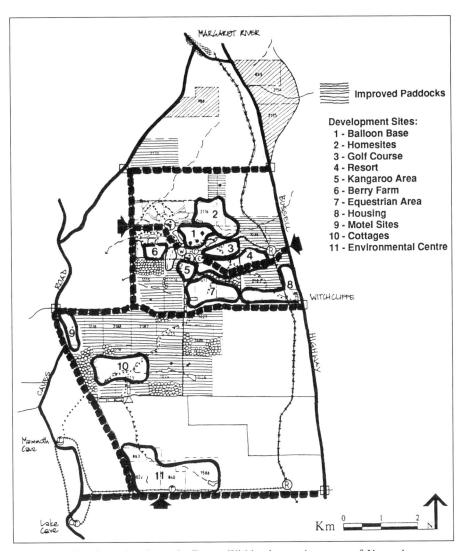

**Figure A-1.** Tourism plan, Leeuwin Estate. Within the tourism zone of Vasse the resource potential of the Leeuwin Estate was researched. Illustrated is a concept for development of attractions, improved access, convention center, and relation to Margaret River (Leeuwin Estate: 1986, 39).

economic feasibility study, adjustment to ongoing plans, more detailed site planning for specific segments of plan, and a staging of work over time.

## Ireland

On September 27, 1985, the *White Paper on Tourism Policy* was published for Ireland. It is a 19-chapter examination of the status and needs regarding policy, stimulated by a weakened economy and an erratic record of tourism in the previous 10 years. This is the first effort toward establishing policy.

The principal Irish tourism agency, the Bord Failte, was established by the Tourist Traffic Acts 1939–1983. In the late 1970s the National Economic and Social Council examined tourism and made recommendations for policy directions. In response, the Department of Industry, Trade, Commerce, and Tourism (ITCT) identified the national objective:

> To optimise the economic and social benefits to Ireland of the promotion and development of tourism both to and within the country consistent with ensuring an acceptable economic rate of return on the resources employed and taking account of:
> —tourism's potential for job creation.
> —the quality of life and development of the community.
> —the enhancement and preservation of the nation's cultural heritage.
> —tourism's contribution to regional development (White Paper: 1985, 8).

Regarding the balance between public and private sector roles, the report emphasized the need for greater acceptance of responsibility by the private sector. "The onus is on commercial operators to seek out viable projects and to take the best advantage of the opportunities available" (White Paper: 1985, 14). Emphasis is placed on creating new ideas to extend the season, new attractions to meet new competition, and innovate for the new markets. Topics held out as promising were: participation in international conferences, educational studies, tapping new markets, establishing new sporting events, providing for longer holidays of retired people, expanding marina development, and innovating restaurant service.

Following is a summary of proposed action for fulfillment of the current policy statement.

Policy: All departments of government with functions related to tourism will take full cognizance of the needs of tourism and consult with the ITCT when proposals impinging upon tourism are being considered.

Importance of tourism: New research should be initiated to create a better measure of employment in tourism.

Role of public and private sectors: The Bord Failte will arrange its promotional activities with joint financing from commercial operators. Grant schemes are to encourage participation by the private sector. A new strategy of brand marketing will be established.

Import tourism: The Bord Failte has been requested to investigate new policy for young people's holiday periods. The interest is to stimulate greater domestic tourism.

Competitiveness: The Bord Failte has been asked to incorporate greater flexibility in its marketing in order for closer response to sudden changes in travel factors, such as currency fluctuation.

General condition: Drops in tourists from Britain, increased age of travelers, and foreign potential are topics to be analyzed. Marketing should be intensified for tourists from Australia, Europe, and the Far East. Special interest holiday trade for young domestic tourists requires new emphasis.

Environment: Much greater emphasis on making areas more attractive and protecting resources is called for. Among the measures needed:

Identify and correct sites of greatest litter.
Initiate computerized mapping of country's physical attractions.
Study and identify national heritage landscapes.
Review water pollution problems.
Intensify cooperation between Bord Failte and planning authorities.
Intensify cooperation with Department of the Environment.
Expand program of awards of excellence.
Establish an Interdepartmental Committee on Tourism (White Paper: 1985, 72).

Internal transport: Improved roadsigning, improved motor coach tours, and better regulations for passenger-carrying boats are to be accomplished.

Tourism amenities and facilities: Several governmental agencies have been requested by the government to expand sports facilities, improve shopping, identify tourist needs in parks and conserved areas, remove restrictions on Office of Public Works to present national monuments to the public, open historic homes to the public, and improve local amenities, especially upgrading sanitary facilities.

Accommodations: Hotel grading will be improved, farm holidays are to be investigated, and regulations for accommodation signing will be reviewed.

Investment: Several changes are decided—that concessionary finance be provided by Industrial Credit Company, that the National Development Corporation is empowered to make investment in tourism projects, and that a tourist development team shall be formed to act as a catalyst to identify tourist project opportunities and matching developers to appropriate projects.

Seasonality: Marketing shall be changed to promote the shoulder seasons, and the feasibility of changing school and work holidays will be investigated.

Promotion and marketing: Bord Failte is charged with monitoring the efficiency of promotion and with investigating new technology to improve information and reservation systems.

Access transport: The Minister of Communications will take tourism into account for harbor proposals; monitor passenger access to Ireland for more self-sustaining services; set up a more competitive aviation policy; promote access from new markets; and establish new feeder air services.

Institutional-public: Expenditure on tourism is to be evaluated by an outside consultant. The Bord Failte is charged with establishing quantifiable objectives and reducing its financial aid to other agencies. An Interdepartmental Policy Coordination Committee shall be established to stimulate full cooperation among all departments related to tourism actions, including: Taoiseach, Agriculture, Communications, Education, Environment, Fisheries and Forestry, Gaeltacht, Health, Justice, Labor, Bord Failte, Shannon Free Airport Development Company, local authorities, Office of Public Works, Fisheries Board, National Institutions of Science and Art, COSPOIR, Bord na gCapall, and CERT.

# Canada

## Saskatchewan

Because Canada at the federal level has delegated tourism development planning to the provinces and territories, the case of Saskatchewan is useful in understanding a provincial tourism strategy. The following is derived from *A Saskatchewan Tourism Development Strategy* (1983?), a consultant's recommendation.

Through research study of the province, the consultant identified several constraints for an expanded tourism industry there:

A negative image of their province by residents.
Lack of industry organization.
Tourist expenditures in Saskatchewan are the lowest in Canada.
High percentage of visitors pass through.
Poor linkage between scenic resources and routes.
Low involvement by cities and province.
Low key marketing efforts.

However, examination of resources revealed many opportunities and assets:

Abundance of wilderness and wildlife.
Diversity of cultural and historic resources.
Many community events.
Regina and Saskatoon are destinations in their own right.
Access is generally available.

The report identifies three goals for a tourism development strategy: (1) greater product development to increase trip expenditures, (2) greater marketing emphasis to increase share, and (3) greater awareness and greater support for industry organizations.

Recommended are three publicly funded program initiatives:

1. Programs to encourage *product development,* including the growth and development of key attractions, both urban and rural, parks and recreation development, scenic travel corridors, community events, and wilderness area development.
2. Increased *program development to enhance a more positive image* of Saskatchewan and to create an awareness in nonresident markets of the travel opportunities available and cooperative support for industry marketing efforts.
3. *Support or assistance for industry* and tourism organizations to develop an integrated, coordinated approach to tourism development at community, regional, and provincial levels.

An expanded version of this report includes recommendations on a heirarchy of destination zones: primary tourism destination areas (urban and nonurban), secondary tourism destination areas including wilderness waterways and outdoor recreation areas, major-minor attractions, tourism service centers (provincial and regional), and transportation corridors (provincial, regional, and other specialized routes). The primary and secondary tourism development zones are listed in Table A-2.

**TABLE A-2**
**PROPOSED DESTINATION ZONES, SASKATCHEWAN**

Study of resource characteristics of the province and potential travel market segments revealed the potential for several primary and secondary tourism destination zones.

| **AREA** | **KEY ATTRACTION** |
|---|---|
| *Primary Zones: Urban:* | |
| 1. Saskatchewan Provincial Service Centre/Urban Destination Area | Major urban themed attractions and events operated year round. |
| 2. Regina Provincial Service Centre/Urban Destination Area | Major urban themed attractions and events operated year round. |
| *Primary Zones: Nonurban:* | |
| 3. Saskatchewan Rivers Heritage Area Historical Destination | Major interpretation of historic resources related to the 1885 Northwest Rebellion and Metis Society themes integrated with recreational facilities and services. |

**TABLE A-2** *(continued)*

| AREA | KEY ATTRACTION |
|---|---|
| 4. Lake Diefenbaker Tourism Destination Area | Four-season destination community associated with the water and land-based recreational opportunities of Lake Diefenbaker. |
| 5. Qu'Appelle Valley Tourism Destination Area | Hospitality training facility and summer resort in the Qu'Appelle Valley. |

*Secondary Tourism Destination Zones*

| | |
|---|---|
| 6. Prince Albert Tourism Destination Area | Four-season destination area development emphasizing intensive development and use of certain recreational resources of Prince Albert Park. |
| 7. Cypress Hills Tourism Destination Area | Four-season destination area development focusing upon the cold and warm weather recreational and historical opportunities associated with the Cypress Hills. |
| 8. Churchill River Reach Wilderness Waterway | Nationally significant candidate Canadian Heritage River for extensive outdoor recreational activities. |
| 9. Northeast Lake District Outdoor Recreation Area | Extended season extensive outdoor family-oriented recreational opportunities centered upon the lakes. |
| 10. Northern Saskatchewan Extensive Recreation Area | Sport fishing and hunting recreation area at remote access and northern road camps. |

(Saskatchewan Tourism: 1983?, 15)

These destination zones are recommended to be developed within a 20-year strategy. They are based upon the concept that the province should concentrate on those geographical areas where the resources and market segments could best be combined with development. These would be areas with high resource capability, potential for critical economic mass, strong market appeal, and, therefore, suitable for funding. Based upon these factors, major tourism destination areas along interprovincial travel corridors and at major urban centers were delineated to be the core attractions to increase resident use year round, to capture the critical pass-through traffic, attract new interprovincial sectors, and have strong enough market appeal for a significant impact on developing and improving Saskatchewan's image as a tourist destination.

In order to establish provincial priority of specific project development within the identified zones, each project was screened by means of the following:

1. Determine the priority of community or tourism destination areas based on an assessment of resource opportunities and weaknesses. The critics included the significance of the resource, existing market, complementary infrastructure and services, and the potential role of the development in the provincial tourism heirarchy.
2. Determine the priority of the proposal in terms of its market potential.
3. Review of industry concerns and objectives in terms of these considerations identified as essential in satisfying the critical tourism market segments as well as industry needs.
4. Develop an assessment of the existing market potential to support proposed developments and latent market reaction to the project.
5. Provide an assessment of the planner's understanding of a community's or area's reception to increased tourism development.
6. Examine the degree to which the proposed development can complement other existing and future community attractions and services and establish the critical mass necessary to be self-supporting.
7. Assess the degree to which the proposal will be able to enhance economic growth and tourism expenditures for all market segments.
8. Determine the degree to which the development will be a springboard for other complementary developments.

## Alberta

In July 1984, the government of Alberta released a White Paper on "Proposals for an Industrial and Science Strategy for Albertans, 1985 to 1990," which included tourism as a major generator of employment and growth (Position and Policy: 1985, 1). Several responses from this paper encouraged the government to appoint a task force to undertake a comprehensive review of tourism in Alberta to include history, development, training, marketing, world position, and future potential. Included also were to be reactions to the federal tourism statements in the publication, *Tourism Tomorrow: Towards a Canadian Tourism Strategy,* released in March 1985.

As a result, Policy Statement No. 1, *Position and Policy Statement on Tourism,* was issued in June 1985. This document contains four sections and appendices that set forth the challenges and suggested strategies for improving tourism. Of particular interest in setting policy and planning for the future is the section, The Roles of the Private Sector and Government, excerpted and summarized as follows (Position and Policy: 1985, 7), including five goals for the public and private sector partners:

1. Establish higher standards in service skills and attitudes throughout the tourism industry.
2. Develop and market the Province of Alberta as a major four-season travel destination.
3. Foster the development and improvement of physical facilities, attractions, and events.
4. Create a greater awareness among Albertans of the province's tourism potential and the social and economic contributions generated by the industry.
5. Encourage meaningful employment opportunities through manpower planning and training in the tourism industry.

For the private sector, five responsibilities were identified:

1. To provide excellence of service in all areas and to monitor performance, particularly for courtesy and maintenance of facilities.
2. To provide tourism facilities and services that meet the requirements and expectations of the visitor.
3. To promote individual businesses and services to their defined markets.
4. To undertake new development as demand requires.
5. To advise government on policies and programs that will assist the industry.

Regarding the role of the Alberta government, the following seven policy actions were listed:

1. To market Alberta as a tourism destination, establishing the image and awareness of the province in the world marketplace.
2. To assist the private sector in its efforts to increase awareness of tourism and the accompanying benefits to Albertans.
3. To assist the private sector in the development of Alberta's tourism markets through the provision of research, market intelligence, and consultation.
4. To assist the private sector to upgrade/improve facilities and services.
5. To minimize regulations affecting tourism services.
6. To assist other levels of government, such as municipalities, to develop and smooth the way for further development of related recreational and tourist opportunities.
7. To provide adequate protection for significant natural and cultural resources.

All provinces and territories of Canada continue to benefit materially from five federal government financial assistance programs. The *Small Business*

*Loans Act* (SBLA) assists proprietors of small business enterprises to obtain term credit for a wide range of business improvement purposes. The *Federal Business Development Bank* (FBDB) has as its function to provide financial services (loans, loan guarantees, and financial planning), investment banking, and management services. The FBDB's financial assistance is provided for the establishment, modernization, expansion, and operation of a business. The *Marina Policy Assistance* is administered through the Small Craft Harbours Program to encourage the development of public facilities for recreational boaters, in particular, those that might be classed as tourist facilities. The *Program for Export Market Development* (PEMD) is a cost-sharing program for exploring foreign markets where companies have not previously entered. The *Tourist Wharf Program* offers assistance for the construction of wharves and/or launching ramps in an area that has tourism potential, or where tourism is an established industry.

As an example, on May 13, 1985, the government of Alberta and the government of Canada signed a five-year 50-50 cost-sharing $56.3 million Tourist Industry Development Subsidiary Agreement. The agreement is for facility and project development, alpine ski facility development, market development, training and profession development, industry and community support, and opportunity analysis and evaluation administration, public information, and evaluation.

Because greater emphasis has been placed on private sector involvement in recent years in Alberta, a special tourism analysis was prepared in 1984. Using the nominal group technique, Ritchie (1984) surveyed the members of the Executives of the Zone and Industry Sector Associations of the Tourism Industry Association of Alberta (TIAALTA). This resulted in identifying 35 themes that described the most significant problems/issues currently facing the private sector of tourism in Alberta. These themes included broad and specific issues and problems within the private sector as well as with government.

From this analysis, TIAALTA recommended actions to deal with the top priority 15 items, grouped into six categories. A condensation of these recommendations follows (Ritchie: 1984, E9).

1. Overall policy/public sector: that the president of TIAALTA establish a task force to increase government recognition of tourism; that another task force be formed to investigate regulations negative to tourism development.
2. Overall policy/private sector: that a TIAALTA task force be formed to identify means for increasing liaison, cooperation, and communication among tourism zones and industry sectors.

3. Overall policy/public-private interface: that a task force study alternative organizational patterns and better private sector input to planning.
4. Development: that a task force address the issue of reducing travel costs in Alberta; that a task force study opportunities for more festivals, events, attractions, and activities.
5. Market/promotion: that a task force determine how promotional funding can be increased, how more effective advertising can be done, and what

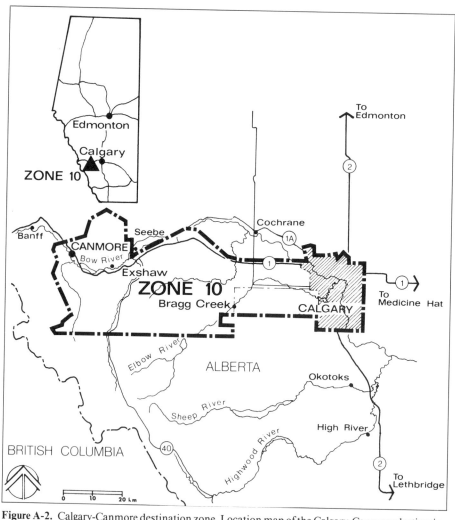

**Figure A-2.** Calgary-Canmore destination zone. Location map of the Calgary-Canmore destination zone study area. The study identified 61 tourism development opportunities, identifying services, attractions, access, and policies needed within seven subzones, including Calgary and Canmore as service centers (Pannell: 1985, iii).

central themes could best promote Alberta; that an information task force be established to improve the quality of information available to visitors.

6. Human resources: that a program of hospitality be established; that programs of tourism awareness be initiated; that an education task force be established to review and recommend needed changes in tourism education.

An example of planning a destination zone in Alberta is contained in the two-volume report, *Calgary-Canmore Tourism Destination Area Study* (Pannell: 1985). Volume one describes the objectives and an inventory of the resources and their assessments. The zone includes some 962,500 acres (385,000 hectares) and 627,000 people (7,000 outside Calgary) (Fig. A-2). The objectives of the area strategy were to increase visitation, increase length of stay, and increase visitor expenditure. Key topics described in volume one are: a biophysical overview, tourism resources (attractions, events, parks, campgrounds, accommodations, food, information centers, and transportation), heritage resources, the economy and human resources, visitor characteristics, plans and policies, and 61 tourism development opportunities. A partial list of these opportunities, their impact, and primary action agent is provided in Table A-3.

### TABLE A-3
### TOURISM DEVELOPMENT OPPORTUNITIES, ALBERTA

Study and analysis of the Calgary-Canmore tourism destination zone revealed many opportunities for tourism development. For each one, potential impacts and funding support are listed in this table.

| Opportunities | Capital or Operations Emphasis | POTENTIAL IMPACTS | | | Cost | Primary Action or Funding Emphasis |
|---|---|---|---|---|---|---|
| | | Social | Economic | Environmental | | |
| GENERAL | | | | | | |
| • Attract World Class Sports Events Before and After the Olympics | O | 2 | 2 | -1 | C | Pr |
| • Develop an Annual Winter Festival | O | 2 | 2 | 1 | C | Pr |
| • Develop a Western Heritage Theme | O | 2 | 2 | 0 | C | Pr |
| • Promote the Scenery by Making Highways "Visitor Friendly" | C | 2 | 1 | 0 | D | P |

TABLE A-3 *(continued)*

| Opportunities | Capital or Operations Emphasis | POTENTIAL IMPACTS | | | Cost | Primary Action or Funding Emphasis |
|---|---|---|---|---|---|---|
| | | Social | Economic | Environmental | | |
| • Capitalize on Linkages to Attractions Outside the Zone | O | 1 | 2 | 0 | A | Pr |
| • Prepare Circle and Exploration Tour Publications | C/O | 1 | 1 | 0 | B | Pr/P |
| • Develop and Promote Major Attractions in Locations Outside of Calgary | C/O | 2 | 2 | -1 | X | Pr |
| • Provide Tax Incentives and Concessions for Tourism Developments | C/O | 1 | 2 | 0 | X | P/M |
| • Develop and Promote Tours of Olympic Facilities | O | 1 | 1 | 1 | B | Pr |
| • Expand the Variety of Fixed Roof Accommodation Available | C | 1 | 2 | -1 | X | Pr |
| • Improve and Develop Recreational Vehicle (RV) Parks | C | 1 | 1 | -2 | B | Pr |
| • Develop Transportation Services to Destinations in Rural Zone 10 | O | 1 | 2 | 0 | C | Pr |
| • Adjust Zone Boundaries to include Cochrane, Highway 1A and Stoney Indian Reserve | O | 0 | 1 | 0 | A | Pr |
| • Develop Adventure Tours | O | 1 | 1 | 1 | B | Pr |
| • Develop Educational and Industrial Tours | O | 1 | 1 | 1 | B | Pr |
| • Improve Visitor Information Services to Provide Accurate and Up-to-the-Minute Information on Attractions, Facilities, Events, and Services Available in Zone | C/O | 1 | 1 | 0 | C | Pr |

| Opportunities | Capital or Operations Emphasis | POTENTIAL IMPACTS | | | Cost | Primary Action or Funding Emphasis |
|---|---|---|---|---|---|---|
| | | Social | Economic | Environmental | | |
| • Develop a Central Reservations System for Zone Accommodation, Tours, Sports, Theatre, Events, etc. | C/O | 1 | 2 | 0 | C | Pr |
| • Improve Signage to Attractions and Services Off Major Transportation Corridors | C | 1 | 1 | 0 | B | P/M |
| • Give Canmore and Rural Zone 10 Higher Profile in Calgary Tourist and Convention Bureau | O | 2 | 2 | 0 | A | Pr |
| • Develop and Deliver Hospitality Training Programs | O | 2 | 1 | 0 | C | P |
| • Improve Operation of Existing Tourist Services and Facilities on Indian Reserves | O | 2 | 2 | 0 | C | P |
| • Develop Public Investment Implementation Strategy | O | 1 | 2 | 0 | A | P |
| • Provide Extra Assistance to Expedite Application Review and Processing | O | 2 | 2 | 0 | B | P |
| • Investigate the Feasibility of Developing First Class Casino in the Zone | O | -1 | 0 | 0 | B | Pr/P |
| CALGARY SUBZONE | | | | | | |
| • Develop Attractions, Facilities, and Services in Character Areas and Target Tourist Awareness Programs at these Areas | C | 2 | 2 | 0 | X | Pr |
| • Coordinate Marketing Activities for Major Attractions | O | 2 | 2 | 0 | A | P/M |
| • Implement More Dynamic and Experiential Programs at Major Attractions | C/O | 2 | 1 | 0 | D | P/M |

| Opportunities | Capital or Operations Emphasis | POTENTIAL IMPACTS | | | Cost | Primary Action or Funding Emphasis |
|---|---|---|---|---|---|---|
| | | Social | Economic | Environmental | | |
| • Organize Nightlife and Entertainment in Character Areas | O | 2 | 2 | 0 | C | Pr |
| • Introduce Design Elements into the Urban Environment to Promote Tourism | C | 2 | 2 | 0 | X | M |
| • Promote Complementary or Alternate Activities for Tourists | O | 2 | 1 | 0 | A | Pr/M |
| • Promote Tours and Tour Packaging | O | 1 | 2 | 0 | A | Pr |
| • Develop Boat Cruise with Entertainment on Glenmore Reservoir | C | 2 | 1 | -1 | X | Pr |
| • Plan and Implement Other World Class Events in Calgary | O | 2 | 2 | 0 | X | Pr |
| CANMORE AND AREA SUBZONE | | | | | | |
| • Promote Canmore and Area as the Staging Area for World Class Outdoor Recreation Events | O | 2 | 2 | -2 | C | Pr |
| • Develop Arts Community in Canmore | C/O | 2 | 2 | 0 | X | Pr/P/M |
| • Develop a Landmark or Major Gateway to Attract People to the Downtown Area | C/O | 2 | 2 | 0 | X | Pr/M |
| • Promote Heritage Design in Canmore | C/O | 2 | 1 | 2 | B | M |
| • Capitalize on the Olympic Facilities and Games in the Area | O | 1 | 2 | 0 | X | Pr |
| • Develop Tours Based on Coal Mining and Forestry Themes | O | 1 | 1 | 0 | B | Pr |
| • Increase Opportunities for Water-based Recreation | C | 1 | 1 | -1 | B | P |

| Opportunities | Capital or Operations Emphasis | POTENTIAL IMPACTS | | | Cost | Primary Action or Funding Emphasis |
|---|---|---|---|---|---|---|
| | | Social | Economic | Environmental | | |
| • Improve Trails, Trail Signage, and Access | C | 1 | 1 | -1 | B | P/M |
| • Attract Major Resort Development to Area | C | 2 | 2 | -2 | X | Pr/P/M |
| • Construct a Gondola or Tramway to a Mountain Viewpoint and Restaurant | C | 1 | 2 | -2 | E | Pr |
| • Improve the Visibility of Facilities and Services Along Highway 1 and Improve Access Awareness Signage | C/O | 1 | 2 | 1 | X | Pr/P/M |
| • Develop and Promote Nightlife and Entertainment in Town | C/O | 2 | 2 | 0 | X | Pr |
| • Promote Canmore as a Service Centre for Surrounding Resort and Recreation Developments | O | 2 | 2 | 0 | A | Pr |
| • Prepare Tourism Action Plan for Canmore | O | 1 | 2 | 0 | A | Pr/P |

BOW VALLEY AND HIGHWAYS CORRIDOR SUBZONE

| Opportunities | Capital or Operations Emphasis | Social | Economic | Environmental | Cost | Primary Action or Funding Emphasis |
|---|---|---|---|---|---|---|
| • Plan and Promote the Development of Service Centres and Attractions at Major Intersections on Highways 1 and 1A | C/O | 1 | 2 | -1 | X | Pr |
| • Improve Signage to Attractions Off Highway 1 | C | 1 | 1 | 0 | A | P |
| • Promote the Development of Resort and Seasonal Residential Developments on Patented Lands South of Highway 1 | C/O | 1 | 2 | -2 | X | Pr/P/M |

KANANASKIS WEST SUBZONE

| Opportunities | Capital or Operations Emphasis | Social | Economic | Environmental | Cost | Primary Action or Funding Emphasis |
|---|---|---|---|---|---|---|
| • Revegetate Canal Embankments | C | 1 | 0 | 2 | E | Pr/P |

| Opportunities | Capital or Operations Emphasis | POTENTIAL IMPACTS | | | | Primary Action or Funding Emphasis |
|---|---|---|---|---|---|---|
| | | Social | Economic | Environmental | Cost | |
| • Develop Campground/ Cabin Area with Nordic Facilities on South End of Spray Lakes Reservoir | C | 1 | 1 | -1 | X | Pr |
| • Improve Signage to Kananaskis Country | C | 1 | 0 | 0 | A | P/M |
| KANANASKIS CENTRAL SUBZONE | | | | | | |
| • Attract Developers for the Ribbon Creek Alpine Village | C/O | 1 | 2 | 0 | X | P/Pr |
| • Develop Campground at Bow Valley Provincial Park | C | 1 | 1 | 0 | E | P/Pr |
| • Establish Horseback Riding and Equipment Rental Concession | C/O | 1 | 1 | 0 | C | P/Pr |
| • Develop Trail and Road Systems to Connect with Resort Developments in the Bow Valley | C | 1 | 2 | -2 | X | P |
| KANANASKIS EAST SUBZONE | | | | | | |
| • Promote East Entrance to Kananaskis Country | O | 1 | 1 | 0 | A | P/Pr |
| • Upgrade Powderface Trail to a Year-Round Road from Sibbald Flats to the Elbow Falls Trail | C | 1 | 1 | -2 | X | P |
| SARCEE SUBZONE | | | | | | |
| • Develop Entrance to Rural Zone 10 and Tourist Facilities and Services on Sarcee Indian Reserve | C/O | 2 | 2 | -1 | X | Pr |

(Source: Pannell Kerr Forster: 1985, ix)

Capital or Operations Emphasis:

C Capital
O Operations

Potential Impacts:

2 Significantly Positive
1 Positive
0 No Impact
-1 Negative
-2 Significantly Negative
NA Not Applicable
X Unknown

1) The impact scale represents an overview assessment by the consultants. Detailed impact assessments may be required prior to development to identify mitigative measures to be implemented.

Cost Potential Order of Magnitude:
(Thousand Dollars)

A Under 50
B 50–99
C 100–499
D 500–999
E Over 1,000
X Unknown

2) Cost potential is based on the first year of operation or the capital cost, whichever is applicable. Costs identified represent minimal amounts as anticipated by the consultants. Actual cost will depend on specific development requirements.

Primary Action or Funding Emphasis:

Pr Private
P Provincial
M Municipal

Volume two provides strategy for the development of these opportunities, including themes, strategy alternatives, the players, strategic direction, and timing and costs of the several development opportunities.

# Zanzibar

Consultant recommendations, including those of specialists from the World Tourism Organization, were prepared for the African island of Zanzibar, part of the nation of Tanzania (Methodology: 1985, 31). Included are the development policy and structure plan.

For the *development policy,* the following postulates were put forward as a basis for planning.

International tourism should be developed as an important means of achieving greater understanding and appreciation of Zanzibar's history, culture, and natural environment by foreigners, and of residents developing some understanding of other people's customs and cultures.

International tourism also should be developed to provide additional employment, income, and foreign exchange for Zanzibar to help diversify the island's economy.

Domestic tourism should be developed as an important means of recreation, increasing understanding by Zanzibaris of their own historical, cultural, and environmental heritage and by mainland Tanzanians of Zanzibar's rich historical and cultural heritage.

Domestic tourism also should be developed as a means of redistributing income within Zanzibar, especially from urban to rural areas and from mainland Tanzania to Zanzibar.

Tourism development should be integrated into the overall development policy, planning, and strategy of Zanzibar, and should receive appropriate priorities and its necessary share of development resources.

Tourism should be developed operationally so that it promotes conservation of archaeological sites and historic places, conservation and revitalization of the desirable aspects of traditional cultural patterns arts and handicrafts, and maintenance of the essence of religious beliefs and practices, all of which represent the historic and cultural heritage of Zanzibar; tourism should be planned, developed and organized so that it does not result in serious social problems or cultural disruptions.

Tourism should be developed in a carefully planned, controlled, and organized manner so that it promotes conservation of the natural environment, especially places of scenic beauty, indigenous flora and fauna, important natural ecological systems, outdoor recreation potential, beaches and underwater environments. Tourism development should not result in any type of serious air, water, noise, and visual pollution.

Tourism development should proceed on a controlled systematic basis, according to a staged program of allocating development resources to specified places. The pace of tourism development should be kept in balance with the number and type of tourist arrivals, the development of infrastructure, and with Zanzibar's economic and social capability of absorbing tourism growth.

Tourism should be developed so that it will serve as a catalyst for increased development of related economic activities, such as handicraft production, agriculture, and fisheries and other related industries, and help in supporting improvements of transport facilities and services and other infrastructure.

Tourism should be planned and developed so that it makes maximum use of existing infrastructure and that improved and new infrastructure should serve general purpose needs as well as tourism.

Tourist accommodations and other facilities should be designed to reflect and represent Zanzibar's distinctive architectural styles, the island's tropical environment, and utilize local building materials to the extent possible. Maximum use should be made of renovated existing buildings, which have architectural and historical significance, for tourist facilities.

All aspects of tourism development and operations should be organized so that tourism functions in an efficient and integrated manner to meet the needs of international and domestic tourists and achieve the objectives of tourism development. Especially important is the coordination of tourism development among the various government agencies and parastatal bodies.

Tourist facilities and services should be designed to meet the range of needs of various tourist market segments, including international and local standard, without being unduly expensive to develop on the one hand and always meeting minimum requirements of sanitation, safety, and comfort on the other.

Emphasis should be placed on employment of local persons in tourism, and persons working in all aspects of tourism should be properly trained to function effectively in their employment and be given maximum opportunity for career development and job satisfaction.

The general public should be educated to understand tourism and its role in Zanzibar's development policy and be given all possible opportunities to use and enjoy tourist facilities and attractions while respecting and maintaining a suitable standard of these facilities and services.

For the *tourism structure plan,* an analysis of Zanzibar, applying several criteria, revealed tourism potential for several destination zones, or "areas". The criteria used were:

One or more important attractions within or near the area.

Suitable and environmentally attractive sites for development of tourist accommodations (or existing buildings suitable for renovation), without creating social or environmental problems, e.g., availability of sites that do not require substantial relocation of people or major economic activities.

Adequate access now or in the future to the area from the tourist entry points (airports and seaports).

Availability of other infrastructure now or in the future, especially of adequate potable water, electricity supply, sanitary sewage disposal, and telephone service. Telex service is desirable but not essential.

Availability of potential to develop the necessary other tourist facilities within the area (often these can be provided in the hotel environment).

Adequate transport links to other development areas and to important tourist attractions elsewhere in Zanzibar.

The existing or proposed development of accommodations in the area.

Interest of local residents in having tourism development in or near their area.

Need in the area for employment and economic activities that can be provided by tourism.

Analysis of both markets and resources was employed in the development of these criteria.

Continuing examination of tourism is recommended through a *monitoring program* containing the following elements:

Collecting and analyzing data on the number and characteristics of movements of persons for tourism and leisure and for duty and obligation.

Conducting special visitor reaction surveys.

Maintaining monthly hotel occupancy statistics.

Field checking of actual progress of tourism development projects.

Maintaining close contact with domestic and international tour operators to determine feedback on the success of tours to and within Zanzibar and general tourism trends.

Obtaining regular information on trends in movements of persons from the Tanzanian mainland, Kenya, and other places in East Africa.

Coordinating development schedules with other agencies having responsibility for certain aspects of tourism.

## Taiwan

In 1971, the Travel and Tourism Bureau of Taiwan, or Republic of China, was established with the Transportation Ministry as the federal agency responsible for tourism (Chu: 1986). It includes several departments, such as planning, business, technique, international relations, instruction, training, and general affairs. Attached are three other functional groups: accounting, human resources, and security.

In recent years Taiwan has experienced considerable growth of tourism, and therefore the government has established policy guidelines for development. These basic policies are (Chu: 1986, 392):

Protect natural scenery and explore the potential tourist resources.

Reward investment in the tourist industry, especially hotels.

Strengthen international publicity and attract more tourists.

Ease travel for local people, set forth proper recreational activities, be courteous to international tourists, and develop Chinese culture.

Chu states that tourism development is hampered by lack of coordination between land use agencies, lack of integration of regulations, excessive land use restrictions, inadequate tourist services, lack of business stimulation, and inadequate research. For solutions, he recommends:

More uniform tourism development regulations.

Planning should be done by competent professionals.

Mountain and beach resources should be opened up.

Better management of restaurants and hotels should be initiated.

Financial aid should be available for local tourism service businesses.

Car leasing, arts and crafts, and other businesses need to be better planned.

Air access should be improved.

Promotion requires better organization and support.

Research by societies and academic institutions needs support.

Further recommendations include the appointment of an ad hoc commission to review present and establish new tourism policies for government including the present tourist agency. Among the objectives would be the identification of special resources requiring protection; linkage between budgets and policies; measurements of social, economic, and political impacts; and feasibility for new education and training in tourism.

Concern over planning for tourism development is further documented in another publication, "The Importance of Development" (Long-Term: 1985). Consideration is given to domestic and international travel growth, which in turn will require care in using the limited extent of resources on the island. Among basic principles set forth are the following.

1. Investigate before planning. Only by means of careful investigation of resources can they be protected at the same time they are partially developed for tourism. Such prior investigation can guide locations to be specifically planned for tourism projects. These areas must be kept beautiful and developed to high standards to avoid misuse of manpower and materials as well as the basic resources.

2. Strengthen linkages of the tourism system. Because of the limited size of the island, a regional approach that ties destination areas together is required. Consideration of transportation linkages and focal points for development can add up to a larger and more cohesive whole. Properly planned, developing the system should add to the number of attractions and extend the visitor's length of stay.

3. Encourage local investment in tourism. In addition to needed public services by government, private investment and initiative can be very productive. Examples of successful private developments are "Birds Coming Through Clouds" Garden, Wild animal zoo of "Lin Fu Village," Dragon Pond Pigwig State, Jar Shaped "Ya Ge" Garaeu, and Valley Dragon Tourist Land. These demonstrate advantages of local private investment—high efficiency, dense capital, collect management, and richness of content. Tourists can enjoy many activities in relatively small areas. These are particularly well adapted to tourist tours.

4. Develop new tourist resources. New areas will be needed to handle greater volumes of visitors, especially during peak seasons. But as new areas are opened up, they must be integrated with existing ones. At the same time, the quality of the resources must be protected.

5. Plan for increased touring circuits. In the past, the focus has been on fixed destinations. As markets change, both domestic and foreign, there is greater need for developing tours that are interesting on water as well as land.

# Israel

For many years Israel has recognized the positive value and also the several costs of developing tourism. Political and physical planning continue to be functions of the Ministry of Tourism resulting in development superior to areas without such planning guidelines (Israel Tourism: 1982 and Tourism Master Plan: 1987).

Several destination zones have been identified since the initial master plan in 1976. These zones are based in part on market studies which show the main purposes of visits to be vacationing, visiting friends and relatives, and pilgrimages. In addition, the use of thermal springs for health has increased contrary to the trend elsewhere. And, with increased industrialization, travel for commerce, trade, conferences and conventions has increased in recent years. Because of the location and relationship to dominant markets, 75 percent of international travelers arrive by air. Most originate from the U.S.A., Europe, and Scandinavia. About 40 percent are Jewish and 60 percent are non-Jewish.

Several zones have been identified and supported by national tourism policy. Each one has its own special resource characteristics and market potential. Some are best suited to international travelers whereas others are better adapted to domestic markets. Following is a sketch of these zones of tourism development potential as designated in the official Israel master plan.

The Upper Galilee and Golan Heights zone's potential lies primarily in its water sites, streams, mountains, and the Sea of Galilee. It is best suited to vacationing recreation market segments and has potential for development of resorts, parks and attractions providing swimming, hiking, fishing, boating, waterskiing, and winter sports. These assets are complemented by several historic sites.

The Tiberias and the Northeastern Shores of the Sea of Galilee zone depends greatly on its seashore assets. Recreation, health spas, water sports as well as religious and historical pilgrimages are main assets to be developed. Especially important are sites of significance to Jewish and Christian tourists.

A Western Galilee zone extends from the Sea of Galilee to the coast of the Mediterranean. The coastal plains and the hills of western Galilee offer a rugged landscape of forests and interesting flora. Already, these resources are protected in parks and reserves. Bathing beaches, archeological sites and scenic vistas are among the tourist development assets. The ancient city of Acre contains remains of the Crusades (11-13th century) period such as khans, fortresses, mosques, and bazaars built by Moslems and Jews.

Haifa and its surroundings form another zone of potential. It is the third largest city of Israel with many attractions already developed. Carmel National Park, many historical and archeological sites, as well as urban attractions can be found here. Plans include expansion of the beach, restoration of historic sites and addition of a country club and scenic overlooks.

The Central Coastal zone encompasses flat sandy beaches from Haifa to Ashquelon. This long-season asset is complemented by cultural assets such as antiquities, art institutions, diamond processing, a 2000-year-old seaport, and many urban attractions. Plans include development favoring markets seeking popular activities suited to these resources.

Tel-Aviv is designated as an urban destination zone. This is the center of trade, industry and suited to conference and convention markets. Plans include expansion of the infrastructure as well as development of more urban attractions.

Jerusalem and its surroundings make up another important zone. The religious, historic and cultural attractions are known worldwide and, as the capital and largest city of Israel, Jerusalem is a major urban attraction zone. Plans include better handling of mass tourism, such as improved interpretation of historic sites, restoration of important structures, better signage and addition of tourist services. The surrounding attractions in Bethlehem, wilderness of Judea and villages add to the importance of this zone.

South of Tel-Aviv about 50 kms. lies the urban beach zone of Ashquelon and its surrounding area. Parks, marinas, commercial centers and artificial lakes are planned for this zone. Ancient cities add to the year-round beach attractions of the zone.

The Negev desert zone is in contrast to the mountains and forests of northern Israel. Health resorts on the Dead Sea and desert touring of scenic and historic areas, such as Beer-Sheba, Shivta and Avdat offer potential for expanding tourism in this zone.

The Dead Sea shore, Judea and Samaria zone offers potential for winter resort development and greater use of the health spas, such as at Ein Nue'eit. Scenic drives and historic sites of Christian and Jewish importance could attract more tourists if infrastructure and services were improved.

At the southern tip of Israel lies the Eilat tourist zone. A port on the Gulf of Aqaba, Eilat offers both desert and water resource assets. With expanded infrastructure, the zone is becoming a convention center as well as a vacation resort.

These zones make up the physical planning directions for tourism growth in Israel. The intent is to improve the supply side so that promotion of new markets will, in fact, expand tourism. Policy includes an extended season, improved air access, better intermodal transportation, better links with Japan and South America, increased development of areas of natural beauty, better compatibility between tourists and host areas, improved adjustment between international and domestic tourists and improved security and safety for tourists. The tourism interests look to government for greater responsibility in tourism—improved cleanliness, better sensitivity to international currency exchange, reduction of strikes, improvement of traveler assistance and a larger budget for tourism. Also recommended is better market research to provide new and more reliable data for planning tourism.

# Bibliography

*Action for Jobs in Tourism.* Department of Employment. London: Central Office of Information, 1986.

*Annual Report for the Year ended March 31, 1986.* London: British Tourist Authority, 1986.

*Annual Report for the Year ended March 31, 1986.* London: English Tourist Board, 1986.

"Canada-Alberta Subsidiary Agreement on Tourism Development," Edmonton, Alberta: Tourism and Small Business, 1985.

Chu, TaRong. *Tourism Policy—Theory and Practice.* Taipei, Taiwan: Wen Yan Printing House, 1986.

Davies, Ednyfed Hudson. "Tourism in British Politics," paper presented at the London Graduate School of Business Studies, March 12, 1979. London: The Tourism Society.

*Development of Tourism Act, 1969,* Chap. 51 (Act of British Parliament creating The Tourist Authority and the Tourist Boards).

*Development Opportunities in Tourism—Highland Region of Scotland.* Inverness, Scotland: Highland Regional Council, n.d.

Edgell, David L. "Model State Tourism Policy," unpublished draft. Washington, DC: US Travel and Tourism Administration, 1986.

Edgell, David L. "Recent US Tourism Policy Trends," *Tourism Management,* 3 (2), June 1982, 121–123.

"Fact Sheet: State Functional Plans," February 10, 1982. Honolulu: Department of Planning and Economic Development.

Gunn, Clare A. "Proposed State Policy Directions for Tourism," Educational Seminar for State Travel Directors, sponsored by National Council for State Travel Directors, Lawrence, KS, July 11, 1984.

*Hawaii State Plan Revised,* Hawaii State Plan Policy Council. Honolulu: Department of Planning and Economic Development, 1986.

"Implementation Strategies for Tourism Development Plans," unpublished paper. Perth: Western Australian Tourist Commission, n.d.

*Israel Tourism Projects Development—Information for the Investor.* Jerusalem: Ministry of Tourism, 1982.

Keith, Kent M. "The Hawaii State Plan Revisited," *University of Hawaii Law Review,* 7 (1), Spring 1985, 29–61.

*Leeuwin Estate Tourism Development Plan.* Prepared by Scenic Spectrums. South Melbourne, Australia, December 1986.

Liu, Juanita C. and Turgut Var. "Resident Opinion on the Effects of Tourism Development in Hawaii," Tourism Research Publications, Occasional Paper No. 8. Honolulu: Univ. of Hawaii, 1984.

*Long-Term Development Plan for Tourism in Taiwan (1986–2000).* Taipei, Taiwan: Travel and Tourism Bureau, 1985.

*Methodology for the Establishment and Implementation of Tourism Master Plans.* Madrid: World Tourism Organization, 1985.

Pannell Kerr Forster. *Calgary-Canmore Tourism Destination Area Study,* Vols. 1 and 2. Edmonton, Alberta: Travel Alberta, 1985.

*Planning for Tourism in England.* London: English Tourist Board, 1978

*Position and Policy Statement on Tourism.* (Policy Statement #1 in Response to the White Paper: An Industrial and Science Strategy for Albertans, 1985–1990).

Price Waterhouse and Clare A. Gunn. *Proposed Tourism Master Plan.* Albany, New York: Department of Commerce, 1986.

*Proceedings, Governor's Tourism Congress, 1984.* Honolulu: Department of Planning and Economic Development, 1985.

"Progress Report on Implementation of the State Tourism Functional Plan," September 1986. Honolulu: Department of Planning and Economic Development.

*Recreation and Tourism Land Use Planning in Australia: An Information Manual.* Hobart: Tasmanian Department of Sport and Recreation, 1987.

Richter, Linda K. "State-Sponsored Tourism: A Growth Field for Public Administration?" *Public Administration Review,* 45 (6), November/December 1985, 832–839.

Ritchie, J.R. Brent. *A Program for Furthering Tourism Development in Alberta.* Phase I, "Getting Our Act Together." Calgary: Tourism Industry Association of Alberta, 1984.

*A Saskatchewan Tourism Development Strategy.* Regina, Saskatchewan, Canada: Saskatchewan Tourism & Small Business, 1983(?).

*South West Region Tourism Development Plan.* Prepared by Coopers & Lybrand, Hassell Planning Consultants PTY LTD, Rob Tonge & Associates, Social & Environmental Assessment PTY LTD. Perth, W.A.: Western Australian Tourism Commission, 1986.

*State of Hawaii Growth Policies Plan: 1974–1984.* Honolulu: Department of Planning and Economic Development, 1974.

*State Tourism Functional Plan.* Honolulu: Department of Planning and Economic Development, 1984.

*Strategic Plan, 1981–1985.* London: British Tourist Authority, 1981.

*Tourism Development Plan, South West Region.* Perth: Western Australian Tourism Commission, 1986.

"Tourism Development Plans," unpublished paper. Perth: Western Australian Tourism Commission, n.d.

*Tourism in Japan 1986.* Tokyo: Japan National Tourist Organization, 1986.

*Tourism in the Heart of England—A Strategy for Growth.* Worcester: Heart of England Tourist Board, 1985.

*A Tourism Strategy for the Thames and Chilterns, 1984.* Abington, England: Thames and Chilterns Tourist Board, 1984.

*Tourism Master Plan.* Jerusalem: Ministry of Tourism, 1987.

"Tourism Statistics for Western Australia," *Touristics,* 5, September 1986.

*Tourism Tomorrow: Towards a Canadian Tourism Strategy.* Ottawa: Minister of State for Tourism, 1985.

*Tourism, the Golden Opportunity.* Perth: Western Australian Tourism Commission, 1985(?).

*Western Australian Tourism Profile.* Perth: Western Australian Tourism Commission, 1986(?).

*West Hawaii.* Hawaii Tourism Impact Plan, Vol. II, Regional. Honolulu: Department of
    Planning and Economic Development, 1972.
*White Paper on Tourism Policy.* Dublin, Ireland: Government Publications Sale Office,
    1985.

# Index

# About the Author

Dr. Clare A. Gunn is professor emeritus, Recreation and Parks Department, Texas A&M University. A pioneer in tourism education, his educational teaching and research career has spanned four decades, and at five universities. He has responded to invitations to present papers and perform consulting work throughout the United States, Canada, and over a dozen countries worldwide. His concepts, presented in lectures, books, and journal articles, have influenced policies and practices, particularly bridging the fields of design, planning, and tourism.

The author has received many honors and awards, including a special citation from Governor Mark White of Texas "for 38 years of inspirational teaching," a special award from the American Society of Landscape Architects for his book, *Vacationscape: Designing Tourist Regions,* and a special commendation by a joint resolution of the Senate and House of the State of Texas. He has been named Fellow of the American Society of Landscape Architects and is licensed in three states.

Dr. Gunn was the first to receive a Ph.D. in Landscape Architecture from an accredited university program, the University of Michigan. He received a master's degree, Land and Water Conservation, and a bachelor's degree, Landscape Architecture, from Michigan State University. He developed the tourism program and all tourism courses at Texas A&M University.

His tourism career began with many years of experience in advising tourist business operators and publishing technical assistance bulletins on tourism design and planning. His research has identified the tourism functioning system and basic planning principles for tourism destination development. His concepts of regional planning are well known and applied throughout the world.

132 A